Springer Series on
Atoms+Plasmas

7

Editor: I. I. Sobelman

E.B. Alexandrov   M.P. Chaika
G.I. Khvostenko

Interference
of Atomic States

With 59 Figures

Springer-Verlag

Berlin Heidelberg New York
London Paris Tokyo
Hong Kong Barcelona
Budapest

Professor Evgeny B. Alexandrov

Russian Academy of Sciences, S. I. Vavilov State Optics Institute,
199034 St. Petersburg, Russia

Professor Maria P. Chaika
Professor Gennadij I. Khvostenko

Institute of Physics, St. Petersburg State University, Universitetskaya nab. 7/9,
199034 St. Petersburg, Russia

Series Editors:

Professor Dr. Günter Ecker

Ruhr-Universität Bochum, Fakultät für Physik und Astronomie,
Lehrstuhl Theoretische Physik I, Universitätsstrasse 150,
W-4630 Bochum 1, Fed. Rep. of Germany

Professor Peter Lambropoulos, Ph. D.

University of Crete, P.O. Box 470, Iraklion, Crete, Greece, and
Department of Physics, University of Southern California, University Park,
Los Angeles, CA 90089-0484, USA

Professor Igor I. Sobelman

Lebedev Physical Institute, Russian Academy of Sciences,
Leninsky Prospekt 53, 117333 Moscow, Russia

Professor Dr. Herbert Walther

Sektion Physik der Universität München, Am Coulombwall 1,
W-8046 Garching/München, Fed. Rep. of Germany

Managing Editor: Dr. Helmut K.V. Lotsch

Springer-Verlag, Tiergartenstrasse 17, W-6900 Heidelberg, Fed. Rep. of Germany

ISBN-13:978-3-642-84444-7   e-ISBN-13:978-3-642-84442-3
DOI: 10.1007/978-3-642-84442-3

Library of Congress Cataloging-in-Publication Data. Aleksandrov, E. B. (Evgeniĭ Borisovich) Interference of atomic states / E. B. Alexandrov, M. P. Chaika, G. I. Khvostenko. p. cm. – (Springer series on atoms & plasmas ; 7) Includes bibliographical references and index. ISBN-13:978-3-642-84444-7 1. Hanle effect. 2. Interference (Light) I. Chaĭka, M. P. (Mariíá   Pavlovna) II. Khvostenko, G.I. (Gennadiĭ Ivanovich) III. Title. IV. Series. QC467.A38 1993   535.8'4 – dc20   92-38510

Typesetting: Macmillan India Ltd., India
54/3140 - 5 4 3 2 1 0 – Printed on acid-free paper

Preface

In this monograph we describe an important and relatively new class of phenomena in the field of high-resolution atomic spectroscopy: the interference effects manifest in the angular distribution and polarization of spontaneous radiation and absorption by atoms. Although the quantum-theoretical description of these interference effects is quite subtle, it turns out – as so often in quantum mechanics – that a simple classical or semi-classical description offers much insight and can even explain quantitative features. In this presentation, however, we attempt to give the full story. Beginning with the simple semi-classical description, we then present the quantum-mechanical analysis based on the density-matrix formalism and the statistical tensor. The remaining two chapters discuss experimental observations and data analysis. A great variety of effects have now been observed and can be used to obtain highly accurate information about hyperfine structure, atomic constants, interaction constants, etc.

The authors have assumed only a basic knowledge of quantum mechanics and electromagnetism, thus making the book accessible to those beginning a graduate studies program. It is also aimed at practising spectroscopists and all researchers for whom atomic spectroscopy is an important tool – for these readers it will hopefully offer some new solutions and ideas for furthering their research.

February 1993

E.B. Alexandrov
M.P. Chaika
G.I. Khvostenko

Preface

Contents

1. Introduction

The origins of the scientific field now called interference of atomic states can be traced back to the 1950s and 1960s. The interference phenomena are in essence very simple and could in principle have been studied much earlier. We note that the Hanle effect – the depolarization of resonance fluorescence in a magnetic field – has been known since the early 1920s [1.1]. It is now clear that this is an interference effect, but at the time of its discovery there were no concepts about the interference of atomic states. The further investigation and application of the Hanle effect thus had to wait some time. One can perhaps attribute this situation to a rather rigid and distorted understanding of the theory of quantum mechanics by the physicists of the time. Quantum mechanics, replacing the traditional universe of continuous processes by the universe of discontinuities, impressed the practical physicists so much that they grasped and familiarized themselves only with the discrete part of quantum mechanics.

Classical physics treats atoms as oscillators whose amplitudes can vary with time, whereas quantum mechanics introduces the concepts of energy levels and jumps between them. In such a way, if we take into consideration the rigorous theory of quantum mechanics instead of the simplified one, then an atom as a wavelike object will possess new properties – for example, it can occupy two energy levels at the same time. Here the classical picture, in which the frequency of harmonic oscillation corresponds to the energy level, is actually more appropriate. If an atom occupies two levels, then this simply means that two oscillations exist simultaneously. As a rule these two oscillations interfere and beat with time, but despite the fact that the fundamental oscillations are at enormously high frequencies – 10^{15} per second or thereabouts – beats are comparatively slow. An excellent model is provided by a ringing crystalline glass, which not only rings, but also sings – this is the beating of two adjacent frequencies.

However, such beats in atoms (or molecules) had not been observed before the early 1960s. It turns out that if atomic levels are suddenly excited, then the atom emits pulsating light – it sings [1.2, 3].

The discovery of beats did much to improve our understanding of atoms. The prevailing physical picture of atoms has been through some dramatic changes over the years. Early this century the resemblance of atoms to the solar system was established – electrons orbit about the nucleus in the same way as planets around the sun. Quantum mechanics seemed to banish motion from atoms. An atom turned out to be a nucleus, surrounded by stationary shells of an electronic cloud that can undergo rare but abrupt changes in form and dimensions. The discovery of beats brought back motion to atoms: waves propagate around the atomic shells and the atom pulsates and rotates like an oscillating drop. These oscillations cease only in exceptional cases, for example, when the frequencies of interfering levels coincide; this is called a level crossing. Thus beats and level crossing are observed in succession, the one being replaced by the other. One therefore sees that these are in fact different manifestations of the same phenomenon, namely interference of states[1].

Now a few words about the terminology "interference of states". In optics interference is a well-known phenomenon. Optical fields and all other electromagnetic fields obey the principle of superposition, where the amplitude A of the field at any point in space equals the sum of the field amplitudes A_i at this location due to the presence of different sources. If a phase relationship exists between the component fields, then these fields are considered to be coherent, and the power or intensity of the resultant oscillating field is not simply the sum of the intensities of the component fields:

$$W = A^2 = \left| \sum_i A_i \right|^2 = \sum_i |A_i|^2 + \sum_{i \neq k} A_i A_k^* \neq \sum_i |A_i|^2 \ .$$

The principle of superposition is also valid in quantum mechanics. If the wavefunction is represented in terms of its expansion over eigenstates of an operator, for example the Hamiltonian, then the amplitude of the resultant state equals the sum of amplitudes of the component states

$$\Psi(x, y, z, t) = \sum_n C_n \Psi_n \ .$$

The entity $\Psi(x, y, z, t)$ is the probability amplitude for finding an electron at a given point in space at a given time instant. The probability itself of this event is given by $|\Psi|^2$:

$$|\Psi|^2 = \sum_{n,k} C_n C_k^* \Psi_n \Psi_k^* = \sum_n |C_n|^2 |\Psi_n|^2 = \sum_n |C_n|^2 \ .$$

Hence, it is obvious that $|C_n|^2$ is the probability that the atom is in the state Ψ_n.

Unlike the electromagnetic fields, where the observable quantity – intensity – is given by

$$I = |A|^2 \ ,$$

an observable quantity in quantum mechanics is described by the matrix

---

[1] It is important to emphasize that the interference of states has been fully treated in quantum mechanics, and in particular that the concept of quantum jumps accompanying the emission of a photon has not been rejected. This is a discrete process, which lasts no longer than the oscillation period of the light. However, the time distribution of these jumps reflects the presence of motion in the atoms, which in turn is described by the interference of wavefunctions.

element of the operator $\hat{T}$. The expectation value of the observable quantity is

$$\langle \Psi | \hat{T} | \Psi \rangle .$$

Thus, in the equation for this quantity one can expect "interference" terms with coefficients $C_n C_k^*$

$$\langle \Psi | \hat{T} | \Psi \rangle = \sum_{n,k} C_n C_k^* \langle \Psi_k | \hat{T} | \Psi_n \rangle .$$

The products $C_n C_k^*$ are analogous to the interference terms $A_i A_k^*$ in the description of an electromagnetic field.

There is a complete analogy between the summation of fields and summation of wavefunctions: the presence of off-diagonal elements is a result of the superposition principle, and a necessary condition for their presence is the phase relation between the components. Otherwise the off-diagonal terms vanish upon averaging over all phase realizations. When summing electromagnetic fields, the off-diagonal elements are called interference or coherent terms. The off-diagonal elements of the density matrix are also called interference or coherent terms and their manifestation during the emission or absorption of light is an interference phenomenon referred to as the interference of quantum states. Interfering states in quantum systems of atoms or molecules, and sometimes in solid-state materials, are produced under the influence of a particular anisotropic perturbation; this may be polarized illumination, directed resonance radiation, or an electron beam. In fact one can use any excitation that is not isotropic.

The interference of states is manifested in spontaneous emission or absorption by quantum systems (as well as during their interaction with rf fields and during the time of collisional interactions with other particles). The interference of states changes the polarization characteristics of emission (absorption) and the related angular distribution of intensity.

The phenomena related to the interference of states can be classified according to their dependence on time as follows:

1. *Quantum beats* or transient phenomena [1.4, 5]. These can be observed when pulse excitation is used to populate coherently two (at least) states with a relatively narrow energy gap between them. The intensity of spontaneous emission decays exponentially with time at a rate given by the damping constant. In any particular direction it also reveals the oscillation. This is what we call a quantum beat. The phase of the intensity oscillation depends on the direction of the emission and, if it increases at a given moment in a particular direction, then there must be another direction in which it decreases. The oscillations disappear when one averages over all directions. The frequency of the intensity oscillations is determined by the energy gap between the eigenstates that interfere.

The other way to induce the beats is to suddenly remove the degeneracy of coherently populated states. In both ways the duration of the pulse must be much less than the period of oscillation. This condition technically restricts the frequency of beats.

2. *Quantum resonances* or stationary quantum oscillation [1.5]. These oscillations arise when either the intensity (or the direction) of excitation or the energy gap between the interfering states is periodically modulated. In the first case the intensity of spontaneous emission is modulated at the same rate as the excitation, and it is important to note that the amplitude of this modulation reaches a maximum when the frequency of excitation coincides with the energy gap between interfering states. In the second case one obtains parametric resonance with a large number of maxima. For each maximum, the ratio of modulation frequency to energy gap is an integer. The height of the maxima depends in a complex way on the amplitude of the energy gap modulation and on the ratio of frequencies.

3. *Time-independent phenomena*, i.e. level crossing and anticrossing [1.6, 7]. These phenomena occur when one uses time-independent excitation and interfering states that are energetically degenerate (at any rate partially). The spontaneous emission is polarized and its intensity varies with direction. When the degeneracy is somehow lifted using a controllable magnetic field, the polarization of radiation and its intensity in a given direction will change. This change versus magnetic field accounts for level-crossing and anticrossing signals. The width of the signals is related to the atomic constants: relaxation time, magnetic moments of the nucleus and electron shells and their interactions.

The interference phenomena can be also classified into groups on the basis of angular dependence and character of polarization of emitted light. These features are connected with so-called atomic polarization moments.

The first polarization moment is called orientation and is produced when circularly polarized light is used for excitation. Physically it corresponds to an externally induced macroscopic magnetic moment in an ensemble of quantum particles. If the vector of orientation does not coincides with the axis of quantization, then there exists a transverse component of orientation attributed to the interference of energy states whose magnetic numbers differ by one. This interference manifests itself in the absorption or emission of circularly polarized light in the direction across the axis of quantization.

The second polarization moment is called alignment. An alignment is produced when one uses linearly polarized light or an unpolarized light beam to excite the atoms and in the case of excitation by an anisotropic collision of any kind. Excitation by circularly polarized light also results in an alignment (provided that the angular moment J of the state obeys to the relation $J \geq 1$). Alignment is described by a tensor of rank two. If the axis of the tensor is perpendicular to the axis of quantization then there exists a transverse alignment which is attributed to the interference between states whose magnetic quantum numbers differ by two (while the longitudinal alignment is attributed to the population difference of these states).

The null polarization moment has no relation to the interference of states. It simply characterizes the total population of the state with a given total angular momentum J.

Polarization moments higher than the second are not produced and not manifested in linear atom–light interactions within the dipole approximation.

Time-independent transverse orientation and alignment disappear when the degeneracy of levels is lifted and in both cases the optical properties of atoms tend to become spherically symmetric (the ideal spherical symmetry corresponds to the absence of any components of any polarization moments with the rank $\kappa \geq 1$). When the degeneracy of levels is lifted the transverse components may exist as transient or stationary time-dependent values. These correspond to the above-mentioned quantum beats and quantum resonances.

Experiments dealing with the interference of states are performed mainly to determine the atomic constants and to measure external fields. The decay constant derived from an analysis of the interference signals characterizes the lifetime of the alignment or the orientation. For free atoms, this time corresponds to the radiation lifetime, but for interacting atoms this identity may no longer hold. An analysis of the interference signals obtained in the presence of foreign gases allows one to measure the cross-section of the depolarizing collisions.

Interference signals also have important application in the area of hyperfine structure measurement of atomic energy levels. The traditional technique for measuring hyperfine structure consists in an investigation of the spectral line profile with the help of a high resolution device, such as a Fabry–Perot interferometer. The limit for such methods is set by the Doppler linewidth. When using the interference of states for the same purpose, the Doppler width plays exactly the opposite role: when the Doppler width is small (compared to the natural linewidth) the signal due to the interference of states can be distorted, whereas when it is large these signals possess a particularly simple form and can easily be processed. The possibility of investigating hyperfine structure using the interference of states is related to the fact that, in the presence of hyperfine structures, the magnetic sublevels are degenerate not only in the absence of a magnetic field, but also in the presence of a non-zero magnetic field. Figure 1.1 shows a sketch of the energy level structure of rubidium $6P_{3/2}$ ($I = 5/2, J = 3/2$) in a magnetic field. The crossing of levels that can be populated coherently and thus give the alignment signals (level crossing signals) are encircled. At the crossing field one will observe particular features in the radiation intensity. Knowing the magnetic field needed to produce the level crossing it is possible to establish the separation between the hyperfine sublevels in a zero magnetic field, and from this one can easily find the hyperfine structure constants.

Additional level degeneracy can arise in an electric field. Such crossing gives information about the electric polarizability tensor.

The measurement of quantum beat frequencies enables one to find the Landé g factor and also to measure directly the hyperfine or even fine structure of atomic energy levels.

The alignment can be produced not only due to excitation of atoms by directed rays of light or particles but also may arise spontaneously as a result of light reabsorption within a restricted volume containing atomic vapour,

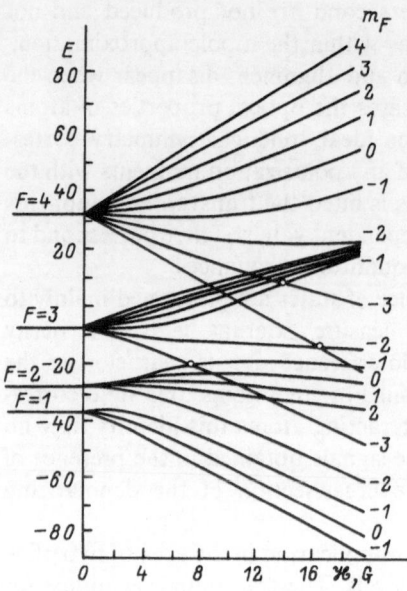

Fig. 1.1. The energy level structure of $^{85}\text{Rb}\,6\,^2\text{P}_{3/2}$ ($I = 5/2$, $J = 3/2$) in a magnetic field

provided its dimension is comparable with the free path length of a photon. Thus, even in an ordinary glowing gas discharge tube one can observe the interference of states. However, the additional intensity related to this phenomenon is small compared to the average intensity of the radiation. Alignment is also revealed in the Sun's corona and prominences because the necessary conditions are fulfilled: the light is anisotropic, and the pressure is sufficiently low that the alignment produced would not be instaneously destroyed by collision. The main reason why astrophysicists are interested in self-alignment in the Sun is because the degree and direction of polarization of the coronal radiation and prominences depend not only on the angular intensity distribution of the resonance light in them, but also on the magnitude and direction of the magnetic field. Measurement of the radiation characteristics thus enables one to draw conclusions about the magnetic field of the Sun. Under terrestrial conditions the determination of a magnetic field on the basis of the behaviour of the polarization moments is an unambiguous and in principle easy problem. The method (Soviet patent due to E.B. Aleksandrov, 1964) will now be described briefly. An orientation or alignment is produced in the ground state atoms inside a cell by using an optical pump. As a result, the lower state acquires an anisotropy of absorption: its absorption depends on the polarization, as well as the direction of the probing light beam with respect to the pump beam. This anisotropy reaches a maximum in zero magnetic field. An unknown magnetic field can be determined by compensating the magnetic field to be measured with the help of Helmholtz coils. The necessary compensating field is fixed using the absorption extremum and its value determined from the current in the compensating coils. The high sensitivity of this technique is related to the narrow linewidth of the lower level: When special precautions are taken, such a magnetometer can attain a sensitivity 10^{-10} G.

At the very outset we claimed that the phenomena of interference between atomic states are simple in their essence. In fact they are so simple that their basic qualitative manifestations can be described in terms of the classical radiating dipole. Therefore, before providing the quantum mechanical description (Chap. 3) we give a demonstration of the interference phenomena on the basis of a classical dipole (Chap. 2). Chapter 4 deals with the experimental observation of interference signals and the last chapter, Chap. 5, describes methods of calculating the signals in order to extract various kinds of information from them.

2. Classical Description of Interference Phenomena in Radiation

2.1 The Classical Oscillator Model of Atomic Emission

The Lorentz model, which describes an atom as a damped harmonic oscillator, can be used successfully to describe the absorption, emission and scattering of light by atomic systems. Hanle used this approach to describe his first experiments on the depolarization of resonance fluorescence by a magnetic field [2.1]. In addition to being instructive and simple, it will be shown here that this model can give a good qualitative picture of many types of interference phenomena. We therefore begin with a consideration of the classical model of an optical emitter. (One has to keep in mind, however, that the atomic structure in this model is far from realistic.)

Such an emitter is considered to be composed of an electron bound by a quasi-elastic force to the residual atom. At equilibrium, the centres of positive and negative charges coincide, and as a result the system will have a zero dipole moment. An electron that is shifted from the equilibrium position under the action of a quasi-elastic force will undergo free harmonic oscillation governed by the equation

$$\ddot{r}(t) + \omega_0^2 \, r(t) = 0$$

[here $r(t)$ is the displacement vector of the electron from its equilibrium position at time t] with a solution

$$r(t) = \text{Re}\{re^{-i\omega_0(t-t_0)}\} \ . \tag{2.1}$$

For brevity we shall henceforth drop the symbol Re, and wherever the initial time reference is immaterial, we put $t_0 = 0$. Since the nucleus is far heavier than the electron, it is possible to neglect the residual atomic motion and consider the ion core to be stationary. When $r \neq 0$ the system will clearly have a dipole moment

$$d = -er \ . \tag{2.2}$$

In this expression we have chosen the direction r from the nucleus to the electron. Together, Eqs. (2.1) and (2.2) determine the variation of the dipole moment with time.

The time-dependent dipole emits an electromagnetic wave, which is characterized at point P at a distance R from the dipole ($R \gg r$) by the electric field strength (in CGS units) [2.2]

$$E_\mathrm{P} = \frac{1}{c^2 R}(\ddot{d}(t) \times n \times n) = \frac{e}{c^2 R}[\ddot{r}(t) - n(\ddot{r}(t) \cdot n)]$$ (2.3)

and the magnetic field

$$\mathscr{H}_\mathrm{P} = \frac{1}{c^2 R}(\ddot{d}(t) \times n) = \frac{-e}{c^2 R}(\ddot{r}(t) \times n)$$ (2.3a)

here n is the unit vector in the direction from the dipole to the point P.

A knowledge of E is quite sufficient to work out the angular intensity distribution of any harmonic oscillator. The energy flow is proportional to the square of the strength of the electric field and is given by the Poynting vector [2.3]:

$$S = \frac{c}{4\pi}(E \times \mathscr{H}) = \frac{c}{4\pi}E^2 n \ .$$ (2.4)

From (2.3), one sees that in order to calculate E_P, it is sufficient to find the second derivative of the vector $r(t)$ with respect to time $- \ddot{r}(t)$. In the general case (for periodic variation of the dipole moment), both the distance $|r|$ between charges, and the direction of the vector r change with time. If the oscillator is linear, then the direction r does not change (except in sign). For a circular oscillator, i.e. a rotator (an electron orbits around the nucleus), the magnitude $|r|$ will remain constant and only the direction of the vector r will vary.

One can take into account all possible oscillations of the dipole, whose direction is given by the angles θ and ϕ (Fig. 2.1), and establish individually the behaviour of the oscillations with respect to each of the cartesian axes. We express r in form

$$r = r_x e_x + r_y e_y + r_z e_z \ ,$$ (2.5)

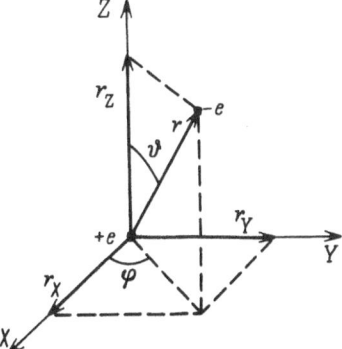

Fig. 2.1. The direction of the dipole in space

where e_x, e_y and e_z are the cartesian unit vectors. The cartesian components r_x, r_y and r_z are defined through the angles θ and ϕ in the following way:

$$r_x = |r| \sin \theta \cos \phi, \qquad r_y = |r| \sin \theta \sin \phi, \qquad r_z = |r| \cos \theta . \tag{2.6}$$

In order to describe the emission from an atom, it is convenient to introduce, besides the cartesian unit vectors, the system of circular basis vectors, which is related to the cartesian coordinate system as follows:

$$e_+ = -(e_x + ie_y)/\sqrt{2} ,$$

$$e_- = (e_x - ie_y)/\sqrt{2} , \tag{2.7}$$

$$e_0 = e_z .$$

As will be seen later (Chap. 3) these same vectors also arise in the formalism of quantum mechanics as the state eigenvectors of atoms.

Expanded in terms of the unit vectors of (2.7), the vector r takes the form

$$r = r_+ e_+ + r_- e_- + r_0 e_0 , \tag{2.8}$$

where

$$r_+ = -(r_x - ir_y)/\sqrt{2} = -|r| \sin \theta \, e^{-i\phi}/\sqrt{2} ,$$

$$r_- = (r_x + ir_y)/\sqrt{2} = |r| \sin \theta \, e^{i\phi}/\sqrt{2} , \tag{2.9}$$

$$r_0 = r_z = |r| \cos \theta .$$

[One can easily derive (2.9) by comparing the expressions (2.5) and (2.8) and making use of (2.6) and (2.7).]

It is not difficult to see that a variation of the vector $r(t)$ containing only a component along e_+ corresponds, for a harmonic oscillation with frequency ω_0, to its rotation in the xy-plane. Actually, from (2.1, 8, 9) we find

$$r(t) = -|r_+|e_+ \exp(-i\omega_0 t - i\phi_+)$$

$$= \frac{|r_+|}{\sqrt{2}} [e_x \cos(\omega_0 t + \phi_+) + e_y \sin(\omega_0 t + \phi_+)] . \tag{2.10}$$

Here the quantity ϕ_+ determines the initial phase of the oscillation

$$\phi_+ = \text{arctg} \, \frac{r_y(t=0)}{r_x(t=0)} .$$

The length of the vector $r(t)$ is constant in time and equal to $|r_+|/\sqrt{2}$; its direction changes with time. It is seen from (2.10) that a rotation in the x–y plane is the result of adding two perpendicular linear oscillations that are shifted in phase by $\pi/2$. This corresponds to a rotator moving in a right handed coordinate system, i.e. rotating in the clockwise direction when observed along the positive z-axis (and counterclockwise if observed in the opposite direction). Similarly, the oscillation

$$r(t) = |r_-| e_- \exp(-i\omega_0 t + i\phi_-)$$

$$= \frac{|r_-|}{\sqrt{2}} [e_x \cos(\omega_0 t - \phi_-) - e_y \sin(\omega_0 t - \phi_-)]$$

is a rotator with a rotation in the anticlockwise direction, and the oscillation $r(t) = r_0 e_0 \exp(-i\omega_0 t - i\phi_0)$ is a linear oscillation along the z-axis. Since any rotator can be represented as a sum of oscillations of two linear and mutually perpendicular dipoles, any linear oscillator can be represented as a sum of two rotators. Let the linear dipole oscillate in the xy-plane. From (2.5, 6) and assuming that the angle θ is equal to $\pi/2$ we find

$$r(t) = (r_x e_x + r_y e_y) e^{-i\omega_0 t} .$$

From (2.8, 9) and with the same condition, this vector can also be written

$$r(t) = \frac{|r|}{\sqrt{2}} (-e^{-i\phi} e_+ + e^{i\phi} e_-) e^{-i\omega_0 t} .$$

The dipole does not change its position in space. In the last expression it is expressed as the sum of two rotators – a right handed and a left handed one. Both rotators have the same amplitude and, of course, the same frequency. The initial phase ϕ is given by the direction of the linear dipole in the xy-plane.

We turn now to the determination of the radiation (intensity) pattern and the polarization of the radiation. We express the vector n in terms of the angles θ_n and ϕ_n

$$n = \sin \theta_n \cos \phi_n e_x + \sin \theta_n \sin \phi_n e_y + \cos \theta_n e_z . \tag{2.11}$$

Substituting (2.1) and (2.11) into (2.13) and considering (2.5, 6) or (2.8, 9) we can obtain the field strength due to an arbitrarily oriented dipole or system of dipoles.

We first consider the radiation pattern of a rotator that rotates in a clockwise direction. The time-dependent vector $r(t)$ of such a rotator is given by the expression (2.10). Substituting (2.10, 11) into (2.3) we find

$$E_P = \frac{e\omega_0^2}{c^2 R} [(r \cdot n)n - r] e^{-i\omega_0 t}$$

$$= \frac{e\omega_0^2 |r_+|}{c^2 R \sqrt{2}} [(-\cos^2 \theta_n \cos \phi_n + i \sin \phi_n) e_x - (\cos^2 \theta_n \sin \phi_n + i \cos \phi_n) e_y$$

$$+ \sin \theta_n \cos \theta_n e_z] \exp(-i\omega_0 t + i\phi_n - i\phi_+) . \tag{2.12}$$

Further, squaring E_P and substituting it into (2.4), we obtain an expression for the Poynting vector, i.e. for the energy flow per unit time through unit area in the direction of n:

$$S = \frac{e^2 \omega_0^4 |r_+|^2}{8\pi c^3 R^2} [\overline{\sin^2(\omega_0 t - \phi_n + \phi_+)}$$

$$+ \cos^2\theta \overline{\cos^2(\omega_0 t - \phi_n + \phi_+)}]n \ . \tag{2.13}$$

Since the average value of $\cos^2(\omega_0 t + \phi)$ and of $\sin^2(\omega_0 t + \phi)$ in a given period is equal to $1/2$, one can write S as

$$S = \frac{e^2 \omega_0^4 |r_+|^2}{16\pi c^3 R^2} (\cos^2\theta + 1)n \ . \tag{2.13a}$$

The dependence on the angle ϕ_n vanished, which is quite natural, since the problem is symmetric with respect to the axis. The radiation pattern in the plane containing the axis of rotation is presented on Fig. 2.2.

For a linear oscillator $r(t) = r_z e_z \exp(-i\omega_0 t + i\phi_0)$ the pattern can be found quite easily. The vector E_P is given by

$$E_P = \frac{e\omega_0^2 r_z}{c^2 R} \sin\theta_n [\cos\theta_n (\cos\phi_n e_x + \sin\phi_n e_y)$$

$$- \sin\theta_n e_z] \exp(-i\omega_0 t + i\phi_0) \tag{2.14}$$

and the Poynting vector (taking the average of E_P^2 in a period $T = 2\pi/\omega_0$) is described by

$$S = \frac{e^2 \omega_0^4 r_z^2}{8\pi c^3 R^2} \sin^2\theta_n \, n \ . \tag{2.15}$$

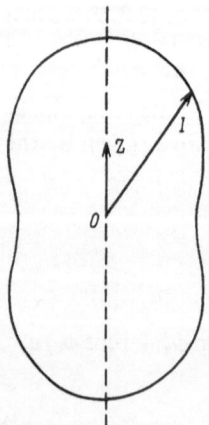

Fig. 2.2. The radiation pattern in the yz plane

The radiation pattern produced by linear dipole is shown in Fig. 2.3. The formulae (2.12) and (2.14) determine the electric field strength of the radiation. Its direction is a very important characteristic of the radiation. The unit vector in the direction of E is called the polarization vector. (In classical literature one finds the polarization of light defined as the direction of the magnetic field, i.e., the direction perpendicular to E.) The direction of the vector E of the linear oscillator, see (2.3), is given by the product

$$e_z \times n \times n .$$

The product $e_z \times n$ is a vector perpendicular to the plane in which the vectors e_z and n lie. After subsequent vector multiplication by n, the resulting vector will be rotated by 90°, but will remain perpendicular to the vector n. Such a vector lies in the OP plane (i.e. in the e_z, n plane).

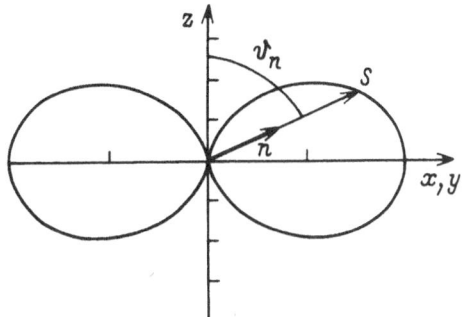

Fig. 2.3. The radiation pattern produced by a linear dipole

The calculation of the polarization of the radiation from a rotator is a rather complex task. It is given by the formula (2.12). In particular, when the point P falls in the xy-plane

$$E_\mathrm{P} \propto (e_y \cos\phi_n - e_x \sin\phi_n)\exp(-\mathrm{i}\omega_0 t + \mathrm{i}\phi_n) . \qquad (2.16)$$

The first factor defines the polarization of radiation. It is linear, lies in the xy-plane and is perpendicular to n. The second factor gives the phase of radiation. As can be inferred from the formula, it depends on the direction of observation, which is determined by the angle ϕ_n. The phase shift is simply equal to this angle.

In the direction $0z$, the rotator emits circularly polarized light

$$E \propto -\frac{e_x + \mathrm{i}e_y}{\sqrt{2}}\exp(-\mathrm{i}\omega_0 t - \mathrm{i}\phi_+) = e_+ \exp(-\mathrm{i}\omega_0 t - \mathrm{i}\phi_+) . \qquad (2.17)$$

In all other directions an elliptically polarized wave will be emitted. The polarization approaches circular for decreasing θ and tends to linear when θ is near to $\pi/2$. The plane of oscillation, of course, is perpendicular to n, and the major axis of the ellipse lies in the xy plane.

Our model has been constructed for a freely radiation dipole. However, a real emitting dipole loses energy, and consequently its amplitude decreases with time

$$E = E_0 \exp(-i\omega_0 t - \Gamma t/2) \ . \tag{2.18}$$

The decay constant Γ can be determined either from the law of conservation of energy or by calculating the retarding influence of the radiated field on the dipole. In both cases we obtain [2.3]

$$\Gamma = \frac{2e^2\omega_0^2}{3mc^3} \ , \tag{2.19}$$

where e is the charge, and m the mass of an electron.

Since the quantity Γ is of the order of $10^9 \, \mathrm{s}^{-1}$ and $\omega_0 \sim 10^{15} \, \mathrm{s}^{-1}$ for optical radiation, we have $\Gamma \ll \omega$ and thus the form of (2.18) is legitimate: The solution of the equation of a damped oscillation[1]

$$\ddot{r} + \Gamma\dot{r} = -\omega_0^2 r$$

can be represented by the product

$$r(t) = r\mathrm{e}^{-i\omega_0 t}\mathrm{e}^{-\Gamma t/2} \ .$$

The quantity Γ is called the damping constant (or reciprocal lifetime) and it determines the time in which the radiation power of the dipole has decreased by a factor of $1/e$.

The experimentally measured value of Γ, as well as the value calculated using the methods of quantum electrodynamics [2.3], show that it depends not only on frequency, but also on the quantum state of the atoms. The decay constants of atomic states often differ from those computed classically, sometimes by a significant amount. However, the relationship $\Gamma \ll \omega_0$ always holds true. We note that in the presence of damping the emitted waves will not be strictly monochromatic. The damped oscillation of the form

$$E = E_0 \exp(-i\omega_0 t - \Gamma t/2)$$

---

[1] Strictly speaking, the term responsible for the friction should be proportional to the third-order derivative. Nevertheless its approximation by the first-order derivative is quite satisfactory for our present purposes.

can be expanded as a Fourier integral

$$E = \int_{-\infty}^{\infty} E_\omega \, d\omega \, e^{-i\omega t} \, , \quad \text{where}$$

$$E_\omega = \frac{E_0}{2\pi} \frac{1}{(\Gamma/2) - i(\omega_0 - \omega)} \tag{2.20}$$

is the partial amplitude of the oscillation of frequency ω. However, due to the fact that $\Gamma \ll \omega$, the observables ω will differ slightly from ω_0.

2.2 A Classical Oscillator in a Magnetic Field

In the previous section we discussed the radiation emitted by a free oscillator on which no external fields are acting. However, it is known that a magnetic field (as well as an electric field) can change the emission characteristics of atoms. This phenomenon was first recognized by *Wood* and *Ellett* [2.4]. It was studied in detail by Hanle for the particular example of resonance fluorescence in mercury vapor [2.1]. Before describing the effects studied by Hanle (and named after him) let us look at the emission of a single dipolar harmonic oscillator, placed in a magnetic field and excited at the time instant $t = t_0 = 0$. The equation of motion of the electron can be written as

$$\ddot{r} + \Gamma\dot{r} + \omega_0^2 r = F/m \, , \tag{2.21}$$

where F is the Lorentz force acting on a moving charge e in a magnetic field:[2] $F = (e/c)(\dot{r} \times H)$. Since the Lorentz force depends on the speed and direction of motion of the charge, its explicit form will depend on the type of dipole oscillator.

We choose the coordinate system in such a way that the magnetic field is directed along the positive z-axis, and write the equation of motion in terms of components of the oscillator along the x-, y- and z-axes:

$$\ddot{r}_x + \Gamma\dot{r}_x + \omega_0^2 r_x = -\frac{e\mathscr{H}}{mc}\dot{r}_y \, ,$$

$$\ddot{r}_y + \Gamma\dot{r}_y + \omega_0^2 r_y = \frac{e\mathscr{H}}{mc}\dot{r}_x \, , \tag{2.22}$$

$$\ddot{r}_z + \Gamma\dot{r}_z + \omega_0^2 r_z = 0 \, .$$

---

[2] Sometimes it is useful to know that the positive direction of the magnetic field is taken to be the direction from north to south.

It is immediately clear that the components of the oscillations along the z-axis are not influenced by the magnetic field; if the oscillations of the dipole were originally directed along z, they remain so upon switching on the magnetic field

$$r_z(t) = r_z(0) \exp(-\mathrm{i}\omega_0 t - \Gamma t/2) . \tag{2.23}$$

In contrast, oscillations along the x- and y-axes are coupled to one another by the magnetic field. In order to solve the two systems of equations (2.22), we multiply the second equation by i and add it to the first

$$(\ddot{r}_x + \mathrm{i}\ddot{r}_y) + \Gamma(\dot{r}_x + \mathrm{i}\dot{r}_y) + \omega_0^2(r_x + \mathrm{i}r_y) = \mathrm{i}\frac{e\mathscr{H}}{mc}(\dot{r}_x + \mathrm{i}\dot{r}_y) . \tag{2.24}$$

This equation can be reexpressed as

$$\ddot{r}_- + \Gamma\dot{r}_- + \omega_0^2 r_- = \mathrm{i}\frac{e\mathscr{H}}{mc}\dot{r}_- . \tag{2.25}$$

We have thus obtained an equation for a rotator that rotates in an anticlockwise direction. Its solution is given by

$$r_-(t) = r_-(0) \exp(-\mathrm{i}(\omega_0 - \Omega)t - \Gamma t/2) , \tag{2.26}$$

where $\Omega = e\mathscr{H}/2mc$ is the Larmour frequency. This solution shows that the magnetic field can reduce the frequency of the rotator, but leaves the polarization of the radiation unchanged.

The second linear combination of the equations, obtained by taking the difference, leads to the equation for $r_+ = -(r_x - \mathrm{i}r_y)/\sqrt{2}$:

$$\ddot{r}_+ + \Gamma\dot{r}_+ + \omega_0^2 r_+ = -\mathrm{i}\frac{e\mathscr{H}}{mc}\dot{r}_+ .$$

Its solution is given by

$$r_+(t) = r_+(0) \exp(-\mathrm{i}(\omega_0 + \Omega)t - \Gamma t/2) . \tag{2.27}$$

This rotator increases its frequency in a magnetic field.

It is thus convenient in this problem to decompose the linear oscillator into three components along the unit vectors e_+, e_- and e_0. In a magnetic field, the projections of the dipole's oscillations on each of these directions will have its own frequency. The component along the z-axis will oscillate at a constant frequency ω_0, and the components along the circular unit vectors will have frequencies

$$\omega_0 + \Omega \quad \text{and} \quad \omega_0 - \Omega .$$

The resultant change of the radius vector of the electron with time is given by the expression

$$r(t) = r_0(0)e_0 \exp(-i\omega_0 t - \Gamma t/2)$$
$$+ r_+(0)e_+ \exp(-i(\omega_0 + \Omega)t - \Gamma t/2)$$
$$+ r_-(0)e_- \exp(-i(\omega_0 - \Omega)t - \Gamma t/2) \ . \tag{2.28}$$

In order to make transparent the behaviour of the vector r with time, we return to its expansion in Cartesian coordinates, using the relations (2.5, 6) (Fig. 2.1):

$$r(t) = |r|\{\cos\theta\, e_z + \sin\theta\, [e_x \cos(\Omega t + \phi)$$
$$+ e_y \sin(\Omega t + \phi)]\} \exp(-i\omega_0 t - \Gamma t/2) \ . \tag{2.29}$$

The time dependence is contained in three factors. The factor $\exp(-i\omega_0 t)$ describes the fast oscillations of the linear dipole at its optical frequency. The term in the square brackets describes slow oscillations with the Larmour frequency $\Omega = eH/2mc$, the frequency at which the projection of the dipole in the xy-plane rotates about the z-axis. As a whole the vector r precesses conically about the z-axis with a full angular width

$$2\theta = 2\arctan\frac{\sqrt{r_x^2 + r_y^2}}{r_z}$$

and with a frequency Ω.

2.3 Emission from an Oscillator in a Magnetic Field

The magnitude and direction of the emitted electric field can be found simply by substituting the expression (2.29) into (2.3). Since $\Omega \ll \omega_0$ and $\Gamma \ll \omega_0$, the time derivatives of the expression (2.29) in the square brackets and of the factor $\exp(-\Gamma t/2)$ can be neglected. Then

$$E_P = \frac{e\omega_0^2}{c^2 R}\, r(t) \times n \times n \tag{2.30}$$

where $r(t)$ is given by the expression (2.28).

From (2.30) we can find the light intensity in any direction. The quantity $|E|^2$ will be proportional to the flux of the electromagnetic field, which corresponds to the light intensity measured in a given direction (provided there are no components of the optical transmission channel that possess polarizing or absorbing properties. It is useful to write $|E|^2$ in the following way:

$$|E|^2 = \frac{\omega_0^4 e^2}{c^4 R^2}\, [(r(t)\cdot n)n - r(t)]^2 = \frac{\omega_0^4 e^2}{c^4 R^2}\, [-(r(t)\cdot n)^2 + |r(t)|^2] \ . \tag{2.31}$$

For convenience let us choose the direction of the x-axis such that the dipole $d = -er(0)$ lies in the xz-plane ($\phi = 0$). The direction of the dipole is determined by the angle θ which it makes with the z-axis, whereas the direction of observation, as before, is given by the angles θ_n and ϕ_n (2.10). We find the expression for intensity $I = |S|$ [see (2.4)] by making use of the equation (2.29) for the vector r. After averaging I over optical oscillations we have

$$(r \cdot n)^2 = \tfrac{1}{2}|r|^2 [\cos\theta\cos\theta_n + \sin\theta\sin\theta_n\cos(\Omega t - \phi_n)]^2 e^{-\Gamma t} .$$

The power of the radiation (the radiative energy transmitted through unit area per unit time) is given by

$$I = \frac{\omega_0^4 e^2}{8\pi c^3 R^2} e^{-\Gamma t}|r|^2 [1 - \cos^2\theta\cos^2\theta_n - \tfrac{1}{2}\sin^2\theta\sin^2\theta_n$$
$$- \tfrac{1}{2}\sin 2\theta\sin 2\theta_n\cos(\Omega t - \phi_n) - \tfrac{1}{2}\sin^2\theta\sin^2\theta_n\cos(2\Omega t - 2\phi_n)] . \quad (2.32)$$

It is a remarkable fact that the time dependence of the intensity contains two frequencies: the Larmour frequency Ω, and twice the Larmour frequency, 2Ω. The presence of the frequencies Ω and 2Ω in the intensity variation can be described as an interference effect: it is precisely the result of interference between radiation from a dipole component along the magnetic field with a frequency ω_0 and the emission of a rotator with frequencies $\omega_0 + \Omega$ and $\omega_0 - \Omega$. Intensity variation at the Larmour frequency exists when two conditions are fulfilled: first, the direction of observation must be at an angle $\theta_n \neq 0$ and $\theta_n \neq \pi/2$ to the magnetic field, and secondly, the dipole itself should form any angle θ, whose value is neither zero nor $\pi/2$, with the magnetic field. The intensity variation at twice the Larmour frequency attains a maximum when the direction of observation is perpendicular to the magnetic field, i.e. under conditions where there is no variation at the Larmour frequency itself. In this case the angle between the dipole and the magnetic field should not be equal to zero, and if it is equal to $\pi/2$, then the amplitude of the intensity oscillation will be maximum. When the dipole is parallel to the magnetic field and there is no variation in intensity, the characteristics of the radiation do not depend on the magnetic field, a point which we already remarked on earlier – the magnetic field has no influence on the dipole, see (2.23).

We can also discuss finer characteristics of the dipole emission, specifically, its polarization, i.e. the direction of the electric field vector E as a function of magnetic field H and direction of observation n. The relationship between the direction of polarization of the radiation and the direction of the dipole is discussed above; see (2.16, 17). The vector E is always perpendicular to the direction of observation n and lies in the same plane as the dipole d, i.e. in the r, n plane. In a magnetic field the direction r changes with time, and the vector E rotates together with r. What is more interesting is to follow the variation of the amplitude of the light transmitted through a polarizer whose axis l is determined by the angles θ' and ϕ' as follows:

$$l = (e_x \sin\theta'\sin\phi' + e_y\sin\theta'\cos\phi' + e_z\cos\theta') , \quad (2.33)$$

where l is a unit vector. We call l the polarization vector. From (2.33) and (2.30) we find the projection of E on the vector l:

$$l \cdot E = \frac{e\omega_0^2}{c^2 R} [(l \cdot n)(r \cdot n) - l \cdot r] \ . \tag{2.34}$$

Since l is always perpendicular to the direction of observation due to the orthogonality of optical waves, i.e. $l \cdot n = 0$, we have

$$l \cdot E = -\frac{e\omega_0^2}{c^2 R} l \cdot r \ . \tag{2.35}$$

Substituting the expression for l (2.33) and for r (2.29) into (2.35), we get

$$l \cdot E = -\frac{e\omega_0^2}{c^2 R} |r| [\cos\theta \cos\theta'$$

$$+ \sin\theta \sin\theta' \cos(\Omega t + \phi - \phi')] \exp(-i\omega_0 t - \Gamma t/2) \ . \tag{2.36}$$

By squaring $l \cdot E$, we find the intensity (power) of the radiation transmitted through a polarizer whose axis is parallel to the vector l. The direction n no longer enters into the formula (2.36). Consequently, in every direction, perpendicular to the polarization vector l the same power will pass through a polarizer whose axis is parallel to l. This condition is very important because it facilitates the choice of experimental schemes.

Let us examine the following special case. Let r $(t = 0)$ be directed along the positive x-axis. According to (2.36), if the polarization axis is parallel to the x-axis and the observation is carried out along any direction in the yz-plane, including the z-axis, the same result should be obtained. We note that the behaviour of the total intensity (i.e. without decomposition using polarizers) does depend significantly on the angle between n and the z-axis, i.e. on the angle θ_n. When making an observation along the y-axis the observer will see approximately the same time variation of the intensity both with and without a polarizer. In contrast, along the z-axis with a polarizer one will observe the same as along the y-axis, and without a polarizer, independent of the applied magnetic field, one will observe no variation in intensity. This can be clearly seen from formula (2.36). Actually, for the special case considered $\theta' = \theta = \pi/2$; $\phi' = 0$; thus,

$$l \cdot E = -\frac{e\omega_0^2}{c^2 R} |r| \cos(\Omega t + \phi) \exp(-i\omega_0 t - \Gamma t/2) \tag{2.37}$$

and the radiated power is given by

$$I = \frac{e^2 \omega_0^k}{16\pi c^3 R^2} |r|^2 [1 + \cos(2\Omega t + 2\phi)] e^{-\Gamma t} \ . \tag{2.38}$$

In the general case, when r and l are arbitrarily oriented in space, the radiated power is given by

$$
I = \frac{e^2 \omega_0^4 |r|^2}{16\pi c^3 R^2} \left[2\cos^2\theta \cos^2\theta' + \sin^2\theta \sin^2\theta' \right.
$$

$$
+ \sin^2\theta \sin^2\theta' \cos 2(\Omega t + \phi - \phi')
$$

$$
\left. + \sin 2\theta \sin 2\theta' \cos(\Omega t + \phi - \phi') \right] e^{-\Gamma t} . \tag{2.39}
$$

The relationships presented here are sufficient to interpret the simplest, but at the same time very important interference phenomena: intensity beats and the Hanle effect.

2.4   Emission from an Ensemble of Oscillators

We described above the radiation from a single classical emitter. In reality, we observe in an experiment the emission from a large number of atoms or molecules, and it is therefore necessary to discuss the radiation from an ensemble of emitters. At this stage of our discussion we will model the ensemble of atoms by a large number of classical dipoles, distributed over some finite volume according to the law of probability. Their relative states will depend on the problem to be discussed.

The simplest situation is that in which all atoms, i.e. all classical dipoles, behave in the same manner. This problem reflects the behaviour of a radiating ensemble of dipoles, all excited identically and simultaneously. As is well known, the radiation from a collection of field emitters at any point will be the sum of the fields emitted by each individual emitter:

$$
E(P) = \sum_i E_i(P) .
$$

To represent the case of free particles, we will assume that our emitters are distributed randomly in space. Consequently, at every point P light from each emitter arrives with a random phase. When we calculate the intensity by squaring $\sum E_i(P)$, all cross-terms of the form $\sum_{i,j} E_i E_j^*$, related to interference between the emission from different atoms, thus average to zero, and we have

$$
|E(P)|^2 = \sum_i |E_i(P)|^2 . \tag{2.40}
$$

Since all the interference terms vanish and only the sum of the powers of individual dipoles remains, the intensity dependence on any of a number of parameters can be given as a sum over individual dipoles

$$
|E|^2 = \sum |E_i|^2 \quad \text{or} \quad I = \sum_i I_i .
$$

In many cases this simplification can facilitate the interpretation of observations and help to predict the effects of changing the experiment conditions.

We pause here to consider a number of assumptions that were implicitly made, but not qualified, in the above discussion. First of all the term "excited identically" must be understood to mean that all charges move in the same direction in space. If the dipoles are linear, then all oscillations are parallel, and in the case of rotators, the axes are parallel. Furthermore, it is possible to describe the emission of all dipoles by a single formula with one and the same parameters, provided these parameters are really identical. The angles determining the direction of oscillation of the dipoles have already been constrained to be identical, but the parameters describing the conditions of the observations (angles θ_n and ϕ_n or θ' and ϕ') must also be the same. This automatically holds true if the linear dimension of the volume occupied by the emitters is small in comparison to the distance, R, to the point of observation. Finally, the last stipulation refers to the density of the emitters. It is assumed to be so small that the mutual influence of the particles, as well as any absorption (or amplification) of the radiation emitted in such a medium, can be neglected.

The requirement of simultaneous excitation of the dipoles also requires some further comments. It is difficult to imagine perfect simultaneity of excitation: Even if the excitation is performed by a light pulse, this has a finite speed of propagation and excites atoms at different locations at different times. Nevertheless, it is perfectly acceptable to retain the formula under the condition that the time difference between excitations, Δt, is less than the damping time:

$$\Delta t < 1/\Gamma \ .$$

Having at one's disposal the radiation pattern, it is not difficult to imagine how the intensity reaching the observer will behave in different cases. To be instructive, we restrict the description to the case of linear dipoles. The observer is assumed to be located far from the ensemble, so that we can consider that all atoms are observed under the same angle in a given coordinate system.

2.5 Beats in Intensity[3]

With the above assumptions, the emission from an ensemble of dipoles will be identical in form to the emission of a single dipole. The radiation intensity of an ensemble of, for example, linear dipoles, which is pulse-excited at time $t = 0$, is given directly by the formula (2.38) or (2.39), multiplied by the number of dipoles. The intensity oscillates in time with a frequency Ω whilst gradually decaying [2.7]. The physical meaning of these oscillations is clear from Fig. 2.4:

---

[3] Experimentally, fluorescence beats were first observed in 1964 by *Alexandrov* [2.5] and independently by *Dodd* [2.6].

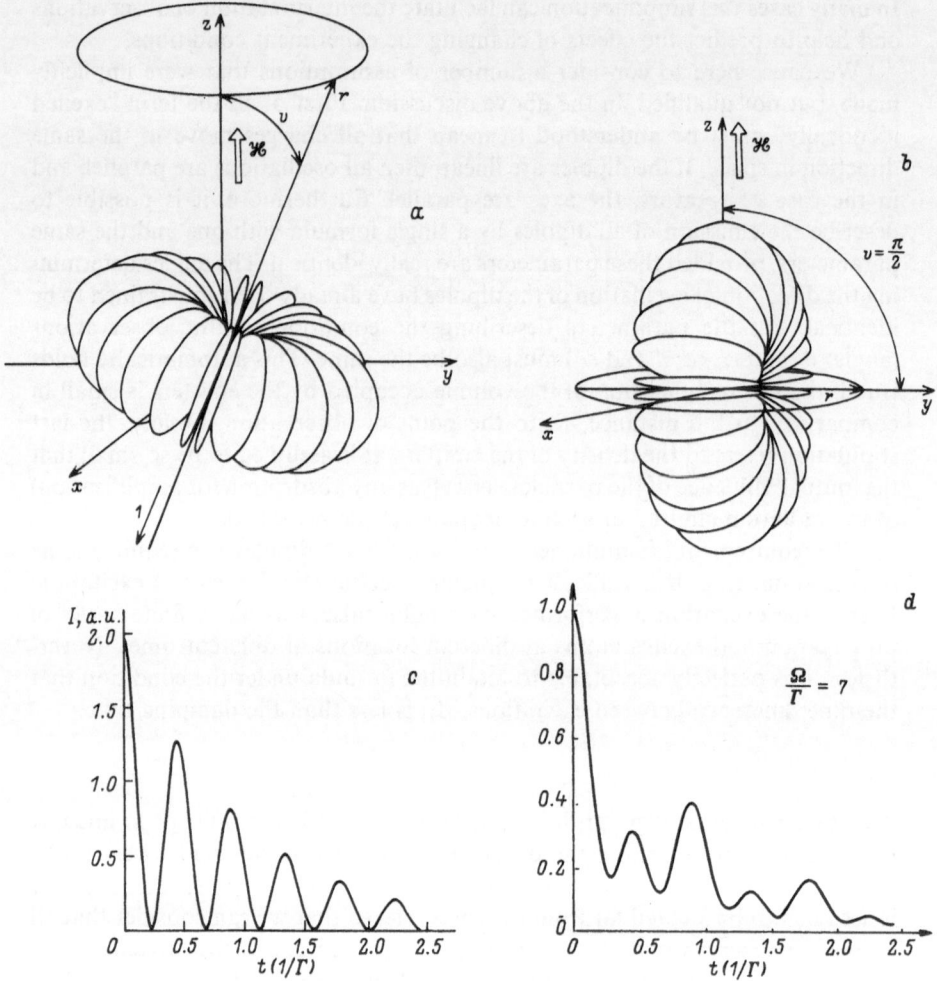

Fig. 2.4. The precession of the dipole in a magnetic field **(a)** $0 < \theta < \pi/2$, **(b)** $\theta = \pi/2$, **(c)** the intensity modulation with the frequency 2Ω **(d)** with 2Ω and Ω simultaneously

When a magnetic field is applied at an angle to the dipole axis, this axis is induced to rotate and the intensity reaching an observer will vary in time, appearing under the envelope of the total damped intensity as a periodic oscillation with frequencies Ω and 2Ω (with periods T and $T/2$). Figure 2.4 illustrates this situation. The arrow ↑ on Fig. 2.4a indicates the axis of the dipole at time t. The double arrow H denotes the direction of the magnetic field. The thin line shows the precession of the dipole about the magnetic field H. Also shown is the radiation pattern for the time instant t; it precess together with the dipole. An observer in direction 1 will register periodic radiation collapse with a period $T = 2\pi/\Omega$, thus confirming the presence of the frequency Ω in the intensity modulation. The modulation with frequency 2Ω is also present. The intensity is modulated with both frequencies Ω and 2Ω in every direction except $\theta_n = \pi/2$ and 0 (Fig. 2.4d). Figure 2.4b represents the other special case where the dipole axis is perpendicular to the magnetic field. An observer at an arbitrary position will now record variations of intensity with a period $T = 2\pi/2\Omega$. The frequency Ω now disappears in the modulation of the intensity. Figure 2.4c shows the intensity modulation at a frequency 2Ω when $\Omega/\Gamma \approx 7$.

2.6 The Hanle Effect

Let all remain the same as in the previous problem, except the simultaneity of the excitation. We assume that the dipoles are excited at random times and we observe their radiation continuously. This is of course a stationary problem and there can be no time variation. There will nonetheless remain a dependence on the magnetic field and on the angle of observation. As before

$$E(P) = \sum_i E_i(P)$$

and, as in the previous problem

$$|E(P)|^2 = \sum_i |E_i(P)|^2 \ .$$

Let us consider the emission from an ensemble of atoms in a magnetic field. The emission from each individual dipole is given by expression (2.39), in which we replace the time t by $t - t_i$, where t is the current time instant and t_i is the time at which the ith dipole starts to emit. The intensity emitted by each dipole is given by

$$I_i = \frac{e^2 \omega_0^4}{16\pi c^3 R^2} |r|^2 \{2\cos^2\theta \cos^2\theta'$$

$$+ \sin^2\theta \sin^2\theta' [1 + \cos 2(\Omega(t - t_i) + \phi - \phi')]$$

$$+ \sin 2\theta \sin 2\theta' \cos(\Omega(t - t_i) + \phi - \phi')\} e^{-\Gamma(t-t_i)} \ , \tag{2.41}$$

and the emission of the ensemble is

$$I = \sum_i I_i \ .$$

Let us replace the formal summation over i by an integral in the first item in the expression (2.41):

$$\sum_i e^{-\Gamma(t-t_i)} = \int e^{-\Gamma(t-t_i)} \, dN \ , \tag{2.42}$$

where dN is the number of oscillators excited during the time from t_i to $t_i + dt_i$. Since we are assuming random excitation, which is equivalent when the number of particles is large to a constant rate of excitation, we can assume that

$$dN = K \, dt_i \ .$$

We then have

$$\sum_i e^{-\Gamma(t-t_i)} = \int_{-\infty}^{t} e^{-\Gamma(t-t_i)} K \, dt_i = K/\Gamma \ . \tag{2.43}$$

The limit of integration from $-\infty$ to t indicates that all the dipoles prepared up to the time of observation t are taken into account.

Making the same replacements in the other two terms in (2.41) we obtain the following expression

$$I = K \frac{e^2 \omega_0^4 |r|^2}{16\pi c^3 R^2} \left[2\left(\cos^2\theta \cos^2\theta' + \frac{1}{2} \sin^2\theta \sin^2\theta' \right) \frac{1}{\Gamma} \right.$$

$$+ \sin^2\theta \sin^2\theta' \frac{\Gamma \cos 2(\phi - \phi') + 2\Omega \sin 2(\phi - \phi')}{\Gamma^2 + 4\Omega^2}$$

$$\left. + \sin 2\theta \sin 2\theta' \frac{\Gamma \cos(\phi - \phi') + \Omega \sin(\phi - \phi')}{\Gamma^2 + \Omega^2} \right] . \tag{2.44}$$

For $\theta = \pi/2$, $\phi = \phi'$ this becomes

$$I = I_0 \left(1 + \frac{\Gamma^2}{\Gamma^2 + 4\Omega^2} \right) ,$$

where $I_0 = K e^2 \omega_0^4 |r|^2 / 16\pi c^3 R^2 \Gamma$.

As anticipated, there is no time dependence. However, the dependence on the magnetic field (via the parameter Ω) remains. The character of the dependence is related to the angles specifying the direction of the observed polarization (θ' and ϕ') and of the dipole at the instant of its formation (θ and ϕ). It is interesting to note that the angles ϕ' and ϕ enter the expression (2.44) in terms of their difference. This is also understandable, since for axially symmetric problems the result should not depend on the particular choice of x- and y-axes within the xy-plane.

We shall analyse formula (2.44) in some detail because it describes the most common among all interference phenomena, namely the Hanle effect. As usual the magnetic field dependence vanishes if the dipole r or the observed linear polarization l is directed along the magnetic field.

In the general case of observation of linear polarization four distinct Ω-dependences contribute to the total: two Lorentzian dependences and two dispersion types of different width (Fig. 2.5). By a suitable choice of experimental conditions it is possible to eliminate one or more of these dependences. Thus, if

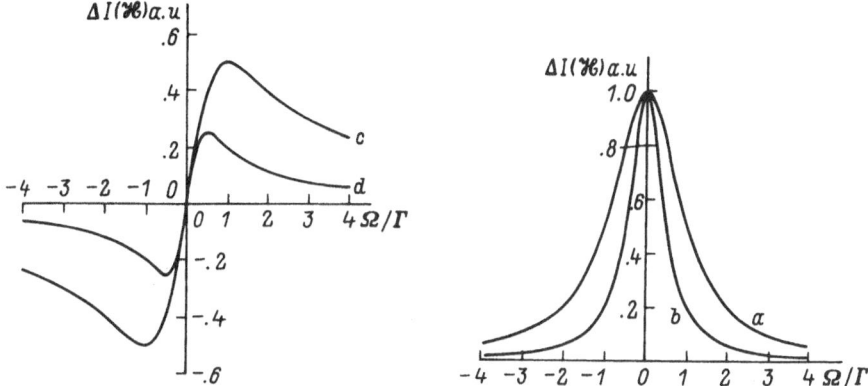

Fig. 2.5. The part of the intensity which depends on the magnetic field

$\phi = \phi'$, the dispersion contours will be removed. However, it is impossible to retain the two dispersion profiles whilst eliminating both the Lorentzian lineshapes, whatever angle is selected. In addition, if either $\theta = \pi/2$, or $\theta' = \pi/2$, i.e. either the dipole or the observed polarization are perpendicular to the magnetic field, then there will remain only a single narrow Lorentzian contour. One dispersion contour can be retained by choosing the difference, $\phi - \phi' = \pi/4$ and simultaneously one of the angles θ or θ' equal to $\pi/2$.

In the following we return again to beats and prove that they can be observed not only during pulse excitation.

2.7 Combination of Hanle Effect and Quantum Beats

Let the time independent excitation be suddenly switched off at the instant t_0 (Fig. 2.6). Before the time t_0 an observer will see the normal Hanle effect. At $t > t_0$ the signal is given by the following expression, in which I_i is given by (2.41)

$$
\begin{aligned}
I(t) = K \int_{-\infty}^{\infty} I_i \, dt_i = \frac{K e^2 \omega_0^4 |r|^2}{16\pi c^3 R^2} \Big\{ & \frac{2\cos^2\theta \cos^2\theta' + \sin^2\theta \sin^2\theta'}{\Gamma} \\
& + \sin^2\theta \sin^2\theta' \\
& \times \frac{\Gamma \cos 2[\Omega(t-t_0) + \phi - \phi'] + 2\Omega \sin 2[\Omega(t-t_0) + \phi - \phi']}{\Gamma^2 + 4\Omega^2} \\
& + \sin 2\theta \sin 2\theta' \\
& \times \frac{\Gamma \cos [\Omega(t-t_0) + \phi - \phi'] + \Omega \sin [\Omega(t-t_0) + \phi - \phi']}{\Gamma^2 + \Omega^2} \Big\} \\
& \times e^{-\Gamma(t-t_0)} \; .
\end{aligned}
\tag{2.45}
$$

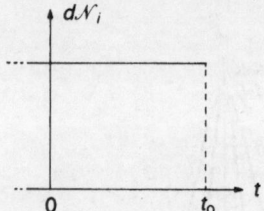

Fig. 2.6. The semi-infinite pulse of excitation

Thus in this case, as was found for pulse excitation, one can observe the sum of damping oscillations at two frequencies. It is possible to exclude the oscillation frequency Ω by selecting appropriate experimental conditions. The only difference from the case of pulse excitation is in the amplitude of the beats: the amplitude now depends on the magnetic field according to the same law that determines the degree of polarization in the Hanle effect.

2.8 Beat Resonances

A particularly interesting case is that of an excitation with a harmonic time dependence:

$$dN_i = K[1 + \varepsilon \cos(\Omega' t + \phi_1)]dt_i \; . \tag{2.46}$$

Making the same substitutions in the initial formula (2.41), and carrying out the same calculations as in the previous problem, one obtains a rather cumbersome expression for the intensity. When $\varepsilon = 0$ it coincides with (2.44), and when $\varepsilon \neq 0$ it has an additional time-dependent term

$$
\begin{aligned}
I(t) = \frac{K e^2 \omega_0^4 |r|^2}{16\pi c^3 R^2} \varepsilon \Bigg\{ & (2\cos^2\theta \cos^2\theta' + \sin^2\theta \sin^2\theta') \\
& \times \frac{\Gamma \cos(\Omega' t + \phi_1) + \Omega' \sin(\Omega' t + \phi_1)}{\Gamma^2 + (\Omega')^2} \\
& + \sin^2\theta \sin^2\theta' \left[\frac{\Gamma \cos(\Omega' t + \phi_1) + (\Omega' - 2\Omega)\sin(\Omega' t + \phi_1)}{\Gamma^2 + (\Omega' - 2\Omega)^2} \right. \\
& \left. + \frac{\Gamma \cos(\Omega' t + \phi_1) + (\Omega' + 2\Omega)\sin(\Omega' t + \phi_1)}{\Gamma^2 + (\Omega' + 2\Omega)^2} \right] \\
& + \sin 2\theta \sin 2\theta' \left[\frac{\Gamma \cos(\Omega' t + \phi_1) + (\Omega' - \Omega)\sin(\Omega' t + \phi_1)}{\Gamma^2 + (\Omega' - \Omega)^2} \right. \\
& \left. + \frac{\Gamma \cos(\Omega' t + \phi_1) + (\Omega' + \Omega)\sin(\Omega' t + \phi_1)}{\Gamma^2 + (\Omega' + \Omega)^2} \right] \Bigg\} \; . \tag{2.47}
\end{aligned}
$$

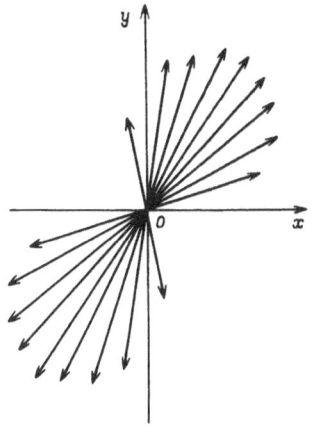

Fig. 2.7. The precession of dipoles that occurs when the light excitation is modulated

The amplitude of the oscillating component has a resonance character centred on the Larmour frequency with two maxima at $\Omega = \Omega'$ and $2\Omega = \Omega'$. It is significant that the amplitude of the resonances does not depend on the relationship between the modulation frequencies of the excitation and the lifetime. This reveals the non-trivial character of the phenomenon of beat resonances.

In addition to resonance terms, (2.47) also contains non-resonance terms, but for modulation frequencies $\Omega' \gg \Gamma$ these turn out to be small.

Based on the model of precessing dipoles one can easily visualize the formation of resonance beats. When the source of excitation is modulated light, the damped dipoles "replenish" synchronously with (in phase with) their precession. As a result one can compare the emitting atoms to the "bundle" of emitting dipoles (Fig. 2.7).

2.9 Parametric Resonance

Parametric resonance can also be described within the scope of the classical model, but here we confine ourselves to a pictorial presentation only. Parametric resonance occurs when the magnitude of the magnetic field is modulated at the Larmour frequency Ω or at the frequency $\Omega_k = k\Omega$, k being an integer. Figure 2.8 illustrates the case $k = 2$. When observing this phenomenon, the pump intensity remains constant with time and the excited dipoles are produced at a constant rate, but the rotation of the dipoles about the magnetic field periodically changes its speed, since the magnitude of the magnetic field changes periodically. When the speed of rotation is small, every successively prepared dipole is close to the preceding one with respect to its direction; upon increasing the speed of rotation their "angular density" decreases. The bundle produced will proceed in its collective motion until the dipoles decay. If the magnetic field variation is synchronized to the rotation of the dipoles, then every successive

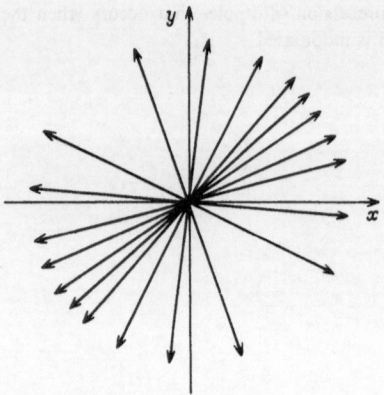

Fig. 2.8. The precession of dipoles occurring when the magnetic field is modulated and the excitation is constant

package will superimpose on the preceding one, which has partially decayed. Due to the non-uniform angular distribution of the dipoles, the light observed in a given direction will be modulated. The amplitude of modulation of the radiation reaches a maximum when the frequency of modulation of the magnetic field exactly coincides with the Larmour frequency.

2.10   Conclusion

All the fundamental interference phenomena mentioned in this chapter have been discussed in terms of a model which considers the emitting atom as a simple dipole. This model is very crude; it cannot reflect the difference between atoms of different elements or different excited states of the same atom, except as variations of the resonance frequency ω_0. For a more adequate description of the polarization characteristics of radiation it is necessary to use the apparatus of quantum mechanics. Quantum mechanics enables certain problems to be solved more easily than in the classical approach. For every atomic state it enables one to find the dependence of the frequency splitting on a magnetic (as well as an electric) field. It describes the case in which the "Larmour" frequency approaches zero for a non-zero magnetic field. As a result of its ability to provide an adequate description, quantum mechanics enables one to find atomic constants and to characterize external perturbations on the basis of the observed behaviour of the radiation, i.e. its variation in time, dependence on the applied magnetic field, etc.

In support of the classical model one has to say that it provides a good description of all the qualitative aspects of the behaviour of the radiation. All angular dependences remain exactly the same. Thus if we generalize the classical model, replacing the linear relationship between the frequencies Ω and the magnetic field H by a more general $\Omega(H)$, then any observable dependence (that can be described using the language of quantum mechanics) can be expressed as the radiation from an ensemble of dipoles:

$$|E|^2 = \sum_i |E_i|^2 \; .$$

In a number of cases this possibility can significantly facilitate the interpretation of observations and experimental predictions.

3. Quantum Mechanical Description of Interference Phenomena

In the previous chapter we treated all the basic interference effects associated with dipole emission using a very crude model. These effects are correctly described by quantum mechanics, and thus the next step of approximation will be to discuss the interaction of an atom that is described quantum mechanically with an electromagnetic field (optical radiation) that is described classically. The next step beyond this is to take into account the quantization of the electromagnetic field. Such an approach will be exploited when proving the impossibility of interference of states with equal angular moments (Sect. 3.9).

The density matrix is the most convenient apparatus for a quantum mechanical treatment of interference phenomena and we shall thus devote significant attention to this concept. For completeness, however, we will also give examples of how interference phenomena can be described in other formalisms; using Schrödinger's equation for the wavefunctions; the scattering matrix; or in the formalism of the statistical tensor – an analogy to the density matrix.

Attention is given to the evidence that the interference terms sum to be zero when averaging the radiation over all directions. For this to be valid for any kind of interference, it is necessary to introduce an anisotropy in the ensemble of atoms.

3.1 The Density Matrix

In quantum mechanics one can specify pure and mixed states. As is well known, quantum mechanics describes averaged microsystems and the concept "pure" or "mixed" state is related to the ensemble: this can be an ensemble of atoms or molecules, or other microsystems, or an ensemble of states of a single particle, but at different instants in time. By a pure state one means a state to which a specified wavefunction – the state vector – can be ascribed. This vector can be resolved with respect to the eigenfunctions of some kind of operator, for instance the Hamiltonian, whose eigenfunctions are stationary states

$$\Psi = \sum_n C_n \Psi_n(0) e^{-i\omega_n t} . \tag{3.1}$$

The assumption that an ensemble is described by a wavefunction is equivalent to the fact that, at each time instant, the values of all the coefficients C_n are known.

Now let us suppose that we have collected into one single ensemble a number of subensembles, each of which can be assigned a wave vector. Moreover, the subensembles are independent and do not interact with one another. Such an ensemble cannot be described by any wavefunction; however, it can be described by the density matrix [3.1].

In order to introduce the density matrix, let us return to the case of pure subensembles. The determination of any physical quantity is equivalent to determining the expectation value of the quantity assigned to the operator $\hat{T}$ corresponding to the quantity in question

$$\int_v \Psi^* \hat{T} \Psi \, dv = \langle \Psi | \hat{T} | \Psi \rangle \equiv T . \tag{3.2}$$

If one substitutes the expansion (3.1) in place of Ψ, then the result will obviously contain the product of coefficients $C_n C_k^*$. The table of multiplication of the coefficients $C_n C_k^*$ is called the density matrix of the statistical operator or simply the density matrix σ. Its matrix elements are

$$\langle \Psi_n | \hat{\sigma} | \Psi_k \rangle = \sigma_{nk} = C_n C_k^* . \tag{3.3}$$

The diagonal elements $\sigma_{nn} = C_n C_n^*$ clearly determine the probability of observing the system in the state Ψ_n, i.e. they describe the population of states. The off-diagonal elements are called interference terms. They describe the phase relationship between different states and are sometimes called coherent terms.

It is possible to show that, for any operator T,

$$\langle \hat{T} \rangle = \mathrm{Tr}\{\hat{\sigma}\hat{T}\} = \sum_{n,k} \sigma_{nk} T_{kn} . \tag{3.4}$$

This equation is sometimes used as a definition of the density matrix. Insofar as the density matrix enables one to determine the physically observable characteristics of a system, one can say that the density matrix determines the state of a system.

The statistical operator of a mixed ensemble, i.e. an ensemble in a mixed state, is given as a sum

$$\hat{\sigma} = \sum_\alpha P_\alpha \hat{\sigma}^\alpha . \tag{3.5}$$

Here P_α indicates the probability that the microsystem belongs to the subensemble α (or, in other words, the relative number of particles in the subensemble) whose state vector and coefficients of expansion are also indicated by the index α. The density matrix of a mixed state is given by

$$\sigma_{nk} = \sum_\alpha P_\alpha \sigma_{nk}^\alpha . \tag{3.6}$$

Such a density matrix describes any mixed ensemble. The ensemble can be a mixture of gases and molecules, a mixture of atoms of different elements, or a mixture of atoms of one element in different stable states. If the system is perturbed by some external action, then it is quite sufficient for subensembles to have, for example, different velocities of thermal motion, in order that the whole ensemble state turns out to be mixed. Since we shall be dealing almost exclusively with ensembles in a mixed state, the fundamental method of description will be the density matrix.

3.2 Derivation of the Density Matrix of Ensembles of Excited States from the Wave Equation

Let us assume that we are dealing with an ensemble of particles, whose density matrix in the initial state is $\sigma_{00} = 1$. This corresponds to an ensemble of identical particles, each in the state Ψ_0.

If the perturbation operator $\hat{V}$ is known explicitly, then the density matrix can be determined by using the Schrödinger equation for the wavefunction. The perturbation $\hat{V}(t)$ changes the state of the subensemble and we find the values of the coefficients C_n when $t > t_0$ by using the methods of time-dependent perturbation theory. This is applicable if the perturbation V is weak, and we will in fact confine ourselves to such cases; strong perturbations will not be discussed in the present treatment. If the coefficients C_n are known, then so is the density matrix:

$$\sigma_{nk}(t) = C_n(t)\,C_k^*(t)\ . \tag{3.7}$$

After switching off all perturbations, the density matrix depends on time harmonically:

$$\sigma_{nk}(t) = C_n C_k^* \exp[\,-\,\mathrm{i}(\omega_n - \omega_k)t\,]\ . \tag{3.8}$$

For the problems that will be discussed in this book, the form of the density matrix can be made rather more concrete. We have already introduced the limit of the linear interaction with light or other excitations. In this limit the atom will have a probability near to unity of being in the lower state Ψ_0, and a small probability of being in the excited state. One can consider the ground state to be time independent and independent of the perturbation. For the coefficient C_n one then has the following equation

$$\mathrm{i}\hbar\dot{C}_n(t) = \sum_k C_k(t)\,V_{nk}(t) + \hbar(\omega_n - \mathrm{i}\Gamma/2)\,C_n(t)\ , \tag{3.9}$$

which is derived from the Schrödinger equation

$$\mathrm{i}\hbar\dot{\Psi} = (\hat{H}_0 + \hat{V})\,\Psi\ .$$

If one substitutes into this the expansion (3.1), then the explicit form of the damping operator is given by

$$\hat{\Gamma} = -\frac{i\hbar}{2}\Gamma\delta_{nk} \ , \tag{3.10}$$

which is equivalent to the introduction of a complex energy $\omega_n \rightarrow \omega_n - i\Gamma/2$. One then multiplies both sides by Ψ_n^* and integrates over all coordinates. The introduction of the damping operator, and the corresponding complex energy generally requires some care, since the whole operator will turn out to be non-Hermitian; however, in the problem under discussion this is permissible.

We assume that the atomic system, at the moment of turning on the perturbation t_0, is in one of the lower states Ψ_0, i.e. the initial conditions are

$$C_{n \neq 0}(t \leq t_0) = 0; \qquad C_0(t \leq t_0) = 1 \ , \tag{3.11}$$

and for simplicity we assume that the energy of this state is equal to zero $\omega_0 = 0$. We then obtain a system of differential equations

$$i\hbar\,\dot{C}_n(t) = \hbar\,\omega_n\,C_n(t) - \frac{i\hbar\Gamma}{2}C_n(t) + V_{n0}(t)\,C_0(t) \ . \tag{3.12}$$

Taking account of the initial conditions the solutions have the following form:

$$C_n(t) = \frac{1}{i\hbar}\int_{t_0}^{t} V_{n0}(t')\,C_0(t')\exp[-i\omega_n(t-t') - \Gamma(t-t')/2]\,dt' \ . \tag{3.13}$$

From these coefficients one can use (3.3) to find the density matrix.

Let the perturbation be given by a monochromatic electromagnetic wave of optical frequency

$$\hat{V} = \boldsymbol{d}\cdot\boldsymbol{\mathscr{E}}_\lambda\exp[-i\omega t - i\phi] + \text{c.c.} \ , \tag{3.14}[1]$$

where $\boldsymbol{d} = -e\boldsymbol{r}$ is the dipole moment operator. Then, according to (3.13), noting that $C_0(t) \approx 1$ and dropping the terms with a sum of optical frequencies in the denominator, we find:

$$C_n(t)$$

$$= \frac{(\boldsymbol{d}\cdot\boldsymbol{\mathscr{E}}_\lambda)_{n0}\,[\exp(-i\omega t - i\phi) - \exp[-i\omega_n(t-t_0) - i\omega t_0 - i\phi - \Gamma(t-t_0)/2]]}{\hbar(\omega - \omega_n + i\Gamma/2)}$$

$$\tag{3.15}$$

---

[1] The interaction of the atom with the electromagnetic wave is described by the perturbation $\hat{V} = -\boldsymbol{d}\cdot\boldsymbol{E} = \boldsymbol{d}\cdot\boldsymbol{E}\exp(-i\pi)$. The minus sign is equivalent to a phase shift by π and is included in the phase ϕ.

and the density matrix elements for $t \gg t_0$ are given by

$$C_n C_k^* = \frac{(d \cdot \mathscr{E}_\lambda)_{n0} (d \cdot \mathscr{E}_\lambda)_{k0}^*}{\hbar^2 (\omega - \omega_n + i\Gamma/2)(\omega - \omega_k - i\Gamma/2)} . \tag{3.16}$$

Here we have introduced the notation

$$(d \cdot \mathscr{E}_\lambda)_{pq} = \langle p | d \cdot \mathscr{E}_\lambda | q \rangle .$$

Hence, when an ensemble of atoms in a pure state is excited by monochromatic light, the density matrix is described by a relatively simple expression. Of course, in the presence of a perturbation the ensemble can also be found in a pure quantum state.

Let us consider now another case of excitation by light consisting of two monochromatic waves with frequencies ω_i and ω_l. We clearly have

$$\hat{V} = \hat{V}_i + \hat{V}_l = d \cdot (\mathscr{E}_i \exp(-i\omega_i t - i\phi_i) + \mathscr{E}_l \exp(-i\omega_l t - i\phi_l))$$

and the density matrix elements are given by

$$\begin{aligned}
C_n C_k^* = {} & \frac{(d \cdot \mathscr{E}_i)_{n0} (d \cdot \mathscr{E}_i)_{k0}^*}{\hbar^2 (\omega_i - \omega_n + i\Gamma/2)(\omega_i - \omega_k - i\Gamma/2)} \\
& + \frac{(d \cdot \mathscr{E}_l)_{n0} (d \cdot \mathscr{E}_l)_{k0}^*}{\hbar^2 (\omega_l - \omega_n + i\Gamma/2)(\omega_l - \omega_k - i\Gamma/2)} \\
& + \frac{(d \cdot \mathscr{E}_i)_{n0} (d \cdot \mathscr{E}_l)_{k0}^* \exp[-i(\omega_i - \omega_l)t - i\phi_{il}]}{\hbar^2 (\omega_i - \omega_n + i\Gamma/2)(\omega_i - \omega_k - i\Gamma/2)} \\
& + \frac{(d \cdot \mathscr{E}_l)_{n0} (d \cdot \mathscr{E}_i)_{k0}^* \exp[i(\omega_i - \omega_l)t + i\phi_{il}]}{\hbar^2 (\omega_l - \omega_n + i\Gamma/2)(\omega_i - \omega_k - i\Gamma/2)} ,
\end{aligned} \tag{3.17}$$

where $\phi_{il} = \phi_i - \phi_l$. It is important to emphasize that time-dependent terms arise here. If there are many fields but each of them is monochromatic, then the density matrix will have the following form

$$C_n C_k^* = \frac{1}{\hbar^2} \sum_{i,l} \frac{(d \cdot \mathscr{E}_i)_{n0} (d \cdot \mathscr{E}_l)_{k0}^* \exp[-i(\omega_i - \omega_l)t - i\phi_{il}]}{(\omega_i - \omega_n + i\Gamma/2)(\omega_l - \omega_k - i\Gamma/2)} . \tag{3.18}$$

Here i and l each run through the same set of values, whose number is equal to the number of monochromatic fields. The time dependence will remain, but it will be quite complex.

We shall also be interested in the case of illumination by a rather broad spectrum, namely one that is broader than $\omega_n - \omega_k + \Gamma$. Let us assume that the intensity of the illuminating light is distributed uniformly in this spectral interval. We model this light by a large number of closely spaced harmonic components. The uniform intensity distribution permits us to assume that $\mathscr{E}_i = \mathscr{E}_l = \mathscr{E}$. Formula (3.18) will then acquire the form:

$$C_n C_k^* = \frac{\mathscr{E}^{2,}}{\hbar^2} (d \cdot e_\lambda)_{n0} (d \cdot e_\lambda)_{k0}^*$$

$$\times \sum_{i,l} \frac{\exp[-i(\omega_i - \omega_l)t - i\phi_{il}]}{(\omega_i - \omega_n + i\Gamma/2)(\omega_l - \omega_k - i\Gamma/2)} . \tag{3.19}$$

where $\mathscr{E} e_\lambda = \mathscr{E}$ and e_λ is the polarization vector of the exciting light. In this sum, the terms with $i \neq k$ have random phase differences ϕ_{il} and upon summing will average to zero. Only the diagonal terms with $i = l$ will remain. The expression (3.19) thus simplifies to

$$\overline{C_n C_k^*} = \frac{\mathscr{E}^2}{\hbar^2} (d \cdot e_\lambda)_{n0} (d \cdot e_\lambda)_{k0}^*$$

$$\times \sum_l \frac{1}{(\omega_l - \omega_n + i\Gamma/2)(\omega_l - \omega_k - i\Gamma/2)} . \tag{3.20}$$

Let us now replace the summation by an integral. In order to do this we introduce the spectral density of the illumination ϱ_ω, which is the energy of the electromagnetic field in a volume 2π per unit frequency interval

$$\sum_l \dots \mathscr{E}_l^2 \to \int \dots \varrho_\omega \, d\omega = \varrho \int d\omega \dots . \tag{3.21}$$

The integration limits can be extended to $-\infty$ and $+\infty$ because of the sharp decrease of the denominator; (2.20) can thus be rewritten as:

$$\overline{C_n C_k^*} = \sigma_{nk}$$

$$= \frac{\varrho}{\hbar^2} (d \cdot e_\lambda)_{n0} (d \cdot e_\lambda)_{k0}^* \int\limits_{-\infty}^{\infty} (\omega - \omega_n + i\Gamma/2)^{-1} (\omega - \omega_k - i\Gamma/2)^{-1} d\omega$$

$$= \frac{2\pi \varrho (d \cdot e_\lambda)_{n0} (d \cdot e_\lambda)_{k0}^*}{\hbar^2} \frac{1}{\Gamma + i\omega_{nk}} = \frac{F_{nk}}{\Gamma + i\omega_{nk}} . \tag{3.22}$$

The term F_{nk} is called the pump matrix. It will be discussed in detail below.

   The formula (3.22) does not contain any time dependence. This is as anticipated, since in the very conditions of the problem there is no time dependence.

   Later on we will come back to the physical phenomena relevant to the formulae just derived, but for now we would like to concentrate on the following situation: When the radiation producing excitation has a continuum spectrum (white light, but not pulse excitation) the coefficients C_n, C_k, are given, in accordance with the formula (3.15), by the expression:

$$\bar{C}_n = \sqrt{\varrho} \, (\boldsymbol{d} \cdot \boldsymbol{e}_\lambda)_{n0}$$

$$\times \int \frac{\{\exp(-i\omega t - i\phi_\omega) - \exp[-i\omega_n(t - t_0) - \Gamma(t - t_0)/2 - i\omega t_0 - i\phi_\omega]\} \, d\omega}{\hbar(\omega - \omega_n + i\Gamma/2)} \, .$$

$$(3.23)$$

This integral, because of the randomness of the phases, is equal to zero, i.e.

$$\bar{C}_n = 0, \qquad \bar{C}_k = 0 \ .$$

However, according to (3.22) $\overline{C_n C_k^*} \neq 0$. Physically this means that the excitation of an ensemble of atoms by light of a broad spectral content produces not a pure but a mixed ensemble of excited atoms. It is impossible to assign any wavefunction to such an ensemble, but it can however be described by the density matrix. This special case demonstrates the merits of the density matrix approach.

Let us consider one more case – pulse excitation at the time instant t_0:

$$\hat{V} = K \boldsymbol{d} \cdot \boldsymbol{\mathscr{E}} \, \delta(t - t_0) \ , \tag{3.24}$$

where the quantity K depends on the choice of δ-function and has the dimension of time. The coefficients are now

$$C_n(t) = \frac{K}{\hbar} \, (\boldsymbol{d} \cdot \boldsymbol{\mathscr{E}})_{n0} \exp[-i\omega_n(t - t_0) - \Gamma(t - t_0)/2] \ . \tag{3.25}$$

The density matrix elements depend on time according to

$$C_n(t) \, C_k^*(t)$$

$$= \frac{K^2}{\hbar^2} \, (\boldsymbol{d} \cdot \boldsymbol{\mathscr{E}})_{n0} \, (\boldsymbol{d} \cdot \boldsymbol{\mathscr{E}})_{k0}^* \exp[-i(\omega_n - \omega_k)(t - t_0) - \Gamma(t - t_0)] \ . \tag{3.26}$$

An important feature for an ensemble of atoms excited by a pulse is that the density matrix contains terms oscillating at the frequency difference $\omega_n - \omega_k = \omega_{nk}$.

Let us return to excitation by a monochromatic light, but let the atoms, in contrast to the case considered up to now, be nonstationary with a distribution of velocities. It is obvious that atoms with the same velocity (both in magnitude and direction) can form pure ensembles, but that the whole system of excited atoms can only be described by the density matrix, which represents an average with respect to the pure ensembles. In the laboratory coordinate system, the characteristic transition frequencies ω_n depend on the velocities of the atoms according Doppler's rule, and they will be distributed over some region of frequencies. For the Maxwell–Boltzmann velocity distribution, the characteristic frequencies are have a Gaussian distribution with a width $\Delta\omega_D$. Making one more assumption, namely, that the frequency width (the Doppler width) is much greater than the natural width Γ, one can readily use the formulae (3.5, 6) to find the density matrix of this mixed state. Let us rewrite (3.6) in the following way

$$\sigma_{nk} = \int C_n^{\alpha} C_k^{\alpha*} \, dP_{\alpha} \ . \tag{3.27}$$

In our case the distribution function over frequencies or over velocities $P(v)$ plays the role of P_{α}. By using the formula (3.6) for the multiplication $C_n^{\alpha} C_k^{\alpha*}$, and taking into account the dependence of $\omega_{n,k}$ on the velocity v we find

$$\sigma_{nk} = \int_{-\infty}^{\infty} \frac{(\boldsymbol{d}\cdot\boldsymbol{\mathscr{E}})_{n0}\,(\boldsymbol{d}\cdot\boldsymbol{\mathscr{E}})_{k0}^*\,P\left(\dfrac{v}{c}\right)d\left(\dfrac{v}{c}\right)}{\hbar^2\left[\omega - \omega_n\left(1 + \dfrac{v}{c}\right) + \dfrac{\mathrm{i}\Gamma}{2}\right]\left[\omega - \omega_k\left(1 + \dfrac{v}{c}\right) - \dfrac{\mathrm{i}\Gamma}{2}\right]} \ . \tag{3.28}$$

Here c is the velocity of light.

The expression under the integral is significant only in a very small region where the value of $\omega - \omega_{n,k}$ is not essentially different from Γ. In that region the value of $P(v/c)$ may be supposed constant and it can be taken out of the integral. Then we find:

$$\sigma_{nk} = \frac{2\pi P(v/c)(\boldsymbol{d}\cdot\boldsymbol{\mathscr{E}})_{n0}\,\boldsymbol{d}\cdot\boldsymbol{\mathscr{E}})_{k0}^*}{\hbar^2\,\omega_n\left[\Gamma\left(\dfrac{\omega_k}{\omega_n} + 1\right)\Big/2 + \mathrm{i}\omega_{nk}\right]} = \frac{F_{nk}}{\Gamma + \mathrm{i}\omega_{nk}} \tag{3.29}$$

because $\omega_n/\omega_k \simeq 1$. The quantity F_{nk} is the pump matrix for monochromatic excitation:

$$F_{nk} = \frac{2\pi P(v/c)(\boldsymbol{d}\cdot\boldsymbol{\mathscr{E}})_{n0}\,(\boldsymbol{d}\cdot\boldsymbol{\mathscr{E}})_{k0}^*}{\hbar^2\,\omega_n} \ . \tag{3.30}$$

The expression (3.29) coincides with the formula for the density matrix of the stationary ensemble of atoms formed under white light excitation (3.22). Obviously, excitation of an ensemble of moving atoms using white light and using illumination with a broad spectrum will both yield a density matrix with the same form.

Until now we have assumed that the ground state Ψ_0 is nondegenerate. For completeness, let us also discuss the case in which degeneracy exists. We now have to expand the wavefunction of the ground state, as is done for excited states, in terms of the eigenfunctions

$$\Psi_0 = \sum_{\alpha} C_{\alpha}\Psi_{\alpha}\mathrm{e}^{-\mathrm{i}\omega_0 t} = \sum_{\alpha} C_{\alpha}\Psi_{\alpha} \tag{3.31}$$

(since $\omega_0 = 0$). Then in the expression (3.12) the subscript "0" in the quantities V_{n0} and $C_0(t)$ must be replaced by the index "α"; it is no longer possible to assume that $C_\alpha(t) = 1$. However, one can assume that C_α is a time-independent quantity. The expression (3.13) can be rewritten as

$$C_n(t) = \sum_\alpha \frac{C_\alpha}{i\hbar} \int_{t_0}^t V_{n\alpha}(t') \exp[-i\omega_n(t - t') - \Gamma(t - t')/2] \, dt' \ . \qquad (3.32)$$

Therefore the factors C_α are preserved in all further intermediate formulae for $C_n(t)$. Clearly, in the expressions for the products $C_n C_k^*$ on the right-hand side there appear terms of the type $C_\alpha C_\beta^*$. In particular, the expression (3.20) will have the following form

$$\overline{C_n C_k^*} = \frac{1}{\hbar^2} \sum_{\alpha, \beta} (d \cdot e_\lambda)_{n\alpha} (d \cdot e_\lambda)_{k\beta}^* \overline{C_\alpha C_\beta^*}$$

$$\times \sum_l \frac{\mathscr{E}_l^2}{(\omega_l - \omega_n + i\Gamma/2)(\omega_l - \omega_k - i\Gamma/2)} \ , \qquad (3.33)$$

which, after integration over frequencies, yields

$$\overline{C_n C_k^*} = \frac{2\pi\varrho}{\hbar^2} \sum_{\alpha, \beta} \frac{(d \cdot e_\lambda)_{n\alpha} (d \cdot e_\lambda)_{k\beta}^* \overline{C_\alpha C_\beta^*}}{\Gamma + i\omega_{nk}} = \frac{F_{nk}}{\Gamma + i\omega_{nk}} \ , \qquad (3.34)$$

where F_{nk} is the pump matrix

$$F_{nk} = \frac{2\pi\varrho}{\hbar^2} \sum_{\alpha, \beta} (d \cdot e_\lambda)_{n\alpha} (d \cdot e_\lambda)_{k\beta}^* \overline{C_\alpha C_\beta^*} \ . \qquad (3.35)$$

3.3 The Equation of Motion of the Density Matrix

We now consider the prepared ensemble of radiating (or absorbing) atoms, and seek a description of the process of its transformation. This can be found when the radiation process can be treated independently of the process of excitation, a circumstance that actually applies in a large number of cases. The emitting ensemble of atoms is created as the result of some "excitation". At this stage, the nature of the external agent providing the excitation is insignificant. We denote the operator of the excitation by $\hat{V}(t)$ and let $V(t < t_0) = 0$, i.e. the excitation is switched on at the time t_0. The complete operator describing the atom under the influence of the perturbation is given by

$$\hat{H} = \hat{H}_0 + \hat{V} \ , \qquad (3.36)$$

where $\hat{H}_0$ is the Hamiltonian, i.e. the energy operator. As is known from quantum mechanics, all operators, including $\hat{\sigma}$, obey the following equation

$$i\hbar\dot{\hat{\sigma}} = [\hat{H}\hat{\sigma}] = \hat{H}\hat{\sigma} - \hat{\sigma}\hat{H} \ . \tag{3.37}$$

Using (3.36), the equation for the density matrix will thus take the form

$$i\hbar\dot{\hat{\sigma}} = \hat{H}_0\hat{\sigma} - \hat{\sigma}\hat{H}_0 + \hat{V}\hat{\sigma} - \hat{\sigma}\hat{V} \ . \tag{3.38}$$

Let us now suppose that each atom is characterized by two energy levels (the lower and the excited) each of which is degenerate. Each of the indices n, k run through all values, corresponding to all possible states of the system of excited levels, and α and β correspond to the lower state. The operator $\hat{H}_0$ will have matrix elements only for states with the same index and it is therefore diagonal.

We now consider the operator $\hat{V}$. As mentioned above, this can take any form, provided it leads to an excitation. However, in order to simplify the problem, we will solve it for the case of optical dipole excitation. In this case the matrix elements of the interaction operator $\hat{V}$ will differ from zero only between the state groups (n, k) and (α, β).

Moreover, the decay is not taken into account in the equation (3.38). It is accepted practice to introduce the decay in terms of a relaxation matrix with elements

$$-i\hbar\Gamma_{pq}\sigma_{pq} \ .$$

As a result, the system of equations (3.33) acquires the form

$$i\hbar\sigma_{nk} = (H_0)_{nn}\sigma_{nk} - \sigma_{nk}(H_0)_{kk} + \sum_{\alpha}(V_{n\alpha}\sigma_{\alpha k} - \sigma_{n\alpha}V_{\alpha k}) - i\hbar\Gamma_{nk}\sigma_{nk} \ ,$$

$$i\hbar\dot{\sigma}_{\alpha\beta} = (H_0)_{\alpha\alpha}\sigma_{\alpha\beta} - \sigma_{\alpha\beta}(H_0)_{\beta\beta} + \sum_{n}(V_{\alpha n}\sigma_{n\beta} - \sigma_{\alpha n}V_{n\beta}) - i\hbar\Gamma_{\alpha\beta}\sigma_{\alpha\beta} \ , \tag{3.39}$$

$$i\hbar\dot{\sigma}_{\alpha n} = (H_0)_{\alpha\alpha}\sigma_{\alpha n} - \sigma_{\alpha n}(H_0)_{nn} + \sum_{k}V_{\alpha k}\sigma_{kn} - \sum_{\beta}\sigma_{\alpha\beta}V_{\beta n} - i\hbar\Gamma_{\alpha n}\sigma_{\alpha n} \ .$$

The diagonal elements $\Gamma_{nn} \equiv \Gamma_n$ and $\Gamma_{kk} \equiv \Gamma_k$ have the meaning of relaxation coefficients for the levels' population. If $|n\rangle$ and $|k\rangle$ are the Zeeman sublevels of a single state, then in the absence of interaction (for instance, collisions) $\Gamma_n = \Gamma_k$. The quantity Γ_{nk}, which reflects the physical decay of the coherence between the Zeeman sublevels is also equal to it, whereas the quantity

$$\Gamma_{\alpha k} = \frac{\Gamma_\alpha + \Gamma_k}{2} \tag{3.40}$$

is half of the natural width of a spectral line. In the amplitude formalism (Sect. 3.2) the quantity $\Gamma_{\alpha k}$ did not appear, because it was assumed that the lower state has zero width, $\Gamma_\alpha = 0$. In the presence of interactions, each of these quantities will have its own value, such that necessarily

$$\Gamma_{pq} \geq \frac{\Gamma_p + \Gamma_q}{2} \ .$$

An optical perturbation can be expressed as a sum of monochromatic waves. For a single monochromatic wave, the operator $\hat{V}$ will have the form (3.14):

$$\hat{V}(t) = \boldsymbol{d} \cdot \mathscr{E} \, e^{-i\omega t} + \text{c.c.} \ ,$$

where $\mathscr{E} = \frac{1}{2} E_0$, see (2.18), characterizes the electric field amplitude of the light wave. The matrix elements of this operator enter into (3.39). For a monochromatic exciting field, they are given by

$$V_{pq} = \mathscr{E}[(\boldsymbol{d} \cdot \boldsymbol{e}_\lambda)_{pq} \, e^{-i\omega t} + (\boldsymbol{d}^* \cdot \boldsymbol{e}_\lambda^*)_{pq} \, e^{+i\omega t}] \ . \tag{3.41}$$

The operator has the form given by (2.14) in a coordinate system linked with the atom, because at any other spatial point, the phase of the field will be different

$$\boldsymbol{E}(t) = \mathscr{E} \exp(-i\omega t + i\omega r/c) + \text{c.c.} \ . \tag{3.42}$$

One thus has to solve the system of equations for each individual particle separately. We will see that only the off-diagonal elements of the density matrix, $\sigma_{n\alpha}$, depend on r, and that the equation for density matrix elements with the indices n, k will no longer depend on r, i.e. for all particles in the volume with equal light intensity, it will be the same. Therefore, one can usually neglect the spatial dependence of the perturbation operator and its phase over the volume.

We will tackle the problem immediately for the general case of excitation by light of any spectral content and with an arbitrary time dependence of the intensity. All the parameters of the radiation are contained in the Fourier representation

$$|\boldsymbol{E}(t)| = \int_\omega a_\omega \exp[-i\omega t - i\phi(\omega)] \, d\omega \ , \tag{3.43}$$

where the coefficient a_ω is the field amplitude at frequency ω. To facilitate solution of the system (3.39) let us replace the integral by a sum and rewrite the perturbation operator in the form

$$\hat{V} = \hat{\boldsymbol{d}} \cdot \boldsymbol{E}(t) = \sum_f \hat{\boldsymbol{d}} \cdot \mathscr{E}_f \exp[-i\omega_f t - i\phi_f] + \text{c.c.} \ . \tag{3.44}$$

Weak perturbations, for which $\sigma_{nn} \ll \sigma_{\alpha\alpha}$, one can neglect the term $\sum V_{\alpha k}\sigma_{kn}$ in the latter equation of the system (3.39). (Note, however, that this term is responsible for stimulated emission.) The equation for $\sigma_{\alpha k}$ now takes the form

$$i\dot{\sigma}_{\alpha k} = (\omega_\alpha - \omega_k - i\Gamma_{k\alpha})\sigma_{\alpha k} - \frac{1}{\hbar}\sum_\beta \sigma_{\alpha\beta} \sum_l \mathscr{E}_l \exp(i\omega_l t + i\phi_l)(\boldsymbol{d} \cdot \boldsymbol{e}_\lambda^*)_{\beta k} \ . \tag{3.45}$$

In this equation the fast oscillating terms, which do not contribute to the intensity, have been dropped. We seek the solution in the form

$$\sigma_{\alpha k} = \sum_l \sigma_{\alpha k}^l \exp(i\omega_l t + i\phi_l) \ . \tag{3.46}$$

Substituting this into (3.45) and making obvious transformations, we find

$$\sum_l \sigma_{\alpha k}^l \omega_l \exp(i\omega_l t + i\phi_l)$$

$$= -(\omega_{\alpha k} - i\Gamma_{k\alpha}) \sum_l \sigma_{\alpha k}^l \exp(i\omega_l t + i\phi_l) + \frac{1}{\hbar} \sum_{\beta,l} \sigma_{\alpha\beta} (\boldsymbol{d} \cdot \boldsymbol{e}_\lambda^*)_{\beta k} \mathscr{E}_l \exp(i\omega_l t + i\phi_l) \ . \tag{3.47}$$

This equation clearly decomposes into l independent equations. The solution of each of the equations for $\sigma_{\alpha k}^l$ will have the form

$$\sigma_{\alpha k}^l = \frac{\sum_\beta \sigma_{\alpha\beta} (\boldsymbol{d} \cdot \boldsymbol{e}_\lambda^*)_{\beta k} \mathscr{E}_l}{\hbar(\omega_{\alpha k} + \omega_l - i\Gamma_{k\alpha})} \ . \tag{3.48}$$

By analogy with (3.46) we seek the solution in the form

$$\sigma_{n\alpha} = \sum_{l'} \sigma_{n\alpha}^{l'} \exp(-i\omega_{l'} t - i\phi_{l'}) = (\sigma_{\alpha n})^*$$

and, as in the case (3.48),

$$\sigma_{n\alpha}^{l'} = \frac{\sum_\beta \sigma_{\beta\alpha} (\boldsymbol{d} \cdot \boldsymbol{e}_\lambda)_{n\beta} \mathscr{E}_{l'}}{\hbar(-\omega_{n\alpha} + \omega_{l'} + i\Gamma_{n\alpha})} \ . \tag{3.49}$$

One can now write the equation for the density matrix elements of the excited state. These are of great interest in all interference phenomena:

$$\dot{\sigma}_{nk} = -i(\omega_{nk} - i\Gamma_{nk})\sigma_{nk}$$

$$-\frac{i}{\hbar^2} \sum_{\alpha,\beta,l,l'} \sigma_{\alpha\beta} \left[\frac{(\boldsymbol{d}\cdot\boldsymbol{e}_\lambda)_{n\alpha}(\boldsymbol{d}\cdot\boldsymbol{e}_\lambda^*)_{\beta k}\mathscr{E}_l\mathscr{E}_{l'}\exp(-i\omega_{ll'}t - i\phi_{ll'})}{\omega_l - \omega_{k\alpha} - i\Gamma_{k\alpha}} \right.$$

$$\left. -\frac{(\boldsymbol{d}\cdot\boldsymbol{e}_\lambda^*)_{\beta k}(\boldsymbol{d}\cdot\boldsymbol{e}_\lambda)_{n\alpha}\mathscr{E}_l\mathscr{E}_{l'}\exp(-i\omega_{ll'}t + i\phi_{ll'})}{\omega_{l'} - \omega_{n\beta} + i\Gamma_{n\beta}} \right] \ . \tag{3.50}$$

We shall replace the summation by an integral and assume that there is no correlation in the spectrum, i.e.,

$$\mathscr{E}_l\mathscr{E}_{l'} = \delta_{ll'}\mathscr{E}_l^2 \ . \tag{3.51}$$

We shall introduce, as previously, the spectral density ϱ_ω. Then, if one can assume that the spectral density is constant within the range of the spectral line, the expression (3.50) becomes

$$\dot{\sigma}_{nk} = -\,i(\omega_{nk} - i\Gamma_{nk})\sigma_{nk} - \frac{i\varrho}{\hbar}\sum_{\alpha,\beta}\int\left[(\omega - \omega_{k\alpha} - i\Gamma_{k\alpha})^{-1}\right.$$

$$\left. - (\omega - \omega_{n\beta} + i\Gamma_{n\beta})^{-1}\right](\boldsymbol{d}\cdot\boldsymbol{e}_\lambda)_{n\alpha}(\boldsymbol{d}\cdot\boldsymbol{e}_\lambda^*)_{\beta k}\,\sigma_{\alpha\beta}\,d\omega\ . \tag{3.52}$$

After integration this becomes

$$\dot{\sigma}_{nk} = -\,i(\omega_{nk} - i\Gamma_{nk})\sigma_{nk} + \frac{2\pi\varrho}{\hbar^2}\sum_{\alpha,\beta}(\boldsymbol{d}\cdot\boldsymbol{e}_\lambda)_{n\alpha}(\boldsymbol{d}\cdot\boldsymbol{e}_\lambda^*)_{\beta k}\,\sigma_{\alpha\beta}\ . \tag{3.53}$$

The last term in (3.53) is the pump matrix of (3.35)

$$F_{nk} = \frac{2\pi\varrho}{\hbar^2}\sum_{\alpha,\beta}(\boldsymbol{d}\cdot\boldsymbol{e}_\lambda)_{n\alpha}(\boldsymbol{d}\cdot\boldsymbol{e}_\lambda^*)_{\beta k}\,\sigma_{\alpha\beta} = F_0\sum_{\alpha,\beta}(\boldsymbol{d}\cdot\boldsymbol{e}_\lambda)_{n\alpha}(\boldsymbol{d}\cdot\boldsymbol{e}_\lambda^*)_{\beta k}\,\sigma_{\alpha\beta}\ . \tag{3.54}$$

The equation of motion of the density matrix can be written as

$$\dot{\sigma}_{nk} = -\,(i\omega_{nk} + \Gamma_{nk})\sigma_{nk} + F_{nk}\ . \tag{3.55}$$

The quantities F_{nk} with different indices can be different and as a whole they constitute the pump matrix. Its diagonal elements reflect the rate of growth of state populations under the influence of a perturbation, whereas the off-diagonal elements describe the production of coherency between different eigenstates.

For any other source of perturbation, the term in (3.39) will be the pump matrix:

$$F_{nk} = \sum_\alpha (V_{n\alpha}\sigma_{\alpha k} - \sigma_{n\alpha}V_{\alpha k})\ . \tag{3.56}$$

The short pulse excitation described by the Dirac delta-function is of particular interest:

$$\hat{V} = K\,\boldsymbol{d}\cdot\boldsymbol{\mathscr{E}}\,\delta(t)\ . \tag{3.57}$$

Using the same techniqe, we find

$$F = F'\,\delta(t)\ . \tag{3.58}$$

Under periodic modulation at frequency Ω', radiation with a continuous spectrum yields

$$I = I_0(1 + \varepsilon\cos\Omega't),\quad \Omega' \ll \omega\ . \tag{3.59}$$

The perturbation operator for such light has the form

$$\hat{V} = \sum_l \boldsymbol{d} \cdot \boldsymbol{\mathcal{E}}_l \{ [(1 - \sqrt{1 - \varepsilon^2})/2]^{1/2} \, e^{-i(\omega + \Omega')t}$$

$$+ [(1 + \sqrt{1 - \varepsilon^2})/2]^{1/2} \, e^{-i\omega t} \} \tag{3.60}$$

and the pump matrix takes the form

$$F = F_{\varepsilon = 0}(1 + \varepsilon \cos \Omega' t) . \tag{3.61}$$

A comparison of the forms of V and F shows that, in all cited examples, the pump matrix depends on time in the same way as the intensity of the excitation process. This enables one to write down the pump matrix for an arbitrary excitation, provided its time dependence and the ratio F_{nk}/F_0 are known.

When the pump matrix is known, the solution of the equation for the density matrix can be found easily. The solution will have the simplest form when the pump is time independent. Then $\dot{\sigma} = 0$ and

$$\sigma_{nk} = \frac{F_{nk}}{\Gamma_{nk} + i\omega_{nk}} \tag{3.62}$$

which matches the expression for σ_{nk} derived from Schrödinger's equation (3.29).

3.4 Spontaneous Emission

The amplitude of an electromagnetic wave of polarization $\boldsymbol{e}_r$, radiated by a quantum system upon its transition from a state Ψ to a state Ψ_μ is well known [3.2] to be proportional to the projection of the matrix elements of the dipole transition on the polarization vector[2] $\boldsymbol{e}_r$

$$E \propto \langle \Psi_\mu | \boldsymbol{d} \cdot \boldsymbol{e}_r | \Psi \rangle . \tag{3.63}$$

Let us assume that the emitting state Ψ is not an eigenstate of the energy operator. We represent the wavefunction in terms of its expansion over the eigenfunctions of the Hamiltonian (3.1)

$$\Psi = \sum_n C_n \Psi_n .$$

---

[2] Comparing the formulae (3.14) and (3.72) of the next section, one can easily see that the identical vectors $\boldsymbol{e}_\lambda$ in (3.14) and $\boldsymbol{e}_r$ in (3.72) correspond to complex conjugate polarization of the light wave.

Then

$$\mathscr{E}_r \propto \sum_n C_n (d \cdot e_r)_{\mu n} \tag{3.64}$$

and the intensity is

$$I(e_r) = K \sum_{n,k} C_n C_k^* (d \cdot e_r)_{\mu n} (d \cdot e_r)_{\mu k}^* . \tag{3.65}$$

If the state Ψ can be decomposed into a series of states Ψ_μ, then one must include yet another summation with respect to μ. In order for this description to approach reality, we must average the quantities $C_n C_k^*$ over all ensembles of radiating atoms. Since this means that the intensities will be averaged, the result will be valid for pure ensembles as well as for mixed ones. The averaging procedure will lead to the density matrix, since $\overline{C_n C_k^*} = \sigma_{nk}$, whereas the terms

$$\sum_\mu (d \cdot e_r)_{\mu n} (d \cdot e_r)_{\mu k}^* = G_{kn} \tag{3.66}$$

constitute the observation matrix. One then rewrites the intensity as

$$I(e_r) = K \sum_{n,k} \sigma_{nk} G_{kn} . \tag{3.67}$$

This formula enables one to determine a number (but not all, as we will see below) of characteristics of the observed radiation. The off-diagonal terms in this formula $(n \neq k)$ determine the interference effects in the radiation. The diagonal terms give its mean intensity.

3.5 Limits of the Density Matrix Apparatus. The Scattering Matrix

We recall that the density matrix concept itself is introduced with the assumption that the excitation process and the process of emission can be treated independently. As a result, the density matrix formalism has certain limits. In order to demonstrate these we shall use the scattering matrix [3.3, 4, 5].

The scattering matrix describes the transition process of a quantum system under the action of an optical field from the lower state $|\mu\rangle$ to the lower state $|\mu'\rangle$, which can coincide with $|\mu\rangle$. There exist two channels for this transition: the first is photon absorption with the system making a transition to the excited state $|n\rangle$ and subsequently falling back into a lower state with the emission of a photon. The second channel begins with the emission of a photon; then follows the photon absorption.

Let us consider the first channel: We wish to find the probability amplitude for the atom to be in state $|\mu'\rangle$ at time t. We denote the function describing the state of the atom in an optical field by Ψ. We will represent it as a superposition of steady-state wavefunctions Ψ_n as has been done previously (3.1):

$$\Psi = \sum_{n'} C_{n'}(t)\, \Psi_{n'} \ ,$$

where n' run through the values μ, n, μ'. The coefficients $C_{n'}(t)$ give the probability amplitude for the atom to be in the state $|n'\rangle$ at the time t. Let us "switch on" the optical field $E = \mathscr{E}_\lambda \exp(-i\omega t) + \text{c.c.}$ at $t = 0$. This determines the initial conditions

$$C_{n'}(0) = \delta_{n',\mu} \ . \tag{3.68}$$

Assuming that the perturbation is weak and that no stimulated emission is present, the quantity C_μ has the form

$$C_\mu = e^{-\gamma t/2} \ . \tag{3.69}$$

The relaxation constant γ depends on the incident light power and on the detuning of its frequency from the resonance frequency of the atomic system. For the weak perturbation considered above, γ is very small (it is determined by the width of the lower state). Let us first find the amplitude of the excited states. Schrödinger's equation for the case in question with a perturbation operator (3.14) has the form

$$i\hbar\, \dot{C}_n = (\boldsymbol{d} \cdot \mathscr{E}_\lambda)_{n\mu} \exp(-i\omega t) C_\mu(t) + \hbar\left(\omega_{n\mu} - \frac{i\Gamma_n}{2}\right) C_n(t) \ , \tag{3.70}$$

where $\hbar\omega_{n\mu} = \hbar(\omega_n - \omega_\mu)$ is the energy difference between the lower and excited states and ω is the frequency of the optical field. The quantity $\boldsymbol{d} \cdot \mathscr{E}_\lambda$ is the optical field perturbation operator, Γ is the decay constant of the excited state $|n\rangle$, and the index λ on $\mathscr{E}$ is introduced in order to distinguish the incident wave from the scattered one $\mathscr{E}_r$. Substituting the explicit form C_μ (3.69) and taking into account the initial conditions, we obtain the solution of equation (3.70) in the form

$$C_n(t)$$

$$= \frac{(\boldsymbol{d} \cdot \mathscr{E}_\lambda)_{n\mu}}{\hbar\left[\omega - \omega_{n\mu} + i(\Gamma_n - \gamma)/2\right]} \left[\exp(-i\omega t - \gamma t/2) - \exp(-i\omega_{n\mu} t - \Gamma t/2)\right] \ . \tag{3.71}$$

We now seek $C_{\mu'}$. For this purpose we will introduce a spontaneous emission operator in a form analogous to (3.14)

$$\hat{V}(\omega') = \boldsymbol{d} \cdot \boldsymbol{\mathscr{E}}_r \, e^{i\omega't} + \text{c.c.} \, , \tag{3.72}$$

where ω' is the frequency of the emitted (scattered) photon.[3] The equation for $C_{\mu'}$ is quite analogous to the equation for C_n, it differs only in the perturbation operator

$$i\hbar \dot{C}_{\mu'}^{(n)}(t) = (V_r)_{\mu'n} C_n(t) = (\boldsymbol{d} \cdot \boldsymbol{\mathscr{E}}_r)_{\mu'n} \, e^{i\omega't} C_n(t) \, . \tag{3.73}$$

Moreover, for simplicity, we put $\omega_{\mu'} = \omega_\mu = 0$ (and thus $\omega_{n\mu} = \omega_{n\mu'}$). Then

$$C_{\mu'}^{(n)}(t) = \frac{1}{\hbar^2} (\boldsymbol{d} \cdot \boldsymbol{\mathscr{E}}_\lambda)_{n\mu} (\boldsymbol{d} \cdot \boldsymbol{\mathscr{E}}_r)_{\mu'n}$$

$$\times \left\{ - \frac{\exp[-i(\omega - \omega')t - \gamma t/2]}{(\omega' - \omega + i\gamma/2)[\omega - \omega_n + i(\Gamma_n - \gamma)/2]} \right.$$

$$+ \frac{\exp[-i(\omega_n - \omega')t - \Gamma_n t/2]}{[\omega - \omega_n + i(\Gamma_n - \gamma)/2](\omega' - \omega_n + i\Gamma_n/2)}$$

$$+ \left. \frac{1}{(\omega' - \omega + i\gamma/2)(\omega' - \omega_n + i\Gamma_n/2)} \right\} \, . \tag{3.74}$$

The states $|n\rangle$ can be numerous and the resultant amplitude is then given by $C_{\mu'}^{(1)}$:

$$C_{\mu'}^{(1)}(t) = \sum_n C_{\mu'}^{(n)}(t) \, .$$

The second channel also contains two stages. The amplitude of the atomic state at the end of the first stage is found from (3.70) by making the corresponding replacement of the indices and by replacing the operator by the spontaneous emission operator. Solving the resulting equation (3.14) gives

$$C_k(t) = \frac{(\boldsymbol{d} \cdot \boldsymbol{\mathscr{E}}_r)_{k\mu} [\exp(i\omega't - \gamma t)/2) - \exp(-i\omega_k t - \Gamma_k t/2)]}{\hbar(\omega' + \omega_k - i(\Gamma_k - \gamma)/2)} \, . \tag{3.75}$$

We substitute this amplitude into the equation obtained from (3.73) by replacing the spontaneous emission operator by the field interaction operator $\boldsymbol{\mathscr{E}}_\lambda$, and the amplitude C_n by C_k. By solving this equation we find the contribution to the

---

[3] Unlike (3.14), the quantity $\boldsymbol{d} \cdot \boldsymbol{\mathscr{E}}$ in the spontaneous emission operator resides in the positive frequency part of the wave.

amplitude of the final state from the second channel:

$$C_{\mu'}^{(2)}(t) = \frac{1}{\hbar^2} \sum_k (d \cdot \mathscr{E}_r)_{k\mu} (d \cdot \mathscr{E}_\lambda)_{\mu' k}$$

$$\times \left\{ \frac{\exp[-i(\omega - \omega')t - \gamma t/2]}{(\omega' - \omega + i\gamma/2)[\omega' + \omega_k - i(\Gamma_k - \gamma)/2]} \right.$$

$$+ \frac{\exp[-i(\omega_k + \omega)t - \Gamma_k t/2]}{(\omega_k + \omega - i\Gamma_k/2)[\omega' + \omega_k - i(\Gamma_k - \gamma)/2]}$$

$$\left. - \frac{1}{(\omega_k + \omega - i\Gamma_k/2)(\omega' - \omega + i\gamma/2)} \right\} . \tag{3.76}$$

The resultant amplitude of the final state is given by

$$C_{\mu'}(t) = C_{\mu'}^{(1)}(t) + C_{\mu'}^{(2)}(t) . \tag{3.77}$$

When $t \to \infty$, i.e. in a steady-state process, the expression for $C_{\mu'}$ simplifies to

$$C_{\mu' t \to \infty} = \frac{1}{\hbar^2} \sum_n \frac{(d \cdot \mathscr{E}_\lambda)_{n\mu} (d \cdot \mathscr{E}_r)_{\mu' n}}{(\omega' - \omega + i\gamma/2)(\omega' - \omega_n + i\Gamma_n/2)}$$

$$- \frac{1}{\hbar^2} \sum_k \frac{(d \cdot \mathscr{E}_r)_{k\mu} (d \cdot \mathscr{E}_\lambda)_{\mu' k}}{(\omega' - \omega + i\gamma/2)(\omega_k + \omega - i\Gamma_k/2)} . \tag{3.78}$$

We note that, because γ is small, the frequency ω' of the scattered light differs slightly from that of the incident light ω.

Resonance Scattering. By resonance scattering one usually understands the scattering of light whose frequency is close to at least one of the resonance frequencies ω_n of an atom. In this case, when $t \to \infty$, one can neglect the second term in (3.78), since its denominator contains the sum of optical frequencies $\omega_k + \omega \gg |\omega_k - \omega|$:

$$C_{\mu' t \to \infty} = \frac{1}{\hbar^2} \sum_n \frac{(d \cdot \mathscr{E}_\lambda)_{n\mu} (d \cdot \mathscr{E}_r)_{\mu' n}}{(\omega' - \omega + i\gamma/2)(\omega' - \omega_n + i\Gamma_n/2)} . \tag{3.79}$$

The only term, or terms, to remain in this sum will be those in which ω_n is close to the frequency of the field ω. In this way the formula coincides with the expression obtained by *Weisskopf* [3.6] and *Heitler* [3.2] from a strict theoretical treatment.

One can associate to every transition from $|\mu\rangle$ to $|\mu'\rangle$ the probability amplitude $S_{\mu\mu'}$ of this process.

$$S_{\mu\mu'} = C_{\mu', t \to \infty} , \tag{3.80}$$

where $C_{\mu'}$ is given for the initial state $|\mu\rangle$. The quantities $S_{\mu'\mu}$ constitute the scattering matrix, which is also called the S-matrix [3.3]. It is a function of the initial and final states of the atoms and of the scattering light.

The probability that an atom that was initially in the state $|\mu\rangle$ will be in the final state $|\mu'\rangle$ due to the absorption from the field of a quantum of frequency ω, and emission of a quantum of frequency ω', is clearly equal to $|S_{\mu'\mu}|^2$.

The polarization characteristics of the exciting field, as well as the spontaneous radiation, are accounted for by the quantities $\mathscr{E}_\lambda$ and $\mathscr{E}_r$ respectively (this relation will be discussed in detail later).

The value $|S_{\mu'\mu}|^2$ is a partial probability of emission of a quantum of frequency ω' and of polarization e_r, $(\mathscr{E}_r = \mathscr{E}_r e_r)$ upon the absorption of light of frequency ω and polarization $e_\lambda (\mathscr{E}_\lambda = \mathscr{E}_\lambda e_\lambda)$.

The intensity (the energy per unit time per unit solid angle) of scattered light of polarization e_r is proportional to the probability of emission of quantum frequency ω' (polarization e_r) and the energy of this quantum. We can write

$$I(\omega, \omega', e_\lambda \cdot e_r)\,d\Omega = \sum_{\mu'} N\gamma\hbar\omega'|S_{\mu'\mu}|^2\,d\Omega \ , \tag{3.81}$$

where N is the number of scattering atoms. Substituting into this the expression for $S_{\mu'\mu}$ and making the legitimate replacements [3.3, 5]

$$\sum_{\omega'} \mathscr{E}_r^2 X(\omega') = \int X(\omega') \frac{\hbar(\omega')^3}{4\pi^2 c^3}\,d\omega' \ ,$$

$$\tag{3.82}$$

$$\sum_{\omega} \mathscr{E}_\lambda^2 Y(\omega) = \int Y(\omega)\varrho_\omega\,d\omega \ ,$$

we find

$$I(e_\lambda, e_r)\,d\Omega = \sum_{\omega, \omega'} I(\omega, \omega', e_\lambda, e_r)\,d\Omega$$

$$= N \int_{\omega, \omega'} \frac{(\omega')^4 \gamma \varrho_\omega}{4\pi^2 c^3 \hbar^2}$$

$$\times \sum_{\substack{n,k \\ \mu'}} \frac{(d \cdot e_\lambda)_{n\mu}\,(d \cdot e_\lambda)_{k\mu}^*\,(d \cdot e_r)_{\mu'n}\,(d \cdot e_r)_{\mu'k}^*\,d\Omega}{\left[(\omega' - \omega)^2 + \dfrac{\gamma^2}{4}\right]\left(\omega' - \omega_n + \dfrac{i\Gamma_n}{2}\right)\left(\omega' - \omega_k - \dfrac{i\Gamma_k}{2}\right)}\,d\omega\,d\omega' \ .$$

$$\tag{3.83}$$

In this summation one encounters terms of two types: those with $n = k$ are the diagonal terms and those with $n \neq k$ are the off-diagonal terms, which describe the interference of states. The intensity radiated by the same ensemble of atoms, but at different frequencies and polarizations will be different and will be given by $|S_{\mu'\mu}|^2$ with another frequency ω' or polarization e_r. One readily arrives at the conclusion that the formula (3.83) describes the probability distribution of the light emitting process with respect to frequencies and polarizations for given characteristics of the exciting field. It also describes this distribution as a function of the characteristics of the excitation.

From an analysis of the expression (3.83), it also follows that the frequency of the emitted photon ω' is close to the frequency of the incident one ω (because γ in (3.86) is very small).

In an experiment, the observed result is usually an average over some kind of frequency region. A particularly interesting case is when the observation spans a frequency region of the order of the spectral linewidth. One can find the relative radiation intensity in the spectral line by integrating the expression (3.83) over the frequency ω'. The integration limits can be extended to $+\infty$ and $-\infty$, since far from the line the contribution to the integral is negligible. As a result of the integration we find an expression describing the absorption line. In the integration we take into consideration that $\gamma \ll \Gamma$:

$$I(\omega, e_\lambda, e_r)\, d\Omega\, d\omega = N\, \frac{\omega^4 \varrho_\omega\, d\omega}{2\pi c^3 \hbar^2}$$

$$\times \sum_{\substack{n,k \\ \mu',\mu}} \frac{(d \cdot e_\lambda)_{n\mu}\, (d \cdot e_\lambda)^*_{k\mu}\, (d \cdot e_r)_{\mu'n}\, (d \cdot e_r)^*_{\mu'k}}{(\omega - \omega_n + i\Gamma_n/2)(\omega - \omega_k - i\Gamma_k/2)} \sigma_{\mu\mu}\, d\Omega\ .$$

$$(3.84)$$

Here we have assumed that the atom possesses a large number of ground states $|\mu\rangle$, the populations of $|\mu\rangle$ are $\sigma_{\mu\mu} = |c_\mu(t = 0)|^2$ and they are defined by the initial conditions. (We recall once more that the quantity ω in this formula is the frequency of the exciting light.) Let us compare the resulting expression with the formulae (3.67) and (3.66). It is quite easy to recognize that the first factor under the summation sign coincides with the density matrix elements derived for excitation by monochromatic light of frequency ω (3.20), and that the last two factors constitute the observation matrix.

When exciting by light of a rich spectral content, one has to integrate the expression (3.84) over frequency.

The density matrix in the form (3.22), which corresponds to excitation by means of a broad spectrum, will be implemented in the considerations to follow. If not stated otherwise, it is to be assumed that the excitation consists of a broad spectrum. The condition of excitation by a broad spectrum is also satisfied by certain other excitation techniques, for example, excitation by collisions.

Nonresonance Scattering. As before, let $\omega_\mu = \omega_{\mu'} = 0$. The field frequency, however, is now far away from all resonance frequencies of the atom

$$\omega - \omega_n \gg \Gamma\ .$$

We return to the general description of the transition probability from the state

$|\mu\rangle$ to $|\mu'\rangle$. The transition probability from (3.78) will be equal to

$$|S_{\mu'\mu}|^2 = \frac{1}{\hbar^4} \frac{1}{(\omega' - \omega)^2 + \gamma^{2/4}}$$

$$\times \left| \sum_n \frac{(\boldsymbol{d} \cdot \boldsymbol{\mathscr{E}}_\lambda)_{n\mu} (\boldsymbol{d} \cdot \boldsymbol{\mathscr{E}}_r)_{\mu'n}}{\omega' - \omega_n + i\Gamma_n/2} - \sum_k \frac{(\boldsymbol{d} \cdot \boldsymbol{\mathscr{E}}_r)_{k\mu} (\boldsymbol{d} \cdot \boldsymbol{\mathscr{E}}_\lambda)_{\mu'k}}{\omega_k + \omega - i\Gamma_k/2} \right|^2 . \qquad (3.85)$$

Analysing the expression (3.85) one can easily see that the two sums are comparable in size and that there is no reason to neglect the second of them, as was done in the case of resonance scattering. The integral intensity of the scattered light with respect to frequencies can be found with the help of (3.81) by integrating over the scattered light frequencies

$$I(\omega, e_\lambda, e_r)d\Omega = \sum_{\omega'\mu'} N\gamma\hbar\omega' |S_{\mu'\mu}|^2 \, d\Omega$$

$$= \int_{\omega'} \frac{N\gamma\hbar^2(\omega')^4}{4\pi^2 c^3} \sum_{\mu'} \frac{|S_{\mu'\mu}|^2}{\mathscr{E}_r^2} \, d\omega' \, d\Omega , \qquad (3.86)$$

where $S_{\mu'\mu}$ is given by the expression (3.85).

Let us consider the frequency dependence of the factors in (3.85). The first factor leads to a sharp maximum of width γ when $\omega = \omega'$ and has a very small value elsewhere. In the region of the maximum, the second factor remains practically constant and it can be taken out of the integral in (3.86), replacing ω' by ω. Moreover, when ω' is close to ω, one can neglect the quantities $i\Gamma/2$ in the denominator, since, according to the condition $\Gamma \ll \omega - \omega_n$ and in the above, $\Gamma \ll \omega + \omega_n$. From the integration, and by making use of the expression (3.82), we find for the intensity:

$$I \, d\Omega = N \int \frac{\varrho_\omega \omega^4 \, d\omega}{2\pi c^3 \hbar^2} \sum_{\substack{n,k \\ \mu,\mu'}} \left| \frac{(\boldsymbol{d} \cdot e_\lambda)_{n\mu}(\boldsymbol{d} \cdot e_r)_{\mu'n}}{\omega - \omega_n} - \frac{(\boldsymbol{d} \cdot e_r)_{k\mu}(\boldsymbol{d} \cdot e_\lambda)_{\mu'k}}{\omega + \omega_k} \right|^2 \sigma_{\mu\mu} d\Omega .$$

$$(3.87)$$

It is clearly different from the resonance case (3.84).

The numerators in (3.87) determine the polarization properties of the scattered light and the denominators characterize the dependence of intensity on frequency detuning. Since we are assuming large frequency detuning, the light scattering turns out to be a weak effect. Nevertheless, it is clearly observable and is known as Rayleigh scattering.

Raman Scattering. Suppose now that the energy of the states $|\mu\rangle$ and $|\mu'\rangle$ is different. In addition to the Rayleigh scattering there will now arise radiation at another frequency. It is known as Raman scattering. Raman scattering is observed primarily in condensed media and serves as a powerful tool for their investigation. The intensity of Raman scattering can be determined from the very same scattering matrix $S_{\mu'\mu}$, but with consideration of the energy difference between the initial $|\mu\rangle$ and final $|\mu'\rangle$ states. This is achieved by making the substitution

$$(\omega' - \omega) \rightarrow [(\omega' + \omega_{\mu'}) - (\omega + \omega_\mu)] = \omega' - \omega - \omega_{\mu\mu'} \; ,$$

where $\hbar\omega_\mu$ and $\hbar\omega_{\mu'}$ are the energies of the atom in the states $|\mu\rangle$ and $|\mu'\rangle$. In the same way as we derived the expression for the Rayleigh scattering intensity (3.87), we now obtain the intensity of Raman scattering as

$$I \, d\Omega = N \int \frac{(\omega + \omega_{\mu\mu'})^4 \varrho_\omega}{2\pi c^3 \hbar^2}$$

$$\times \sum_{\substack{n,k \\ \mu,\mu'}} \left| \frac{(\boldsymbol{d} \cdot \boldsymbol{e}_\lambda)_{n\mu} (\boldsymbol{d} \cdot \boldsymbol{e}_r)_{\mu'n}}{\omega - \omega_{n\mu}} - \frac{(\boldsymbol{d} \cdot \boldsymbol{e}_r)_{k\mu} (\boldsymbol{d} \cdot \boldsymbol{e}_\lambda)_{\mu'k}}{\omega + \omega_{k\mu}} \right|^2 \sigma_{\mu\mu} \, d\omega \, d\Omega \; . \qquad (3.88)$$

This formula shows that when the frequency of the illuminating light approaches the resonance frequency the intensity of scattering increases; the closer the frequency to resonance, the greater will be this increase. An experimental observation of such an increase was reported in [3.7]. The expression (3.86) is valid only for the off-resonance case. However, when the frequency detuning approaches zero, the lifetime of the excited state [expression (3.88)] will have an effect on the expression for the intensity, which is valid only for the off-resonance case, since there certain quantities are dropped and the contribution of the second term is small. In some detuning regions, none of the terms in (3.87) can be neglected. Upon squaring, interference terms arise and in the spectrum of the scattered light new maxima appear [3.8, 9], see also Chap. 4. As the frequency gets closer to the resonance one can neglect the second term in (3.87) and we must take into account the damping constant Γ. Then we shall arrive at the description of resonance fluorescence (3.84).

The expressions (3.87, 88), unlike (3.84), are not transformed in the ordinary way for the density matrix formalism, because neither in the case of Rayleigh scattering, nor in the case of Raman scattering, can the spectrum of the illuminating light approaching the resonance frequency of the atom be regarded as broad.

3.6 Interference Signals

The presence of interference terms in the density matrix is associated with some special features of emission (and, of course, absorption) of light. As will be seen below, these peculiarities are related to the energy gap between the interfering states. It is this dependence of the radiation characteristics on the energy gap which we will henceforth call the interference signals. Most frequently this will be the radiation intensity observed in a given direction and with a specified polarization. Inasmuch as the description is limited (except in Sect. 3.9) by the scope of density matrix theory, one assumes that the observation of the radiation includes a broad spectral range, broader than the Doppler and natural width of the spectral line. None of the spectral line profile variations of the radiation will be included in this class of signals. Moreover, nonresonance excitation will not be treated, since careful analysis shows that the interference signals in this case are very weak. Since the density matrix formalism largely reflects the experimental situations and at the same time possesses appreciable simplicity and convenience, it is quite natural to base the discussion primarily on this formalism and on the equivalent formalism of the statistical tensor (Sects. 3.10–12).

Let us consider the radiation in a given direction and polarization. In Sect. 3.3 we derived an expression for the intensity of the spontaneous emission through the density matrix of the radiating state and the observation matrix (3.67):

$$I = K \sum_{n,k} \sigma_{nk} G_{kn} \ .$$

In general σ_{nk} may depend on time and we shall now consider a variety of different possible interference phenomena.

Hanle Effect and Level Crossing. Let us discuss first the special case of time-independent excitation and let $\Gamma_n = \Gamma_k = \Gamma$. When the optical excitation is time independent, an expression for the density matrix is given by (3.34). At this juncture it is necessary to note that σ does not depend on time, and thus $\dot{\sigma} = 0$. Then, from (3.55), we find

$$\sigma_{nk} = \frac{F_{nk}}{\Gamma + i\omega_{nk}} \ , \tag{3.89}$$

where ω_{nk} is the frequency separation between the energy levels of the states $|n\rangle$ and $|k\rangle$

$$\omega_{nk} = \frac{1}{\hbar}(E_n - E_k) \ .$$

Substituting (3.89) into (3.67), we find

$$I = K \sum_{n,k} \frac{F_{nk} G_{nk}}{\Gamma + i\omega_{nk}} \ . \tag{3.90}$$

In this expression the terms under the sum can be divided into two groups: terms with identical indices – the diagonal terms – and those with different indices – the off-diagonal or interference terms:

$$I = K \sum_{n} \frac{F_{nn} G_{nn}}{\Gamma} + K \sum_{n,k \neq n} \frac{F_{nk} G_{kn}}{\Gamma + i\omega_{nk}} \ . \tag{3.91}$$

The latter group of terms describe the interference signal, i.e. the dependence of the observed intensity on the quantities ω_{nk} (if, of course, there exists a means of changing the frequency separation ω_{nk} continuously by some external action). An interference signal of this type is produced upon level crossing or in the Hanle effect, when two or more levels are degenerate, but the energy levels change continuously upon varying the external magnetic field. Let us add the terms in (3.91) in pairs with indices nk and kn. Noting that

$$F_{nk}G_{kn} = (F_{kn}G_{nk})^* ,$$

we find for each pair of terms

$$I_{kn} + I_{nk} = 2K\left(\frac{\Gamma\,\mathrm{Re}\{F_{nk}G_{kn}\} + \omega_{nk}\,\mathrm{Im}\{F_{nk}G_{kn}\}}{\Gamma^2 + \omega_{nk}^2}\right). \qquad (3.92)$$

It is rather interesting to work out the form of the dependence of I on ω_{nk}. Let the sum in (3.91) contain only two terms, or, equivalently let all ω_{nk} be identical. This corresponds to the case most frequently encountered in experiments. In the general case, the product $F_{nk}G_{kn}$ is complex, and therefore one has to consider three cases: the imaginary part of the product is equal to zero, its real part is equal to zero and both parts are different from zero. In the first case only the first term within the brackets will remain and the intensity has a Lorentzian dependence on the energy separation between the states. The intensity of the light (3.91) will have a maximum if the sign of the real part of (3.92) is positive and a minimum if it is negative. These signals are depicted in Fig. 3.1. If $\mathrm{Re}\{F_{nk}G_{kn}\} = 0$, then the signal will be given by the second term and, as can be seen easily, it has an antisymmetric form (Fig. 3.1, curve b). If, however, the product is complex, then the signal will have an intermediate form and will be a sum of components of the types a and b with different weights (Fig. 3.1, curve c).

The ratio of the real and imaginary parts in the product FG depends on the polarization directions of the incident and observed light.

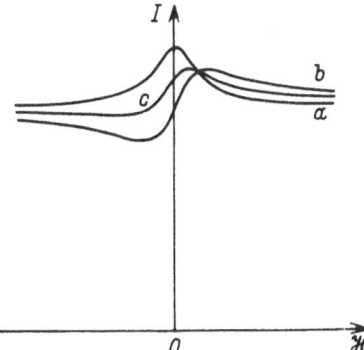

Fig. 3.1. The forms of the Hanle signal: (a) $\mathrm{Im}\{F_{nk}G_{kn}\} = 0$; (b) $\mathrm{Re}\{F_{nk}G_{kn}\} = 0$; (c) the general case

The relationship between the conditions of excitation and observation and the signal shape deserves a rather detailed discussion. This relationship was treated in Chap. 2 within the limits of the classical approximation. In the quantum mechanical approach, the characteristics of excitation and observation are reflected in the pump and observation matrices. The projections of the matrix elements of F and G determine the relationship of the imaginary and real parts of their product, which in turn determines the signal symmetry. In order to demonstrate this relationship, let us consider a simple special case, namely, excitation by linearly polarized light with a vector polarization e_λ perpendicular to the quantization axis. The observation is likewise of linearly polarized light with a vector e_r, which is also perpendicular to the quantization axis, but makes some angle ϕ_0 with the polarization vector of the exciting light. The exciting light is time independent and occupies a broad spectral range. The density matrix will then have the form

$$\sigma_{nk} = \frac{F_{nk}}{\Gamma + i\omega_{nk}} = F_0 \frac{(d \cdot e_\lambda)_{n\mu}(d \cdot e_\lambda)^*_{k\mu}}{\Gamma + i\omega_{nk}} . \tag{3.93}$$

The vector e_λ is determined by the polarization of the exciting light. Let us take the x-axis along this direction. Then

$$\sigma_{nk} = F_0 \frac{(d \cdot e_x)_{n\mu}(d \cdot e_x)_{k\mu}}{\Gamma + i\omega_{nk}} .$$

The expressions for the matrix elements are given later in (5.81–83). We assume for the sake of brevity that the magnetic quantum numbers of states $|n\rangle$, $|k\rangle$, $|\mu\rangle$ are equal to the indices n, k, μ. Then for $n - \mu = 1$

$$(d \cdot e_x)_{n\mu} = d_{n\mu} e^*_{v = n - \mu = 1} e_x .$$

The form of signal is determined by the product of vectors: $e^*_1 \cdot e_x \equiv e^*_+ \cdot e_x = -1/\sqrt{2}$. Recall from (2.7) that $e_+ = (-1/\sqrt{2})(e_x + ie_y)$ and $e_- = (1/\sqrt{2})(e_x - ie_y)$. If $k \neq n$ then $k = \mu - 1$, and $e^*_- \cdot e_x = 1/\sqrt{2}$, and for the density matrix we get the expression

$$\sigma_{nk} = -F_0 \frac{d_{n\mu} d_{k\mu}}{2(\Gamma + i\omega_{nk})} .$$

The observation matrix is also determined by matrix elements

$$G_{kn} = (d \cdot e_r)_{\mu n}(d \cdot e_r)^*_{\mu k}$$

where e_r is the polarization vector of the light reaching the observer. Being perpendicular to the z-axis, the vector e_r may be expressed in the form

$$e_r = e_x \cos \phi_0 + e_y \sin \phi_0$$

$$(d \cdot e_r)_{\mu n} = d_{\mu n} e^*_{v = \mu - n = -1} \cdot e_r = d_{\mu n} e^{i\phi_0}/\sqrt{2} .$$

For the second matrix element we have the following:

$$(d \cdot e_r)^*_{\mu k} = -d_{\mu k} e^{i\phi_0}/\sqrt{2} .$$

substituting in the above equation for the observation matrix elements, we have

$$G_{kn} = -d_{\mu n} d_{\mu k} e^{2i\phi_0}/2 .$$

The product $F_{nk} G_{kn}$ thus becomes

$$F_{nk} G_{kn} = \text{const } e^{2i\Phi_0} .$$

We can now relate the type of signal to the conditions of the experiment: If $\phi_0 = n\pi/2$ (where n is an integer), the product $F_{nk} G_{kn}$ is real valued and the signal has the Lorentz form (Fig. 3.1a). If $\phi_0 = n\pi/2 + \pi/4$, the product $F_{nk} G_{nk}$ is imaginary, and the form of a signal is assymmetric. In other cases the signal is a mixture of the above two forms.

As was mentioned above, under constant excitation one can observe the Hanle effect and the level-crossing effect. The Hanle effect is in fact merely the special case of level crossing at zero magnetic field. Now let us pass on to the case of time-dependent excitation.

Quantum Beats. Excitation by a short pulse, whose duration is shorter than the lifetime of the excited state in question, is of particular interest. The pump function is then a δ-function in time

$$F(t) = F'\delta(t - t_0) . \tag{3.94}$$

Substituting (3.94) into (3.55) and solving, we find

$$\sigma_{nk} = F'_{nk} \exp\left[-\Gamma(t - t_0) - i\omega_{nk}(t - t_0) \right] . \tag{3.95}$$

If the excitation is provided by a light pulse, then (3.94) will acquire the form (3.26). Using the (3.67) for the intensity, we have

$$I(t) = K \sum_{n,k} F'_{nk} G_{kn} \exp\left[-(\Gamma + i\omega_{nk})(t - t_0) \right] . \tag{3.96}$$

In the same way as before, one can distinguish the interference terms

$$\begin{aligned}
I_{nk} + I_{kn} &= 2K[\text{Re}\{F'_{nk} G_{kn}\} e^{-\Gamma(t-t_0)} \cos \omega_{nk}(t - t_0) \\
&\quad + \text{Im}\{F'_{nk} G_{kn}\} e^{-\Gamma(t-t_0)} \sin \omega_{nk}(t - t_0)] \\
&= K'_{nk} e^{-\Gamma(t-t_0)} \cos[\omega_{nk}(t - t_0) + \phi_{nk}] .
\end{aligned} \tag{3.97}$$

The above expression shows that the interference term is a damped harmonic oscillation. The presence of a periodically varying term in the fluorescence intensity constitutes the essence of the phenomenon known as quantum beating

[3.10, 11]. The form of the total signal, including the diagonal terms, for the excitation of only two levels, is presented in Fig. 3.2. It is clear that the amplitude and phase of the beats will depend on the form of the pump and observation matrices F and G, i.e. on the conditions of excitation and observation.

The manifestation of interference just discussed is also known as free quantum beats. This terminology underlines the fact that the coherent effects are observed in the emission of free atoms, i.e. atoms not interacting with radiation. This means that the weak intensity requirement is not imposed on the excitation process itself, provided the observation is carried out after its termination. The pulse excitation itself can be accompanied by nonlinear effects. After the termination of the pulse, when an observation is conducted, all nonlinear interactions vanish and the radiating atom is "free"[4].

Resonance beating is observed when a modulated source of excitation is used [3.13]. Resonance occurs when the frequency of modulation coincides with the energy gap between the interfering states. Let the spectrum of the excitation be sufficiently broad that it excites both states with their different energies simultaneously. In this case the pump matrix can be rewritten in the form (3.61):

$$F = F_{\varepsilon=0}[1 + \varepsilon\cos(\Omega't + \phi_0)] \, , \tag{3.98}$$

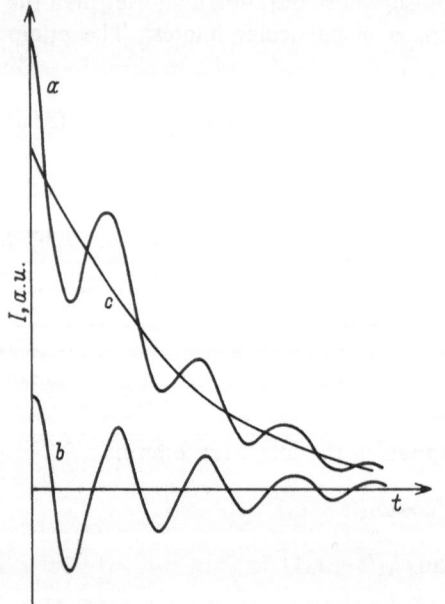

Fig. 3.2. (a) The damping of fluorescence, (b) the curve of the interference signal and (c) the exponential curve describing the damping of population

---

[4] Strictly speaking, this is valid only when the concentration of atoms is low. If the opposite is true, one has to take into account collective effects [3.12].

where ε is the modulation amplitude. The phase ϕ_0 can be set to zero by a suitable choice of the reference time t_0. The solution of (3.55) for this kind of pump can be written

$$\sigma_{nk} = \exp[-(\Gamma_{nk} + i\omega_{nk})t]$$

$$\times \int_{-\infty}^{t} (F_{\varepsilon=0})_{nk}\left[1 + \frac{\varepsilon}{2}(e^{i\Omega't'} + e^{-i\Omega't'})\right]\exp[(\Gamma_{nk} + i\omega_{nk})t']\,dt'$$

$$= (F_{\varepsilon=0})_{nk}\left[\frac{1}{\Gamma_{nk} + i\omega_{nk}} + \frac{\varepsilon}{2}\frac{e^{i\Omega't}}{\Gamma_{nk} + i(\omega_{nk} + \Omega')} + \frac{\varepsilon}{2}\frac{e^{-i\Omega't}}{\Gamma_{nk} + i(\omega_{nk} - \Omega')}\right].$$

$$(3.99)$$

The diagonal terms in the expression for the intensity ($n = k$, $\omega_{nk} = 0$) can be written separately

$$I_d = K\sum_n \sigma_{nn}G_{nn} = K\sum_n (F_{\varepsilon=0})_{nn}G_{nn}$$

$$\times \left[\frac{1}{\Gamma_n} + \frac{\varepsilon}{2}\frac{\Gamma_n(e^{i\Omega't} + e^{-i\Omega't}) - i\Omega'(e^{i\Omega't} - e^{-i\Omega't})}{\Gamma_n^2 + (\Omega')^2}\right].$$

$$(3.100)$$

Assuming that the decay time $1/\Gamma_{nk}$ is the same for all density matrix elements, we find

$$I_d = K\left(\frac{1}{\Gamma} + \frac{\varepsilon\cos(\Omega't - \phi)}{\sqrt{\Gamma^2 + (\Omega')^2}}\right)\sum_n (F_{\varepsilon=0})_{nn}G_{nn} ,$$

$$(3.101)$$

$$\tan\phi = \Omega'/\Gamma .$$

$$(3.102)$$

The component I_d of the fluorescence is modulated as a result of the modulation of the total population of the excited state. The fluorescence modulation amplitude given by

$$\varepsilon\Gamma/\sqrt{\Gamma^2 + (\Omega')^2}$$

is lower than the modulation amplitude of the pump and it decreases with increasing frequency. This is simply a consequence of the inertia of the spontaneous radiation. Because of this, the phase of the intensity oscillation of the spontaneous radiation is delayed relative to the pump. Moreover, the phase shift, like the amplitude, is a function of time. The dependence of the fluorescence phase modulation on the frequency of modulation of the excitation has been used for a long time in order to measure the short lifetime of molecules in condensed matter (the phase fluorometry technique). In atomic spectroscopy this technique was exploited by *Demtröder* in 1962 [3.14].

Let us now write down the intensity attributed to the off-diagonal terms of the density matrix

$$I = K \sum_{k,n} \frac{1}{2} \text{Re}\left\{(F_{\varepsilon=0})_{nk} G_{kn}\right\} \left[\frac{2\Gamma}{\Gamma^2 + \omega_{nk}^2} \right.$$

$$\left. + \frac{\varepsilon \cos(\Omega't - \phi_1)}{\sqrt{\Gamma^2 + (\omega_{nk} + \Omega')^2}} + \frac{\varepsilon \cos(\Omega't + \phi_2)}{\sqrt{\Gamma^2 + (\omega_{nk} - \Omega')^2}} \right]$$

$$+ K \sum_{k,n} \frac{1}{2} \text{Im}\left\{(F_{\varepsilon=0})_{nk} G_{kn}\right\} \left[\frac{2\omega_{nk}}{\Gamma^2 + \omega_{nk}^2} \right.$$

$$\left. - \frac{\varepsilon \sin(\Omega't - \phi_1)}{\sqrt{\Gamma^2 + (\omega_{nk}^2 + \Omega')^2}} + \frac{\varepsilon \sin(\Omega't + \phi_2)}{\sqrt{\Gamma^2 + (\omega_{nk} - \Omega')^2}} \right] ;$$

where

$$\tan \phi_1 = (\omega_{nk} + \Omega')/\Gamma; \quad \tan \phi_2 = (\omega_{nk} - \Omega')/\Gamma . \tag{3.103}$$

The first term in the square bracket accounts for the level crossing, i.e. the Hanle effect, the second term is negligibly small at high modulation frequencies and the third is actually the "resonance beating". This part of the intensity will attain a maximum when $\omega_{nk} = \Omega'$. The maximum value of the signal does not depend on the frequency of modulation.

Parametric Resonance. The phenomenon of parametric resonance occurs when the atomic resonance frequencies coincide with the frequency of modulation of the external magnetic or electric fields [3.11]:

$$\mathscr{H} = \mathscr{H}_0(1 - \varepsilon' \cos \Omega't) . \tag{3.104}$$

The variation in magnetic field will also cause the magnetic sublevels to vary. In a first-order approximation, one can assume that the additional change of the energy level linked with the term $\mathscr{H}_0 \varepsilon' \cos \Omega't$ is linear over the whole range of variation of the magnetic field. The differences of the characteristic frequencies of the atom will then be functions of time of the form

$$\omega_{nk} = \omega_{nk}^0(1 - \varepsilon \cos \Omega't) \tag{3.105}$$

(with a linear relationship between ω_{nk} and $\mathscr{H}$, $\varepsilon' = \varepsilon$). The equation of motion of the density matrix will have the following form

$$\dot{\sigma}_{nx} = - \Gamma\sigma_{nk} - i\omega_{nk}^0(1 - \varepsilon \cos \Omega't)\sigma_{nk} + F_{nk} . \tag{3.106}$$

The pump F_{nk} is assumed to be constant in time. The solution of this equation is

$$\sigma_{nk} = \exp\left(- \int [\Gamma + i\omega_{nk}^0(1 - \varepsilon \cos \Omega't)] \, dt\right)$$

$$\times F_{nk} \int \exp\left(\int [\Gamma + i\omega_{nk}^0(1 - \varepsilon \cos \Omega't)] \, dt\right) dt . \tag{3.107}$$

Since

$$\int [\Gamma + i\omega_{nk}^0(1 - \varepsilon\cos \Omega' t)] \, dt = \Gamma t + i\omega_{nk}^0 t - i\frac{\omega'}{\Omega'} \sin \Omega' t$$

(where the frequency deviation $\varepsilon\omega_{nk}^0$ is denoted by ω') and the function $\exp[-i(\omega'/\Omega')\sin \Omega' t]$ can be written as a series of Bessel functions

$$\exp[-i(\omega'/\Omega')\sin \Omega' t] = \sum_{l=-\infty}^{\infty} J_l\left(\frac{\omega'}{\Omega'}\right) e^{-il\Omega' t} \ ,$$

the solution can be reexpressed in the following way

$$\sigma_{nk} = \exp[-(\Gamma + i\omega_{nk}^0)t] F_{nk} \sum_{p=-\infty}^{\infty} J_p\left(\frac{\omega'}{\Omega'}\right) e^{ip\Omega' t}$$

$$\times \left\{ \int \exp[(\Gamma + i\omega_{nk}^0)t] \sum_{l=-\infty}^{\infty} J_l\left(\frac{\omega'}{\Omega'}\right) e^{-il\Omega' t} \, dt \right\} , \tag{3.108}$$

which, after integration, gives

$$\sigma_{nk} = F_{nk} \sum_{p=-\infty}^{\infty} \sum_{l=-\infty}^{\infty} \frac{J_p(\omega'/\Omega')J_l(\omega'/\Omega')\exp[i(p-l)\Omega' t]}{\Gamma + i(\omega_{nk}^0 - l\Omega')} . \tag{3.109}$$

The observed signal is

$$I = \frac{k}{\Gamma}\sum_n F_{nn} G_{nn} + K \sum_{k,n} \sum_{p,l=-\infty}^{\infty} \frac{J_p(\omega'/\Omega')J_l(\omega'/\Omega')}{\sqrt{\Gamma^2 + (\omega_{nk}^0 - l\Omega)^2}}$$

$$\times [\mathrm{Re}\{F_{nk} G_{kn}\} \cos[(p-l)\Omega' t - \phi_l]$$

$$- \mathrm{Im}\{F_{nk} G_{kn}\} \sin[(p-l)\Omega' t - \phi_l]] , \tag{3.110}$$

where

$$\tan \phi_l = \frac{\omega_{nk}^0 - l\Omega'}{\Gamma} .$$

The term with $p = 0$ and $l = 0$ corresponds to the level crossings.

A parametric resonance consists of an infinite series of resonances that rapidly decreases in amplitude. Their widths do not depend on the amplitude of the rf field. However, a high modulation amplitude ε is required in order to get signals of noticeable size, and this limits the applicability of the method.

All types of interferences of quantum states lead to changes in the emitted light of the type derived qualitatively in Chap. 2. Here we presented a classical treatment of the emission process of an atom as a radiating dipole. However, a correct quantitative description can be obtained only in the framework of quantum mechanics.

3.7 The Radiation Pattern and Polarization for Transitions Between Eigenstates of the Angular Momentum Operator

Before turning to a discussion of the angular dependences in spontaneous radiation, which are highly important for the understanding of interference phenomena, we briefly recap on the description of emission from quantum systems.

The matrix element of the dipole transition operator is considered to be a vector. This vector coincides in direction with one of the circular unit vectors e_0, e_+, e_-, depending on the magnetic quantum numbers of the coupled states. The matrix element of the dipole transition between an initial state of quantum number m and a final state of quantum number $m - 1$ is directed along the unit vector

$$e_+ = - \frac{e_x + i e_y}{\sqrt{2}} . \tag{3.111}$$

Upon transition to the state $m + 1$ it is directed along the unit vector

$$e_- = \frac{e_x - i e_y}{\sqrt{2}} . \tag{3.112}$$

The third possible case is conservation of magnetic quantum number. This corresponds to a matrix element along the unit vector

$$e_0 = e_z . \tag{3.113}$$

This will be shown in Sects. 5.5, 6 when we calculate the matrix elements of the dipole transitions. The polarization of the radiation e^5 upon a transition from $|m\rangle$ to $|\mu\rangle$ is determined by the projection of the eigenvector (Sect. 5.6) of the matrix element $d_{m\mu}^*$ onto a plane perpendicular to the propagation direction of the light. Let the light propagate in the direction Oz'. We fix the coordinate system $x'y'z'$ given by the Euler angles α, β, γ to this direction. Let us denote the polarization vector of the light emitted upon a transition from the state $|m\rangle$ to $|\mu\rangle$ and propagating in the direction of Oz' (this vector lies in the $x'y'$-plane) by $l_{m\mu}$. One can prove that it is equal to ($\mu = m, m \pm 1$):

$$l_{mm-1} = (e'_+ D^*_{+1+1} + e'_- D^*_{+1-1}) d_{mm-1} ;$$

$$l_{mm} = (e'_+ D^*_{0+1} + e'_- D^*_{0-1}) d_{mm} ; \tag{3.114}$$

$$l_{mm+1} = (e'_+ D^*_{-1+1} + e'_- D^*_{-1-1}) d_{mm+1} .$$

---

[5] Strictly speaking in quantum mechanics there is no such concept as polarization of radiation; one can refer only to the quantity $\langle \mu | d \cdot e_r | m \rangle$, which has the meaning of probability amplitude for the emission of a quantum of light with a polarization e_r^* upon a transition from the state $|m\rangle$ to state $|\mu\rangle$.

Table 3.1. The elements of the matrix $D_{ik}^{(1)}$

i	$k = +1$	$k = 0$	$k = -1$
$+1$	$e^{-i\alpha}\dfrac{1+\cos\beta}{2}e^{-i\gamma}$	$-e^{-i\alpha}\dfrac{\sin\beta}{\sqrt{2}}$	$e^{-i\alpha}\dfrac{1-\cos\beta}{2}e^{i\gamma}$
0	$\dfrac{1}{\sqrt{2}}\sin\beta\,e^{-i\gamma}$	$\cos\beta$	$-\dfrac{1}{\sqrt{2}}\sin\beta\,e^{i\gamma}$
-1	$e^{i\alpha}\dfrac{1-\cos\beta}{2}e^{-i\gamma}$	$e^{i\alpha}\dfrac{\sin\beta}{\sqrt{2}}$	$e^{i\alpha}\dfrac{1+\cos\beta}{2}e^{i\gamma}$

Here $e_{\pm}$ are circular unit vectors in the primed cartesian coordinate system, and D_{ik} are the generalized spherical functions. They are known also as the Wigner functions [3.15]. The quantities D_{ik} constitute the so-called rotation matrix D. Their explicit form is given in Table 3.1 by the entries $D_{ik}^{(1)}$.

Let us examine a few specific cases. The simplest is the π-transition ($\mu = m$). The polarization l_{mm} is linear in this case and is given by

$$l_{mm} = (-e'_x\sin\beta\cos\gamma + e'_y\sin\beta\sin\gamma)d_{mm} \ .$$

It lies in a plane passing through the z-axis and makes an angle $(\pi/2 - \beta)$ to this axis. It is not difficult to find the radiation intensity for the π-transition and its dependence on the angle β:

$$I = K|l_{mm}|^2 = K\sin^2\beta\, d_{mm}^2 \ .$$

This same intensity pattern of a linear dipole was also obtained from the classical treatment (2.15). It is obvious that in the direction $\beta = 0$ the π-radiation is absent; light in this direction is determined by the vectors e_+ and e_-. The transition $m \rightarrow m - 1$ i.e. the σ_+ transition corresponds to the former, and the σ_- transition to the latter.

Let us find the polarization of the light corresponding to the σ_- transition

$$l_{mm+1} = [(\cos\beta\cos\gamma - i\sin\gamma)e'_x$$

$$- (\cos\beta\sin\gamma + i\cos\gamma)e'_y](e^{-i\alpha}/\sqrt{2})d_{mm+1} \ .$$

It can be seen that the linear components e'_x and e'_y are shifted in phase by $-\pi/2$ and have different amplitudes. This corresponds to an elliptic polarization of the light. It reduces to linear polarization when $\beta = \pi/2$.

The radiation pattern is determined by the square of the absolute value of the vector l

$$I = K|l_{mm+1}|^2 = \tfrac{1}{2}K(\cos^2\beta + 1) \ .$$

As anticipated, both the polarization and the intensity diagram are the same as those obtained from the classical approach; see (2.13).

The absorption is described by the corresponding complex conjugate quantities. If the magnetic quantum number of the upper state is greater by one than that of the lower state, then light of one and the same polarization e_{+1} will be emitted and absorbed. Note, however, that the initial states for emission and absorption will be the upper and lower states, respectively. The diagram of the absorption coefficient for different polarizations is thus determined by expressions that are complex conjugates of (3.114).

3.8 Influence of Interference Between States on the Polarization of Spontaneous Radiation

The existence of coherency in an excited state changes the polarization characteristics of spontaneous radiation. If one knows the density matrix of the excited state, then the radiation intensity in any direction and any polarization can be found with the help of (3.67). For this purpose, and to offer a feeling for the character of the variation of the radiation in the presence of coherency, we will discuss a specific example.

Let the excited state be three-fold degenerate with respect to the magnetic quantum number, i.e. $J = 1$. The spontaneous radiation leads the atom to the state with $J = 0$. Furthermore, let the excited state be created from the lower state (with $J = 0$) by the absorption of light with a polarization e:

$$e = a_0 e_0 + a_+ e_+ + a_- e_- .\tag{3.115}$$

Since the numerical value of the matrix elements for all components of this transition are identical, denoting them by d, one can write the density matrix in the form

$$\sigma = \frac{d^2}{\Gamma} \begin{pmatrix} |a_-|^2 & a_- a_0^* & a_- a_+^* \\ a_0 a_-^* & |a_0|^2 & a_0 a_+^* \\ a_+ a_-^* & a_+ a_0^* & |a_+|^2 \end{pmatrix} .\tag{3.116}$$

The density matrix elements do not contain the quantity ω, since we have assumed degenerate levels and no perturbation is applied to lift the degeneracy.

We will show that the spontaneous emission due to the chosen transition is completely polarized. We choose the coordinate system with the z-axis along the direction of observation n. The detection of completely polarized light, whose vector polarization is denoted by e_r, is equivalent to confirming that the intensity in the orthogonal polarization is equal to zero.

The polarization of the incident light is denoted by e_λ. Its expansion in terms of circular unit vectors in our coordinate system (3.115) is given by

$$e_\lambda = a_0^\lambda e_0 + a_+^\lambda e_+ + a_-^\lambda e_- .$$

Let us suppose that the polarization of the observed light e_r corresponds to the projection of e_λ on the xy-plane, i.e. perpendicular to n

$$e_r = a_+^\lambda e_+ + a_-^\lambda e_- .$$ (3.117)

We shall prove that the intensity in the orthogonal polarization, which we call e_p, is zero. Since e_r and e_p are orthogonal

$$e_r e_p^* = 0 = a_+^\lambda (a_+^p)^* + a_-^\lambda (a_-^p)^* .$$ (3.118)

In accordance with (3.67):

$$I(e_p) = K \sum_{i,k} \sigma_{ik}(e_\lambda) G_{ki}(e_p) .$$

At the same time we have from (2.116)

$$\sigma_{ik} = \frac{d^2}{\Gamma} a_i^\lambda (a_k^\lambda)^*$$

and G_{ki} from (2.66), but in the new notation:

$$G_{ki} = d^2 a_k^p (a_i^p)^* .$$

Then

$$I(e_p) = K \frac{d^4}{\Gamma} \sum_{i,k} a_i^\lambda (a_k^\lambda)^* a_k^p (a_i^p)^*$$

$$= K \frac{d^4}{\Gamma} \left[\sum_i a_i^\lambda (a_i^p)^* \cdot \sum_k (a_k^\lambda)^* a_k^p \right]$$

$$= K |(a_+^\lambda (a_+^p)^* + a_-^\lambda (a_-^p)^* + a_z^\lambda \cdot 0|^2 .$$

According to (3.118) the sum in the bracket is equal to zero, i.e. $I(e_p) = 0$. In addition, using (3.117) we have

$$I(e_r) = K \frac{d^4}{\Gamma} |a_+^\lambda (a_+^\lambda)^* + a_-^\lambda (a_-^\lambda)^*|^2 \neq 0 .$$

Consequently, the polarization of spontaneous radiation in the present case is determined by the projection of the polarization of the incident excitation on the plane perpendicular to n (3.117).

Let us now lift the degeneracy of the levels. This can be done by applying a magnetic field. Let us direct it along the z-axis. This does not relax the generality of our discussion since the vector e_λ is chosen arbitrarily. One has to rewrite the density matrix as

$$\sigma_{ik} = \frac{d^2}{\Gamma + i\omega_{ik}} a_i^\lambda (a_k^\lambda)^* .$$ (3.119)

In a strong magnetic field when $\omega \to \infty$, the off-diagonal terms of the density matrix become zero and only the diagonal, incoherent terms remain. The emission from such a system is determined by the population of the sublevels, i.e. by the sum of the intensities radiated from π and σ transitions. The radiation pattern for each of these transitions is symmetric with respect to the z-axis, i.e. with respect to the magnetic field direction. This conclusion also follows from the classical treatment: the magnetic field "twists" all emitters about its direction and the whole system acquires axial symmetry.

However, complete agreement between the conclusions of the classical and quantum mechanical descriptions exists only for the classical Zeeman triplet. For other transitions, excitation by light of any polarization leads to spontaneous emission that can be assumed to consist of two parts: unpolarized radiation with spherical symmetry and a part that is polarized anisotropically. It is convenient to prove this using the formalism of the statistical tensor, with which we shall become acquainted in Sect. 3.10. Let us now show that interference phenomena observed without spectral decomposition are necessarily accompanied by an intensity redistribution over angles.

3.9 Redistribution of Radiated Energy Due to the Interference of Quantum States

Like any other interference the interference of atomic states causes only the redistribution of intensity in space and has no influence upon integral intensity of radiation.

This is true also for all cases of interference which depends upon time. All cases of beats may be regarded as a result of the rotation and oscillation of the radiation pattern of a dipole under the action of an intra-atomic magnetic field or external magnetic and electric fields. The description in Chap. 2 gave the same interpretation of beats.

The influence of the interference of states on emitted radiation depends on the properties of the interfering states, i.e. on their quantum numbers. It is convenient to divide all possible cases of interference into three groups. The first group is the interference of states with different angular momenta F (5.21) or J (5.2). The second group comprises states of equal total angular momenta, but differing in magnetic quantum numbers. The third group has identical total angular momenta and magnetic quantum number as well and the states differ only in the principal quantum number n.

The interference of states with different total angular momenta F or J are necessarily accompanied by an anisotropy of the radiation. Upon averaging the radiation over all directions at any instant in time, all the interference effects disappear [3.16–18]. In order to prove this, we consider the formula for the intensity (3.67) [taking account of (2.66)]

$$I(e_r) = K \sum_{n, k, \mu} \sigma_{nk} (d \cdot e_r)_{\mu n} (d \cdot e_r)^*_{\mu k} . \tag{3.120}$$

From the projections of the matrix elements of the dipole transitions between the interfering and lower states, we isolate the Klebsch–Gordon coefficients [3.19, 15] (Sect. 5.6) and the product of unit vectors defining the characteristic polarization of the transitions e_v and of the observed light e_r. The remaining factor (α) will be the same for all terms with the same F_0. One can thus take it out of the summation sign and it will have no significance:

$$I(e) = K \sum_{m_{F_i}, m_{F_k}} \sigma_{m_{F_i} m_{F_k}} \alpha \sum_{m_{F_0}} C^{F_k m_{F_k}}_{F_0 m_{F_0} 1 v'} C^{F_i m_{F_i}}_{F_0 m_{F_0} 1 v} (e_r \cdot e_{v'})^* (e_r \cdot e_v) \ . \tag{3.121}$$

We will show that, in each of the interference terms, the sum over m_{F_i} and m_{F_k} becomes zero when one integrates over all directions and polarizations. For this it will follow that the entire interference, as well as the interference between pairs of substates, vanishes upon averaging with respect to the angles. This demonstrates that the interference of states leads to the redistribution of intensity over polarizations and, consequently, over angles. An integration of the intensity over all directions is equivalent to an integration of the radiation over all polarizations[6], which can be taken to be linear and uniformly distributed over angles. For every term of the sum over m_{F_i}, m_{F_k}, which we denoted by $I(m_{F_i} m_{F_k})$ the integral is written as

$$\int I(m_{F_i} m_{F_k}) \, d\Omega$$

$$= \sum_{m_{F_0}} C^{F_k m_{F_k}}_{F_0 m_{F_0} 1 v'} C^{F_i m_{F_i}}_{F_0 m_{F_0} 1 v} \int (e_r \cdot e_{v'})^* (e_r \cdot e_v) \sin \theta \, d\theta \, d\phi \ . \tag{3.122}$$

We calculate the integral over angles appearing on the right-hand side. The vector e_r can be written in the form

$$e_r = e_z \cos \theta' + e_x \sin \theta' \cos \phi' + e_y \sin \theta' \sin \phi'$$

and the characteristic polarization of the transition is given by

$$e_{v=+1} \equiv e_+; \qquad e_{v=-1} \equiv e_-; \qquad e_{v=0} \equiv e_0 \ .$$

Then

$$e_0 \cdot e_r = \cos \theta' = Y_{10} \sqrt{\frac{4\pi}{3}} \ ,$$

$$e_+ \cdot e_r = \frac{-1}{\sqrt{2}} \sin \theta' e^{i\phi'} = Y_{11} \sqrt{\frac{4\pi}{3}} \ , \tag{3.123}$$

$$e_- \cdot e_r = \frac{1}{\sqrt{2}} \sin \theta' e^{-i\phi'} = Y_{-1-1} \sqrt{\frac{4\pi}{3}} \ ,$$

---

[6] This is evident, for example, from (2.39).

where Y are spherical harmonics. It is known [3.15] that the integral of a product of spherical harmonics over the total solid angle is equal to unity if they are identical and is zero if they are different

$$\int Y_{1v'} Y_{1v}^* \, d\Omega = \delta_{vv'} \, . \tag{3.124}$$

In the simplest case, namely, the interference of magnetic sublevels, the difference $v - v' = \pm 1$. For these cases, the integral (3.124) will be equal to zero, i.e. the interference terms vanish upon averaging over direction. This indicates that the interference of states leads to the redistribution of intensity over directions and polarizations. This result does not depend on the character of the density matrix, and, if the latter depends on time, remains valid at every instant in time. Consequently, the quantum beats must be regarded as a time-dependent redistribution of energy over directions (and polarizations). If an atom possesses no hyperfine structure, and if one considers the interference of states with different J (but not F), then the result of averaging will be the same, because the formulae remain of the same type; the total moment F and its projections are merely replaced by the total moment J and its projections. The same result will be obtained for the interference of atomic states that are perturbed by external fields such that F and J are no longer good quantum numbers. This is because the factor containing these characteristics of the wavefunctions will become zero.

The case $v = v'$ signifies that the interfering states (3.124) will have the same magnetic quantum numbers. When $v = v'$ the integral (3.124) is equal to unity, but then, according to the properties of the Klebsch–Gordon coefficients, the sum over m_{F_0} will become zero:

$$\sum_{m_{F_0}} C_{F_0 m_{F_0} 1v}^{F_k m_{F_k}} C_{F_0 m_{F_0} 1v}^{F_i m_{F_i}} = \delta_{F_k F_i} \delta_{m_{F_k} m_{F_i}} \, . \tag{3.125}$$

In the formula (3.125) $\delta_{m_{F_k} m_{F_i}} = 1$, because $m_{F_k} = m_{F_i}$. However, the total angular momenta can be different and then $\delta_{F_k F_i} = 0$. If they are different, then the interference of states in the case of equal magnetic quantum numbers will lead to the angular redistribution of radiation.

If the wavefunctions describing the interference of states are not eigenfunctions of the momentum operator, i.e. if the quantum numbers F and J are not good quantum numbers (as, for instance, in an atom perturbed by external fields), then one can represent the matrix elements of the electric dipole transitions through the matrix elements of the dipole transitions of the unperturbed states. Then, in the expression (3.121), every term in the sum over m_{F_0} will be replaced by a set of terms, each of them resembling the product (3.122). As a result of the conditions (3.124) and (3.125) each of these terms yields zero after integration over angles.

The interference of states with identical m, involves transitions to other states that all have the same polarization and the same angular redistribution. It is thus interesting to monitor the form of this radiation. For this purpose, it is quite sufficient to compare the intensity in the characteristic polarizations $e_0(\pi)$ and

$e_\pm(\sigma)$. In accordance with (3.121), for $m_{F_0} = m_{F_k} = m_{F_i}$ the interference term

$$I(e_0) = K\alpha \sum_{m_{F_i}} \sigma_{m_{F_i} m_{F_k}} C_{F_0 m_{F_0} 10}^{F_i m_{F_i}} C_{F_0 m_{F_0} 10}^{F_k m_{F_k}} \neq 0$$

is different from zero. However, the total interference part of the intensity is given by

$$I(e_0) + I(e_+) + I(e_-) = 0$$

and consequently

$$I(e_0) = -[I(e_+) + I(e_-)] . \tag{3.126}$$

This means that quantum beats correspond to the periodic variation of the intensity in the π-polarization and the same variation but in antiphase in the σ-polarization. At level crossings, interferences will redistribute the radiation between π- and σ-transitions. The sign of the interference term depends on the sign of the Klebsch–Gordon coefficients, as well as on the sign of the corresponding density matrix element $\sigma_{m_{F_i} m_{F_k}}$, which depends on the technique used to produce the ensemble.

Since the problem is truly symmetric with respect to emission and absorption, it is quite evident that the coherency of quantum states for an ensemble prepared by optical means will be achieved only if one illuminates anisotropically. Of course, any other method used to produce the ensemble in order to observe interference phenomena must also possess anisotropy.

When the quantum numbers m and F are both identical, none of the factors in (3.122) will turn out to be zero, i.e. these factors will give no information about the disappearance of the interference terms upon integrating over all directions. However, the equality of m and of F indicates that the levels must differ by some other quantum number. A possible difference is in the principal quantum number n, and this will be discussed below. But it is quite feasible that there may be a difference in the orbital moment J and this means that the interfering states belong to different components of the fine-structure multiplet, i.e. they are quite far apart from one another in energy. Because the fine-structure splitting is usually greater than the hyperfine splitting, it is thus reasonable to assume that the intensities of all the hyperfine components are observed simultaneously, i.e. the observation offers no spectral distinction with respect to the hyperfine components. Let us rewrite the formula (3.121) in another form, concentrating now on the quantity α (Sect. 5.6) that was not interesting before. The quantity α is a function of the quantum numbers F_0, J_0 of the lower states and $F_{i,k}, J_{i,k}$ of the upper states

$$I = K' \sum_{F_0} (2F_0 + 1)\sqrt{(2J_k + 1)(2J_i + 1)} \begin{Bmatrix} I & J_0 & F_0 \\ 1 & F_i & J_i \end{Bmatrix} \begin{Bmatrix} I & J_0 & F_0 \\ 1 & F_k = F_i & J_k \end{Bmatrix}$$

$$\times \sum_{m_{F_0}} C_{F_0 m_{F_0} 1v}^{F_i = F_k m_{F_i} = m_{F_k}} \int (e_r \cdot e_v)^2 \, d\Omega . \tag{3.127}$$

We have just shown that the sum over m_{F_0} in the case considered is equal to one. But the 6_j symbols have the property [Ref. 3.15, Chap. 9, eq. (9)]

$$\sum_{F_0} (2F_0 + 1)\sqrt{(2J_k + 1)(2J_i + 1)} \begin{Bmatrix} I & J_0 & F_0 \\ 1 & F & J_i \end{Bmatrix} \begin{Bmatrix} I & J_0 & F_0 \\ 1 & F & J_k \end{Bmatrix} = \delta_{J_i J_k} \ .$$

(3.128)

Because the quantum numbers J are different in this case, the sum over F_0 reduces to zero, i.e. under these conditions of observation, the interference term also vanishes.

In the same way one can show that in the interference of excited states with different l but with the remaining quantum numbers equal, the interference term in the intensity disappears upon averaging over all directions and summing with respect to all lower states.

Let us now consider the case in which all the quantum numbers of the interfering states are equal except the principal quantum numbers n. Such states differ from one another only in the radial part of the wavefunction, which is always positive. From the formulae of the type (3.121), derived on the basis of density matrix theory (Sect. 3.2), it follows that the interference term is conserved upon averaging over directions. Actually, for equal J, the condition (3.128) reduces the sum of Racah coefficients to one; the radial parts are always positive and they cannot turn the product (3.127) to zero (independent of the presence or absence of hyperfine structure). At first sight it seems that an interference of states not depending on angle is possible. However, a detailed analysis within the framework of quantum electrodynamics shows that such states are not interfering.

The reason behind this is that quantum electrodynamics takes into account the interaction of states through the vacuum. For states with different angular momenta, it leads to the same expression for the scattered light intensity as was derived from the density matrix formalism. Hence, for states with equal angular momenta yet another term will arise: it is essentially equal to the interference term, but has the opposite sign. It thus cancels the interference term with the result that no trace of interference remains in the complete expression for the intensity.

Let us consider the simplest case of a three level atom with the lower state $|\mu\rangle$ and excited states $|i\rangle_0$ and $|k\rangle_0$ with decay constants Γ_i^0 and Γ_k^0, respectively. We will see how this is approached in quantum electrodynamics, by considering atomic system + field. The initial state of this system is an atom in the state $|\mu\rangle$ and a field that consists of a single photon of frequency ω and polarization e_λ. We denote this state by $|0\rangle$. The final state, comprising an atom in the state $|\mu\rangle$ and photon of frequency ω' and polarization e_r, is given by $|0'\rangle$. All other states of the system with an atom in the state $|\mu\rangle$ will be denoted by $|0''\rangle$, $|0'''\rangle$, etc. When the atom is in an excited state, there is no photon in the field and the state will be denoted in the same way as the atomic state. The lower state width is denoted by γ and we assume it to be sufficiently narrow.

The intensity of the scattered light can be found by making use of the scattering matrix.

Let us use the Feynman light scattering diagram [3.3] to represent our atomic system (a single atom), in which ω and ω' are the frequencies of the absorbed and emitted photons, respectively:

$$\langle 0', \omega' | S | 0, \omega \rangle \equiv S_{0'0} = \overset{\omega}{\underset{|0\rangle}{\rightsquigarrow}} \left[\underset{|i\rangle_0}{\bullet\!\!-\!\!-\!\!\bullet} + \underset{|k\rangle_0}{\bullet\!\!-\!\!-\!\!\bullet} \right.$$

$$+ \underset{|i\rangle_0 \quad |0''\rangle \quad |k\rangle_0}{\bullet\,\overset{\omega''}{\frown}\,\bullet} + \underset{|k\rangle_0 \quad |0''\rangle \quad |i\rangle_0}{\bullet\,\overset{\omega''}{\frown}\,\bullet}$$

$$+ \underset{|i\rangle_0 \; |0''\rangle \; |k\rangle_0 \; |0'''\rangle \; |i\rangle_0}{\bullet\,\overset{\omega''}{\frown}\,\overset{\omega'''}{\frown}\,\bullet}$$

$$\left. + \underset{|k\rangle_0 \; |0''\rangle \; |i\rangle_0 \; |0'''\rangle \; |k\rangle_0}{\bullet\,\overset{\omega''}{\frown}\,\overset{\omega'''}{\frown}\,\bullet} + \cdots \right] \overset{\omega'}{\underset{|0'\rangle}{\nearrow}} . \qquad (3.129)$$

On this diagram, points incidate the interactions leading to the transitions of the atomic system + field, which itself is represented by the solid lines. The state symbol is given below the lines. The wavy lines indicate the presence of a free photon in the system. The diagram shows explicitly only the scattering matrix terms (partial amplitudes) of low order. Upon summation, we take into account an infinite series of terms similar to those shown. However, we will not take into consideration the non-resonance terms.

We note that, for ordinary interference of states, after summing over intermediate states $|0'', ''' \cdots\rangle$ and integrating over all angles and frequencies, all terms, except the first two, have zero contribution. The terms from the third one onwards differ from zero only in the exceptional case, which we discuss now, in which the states have different angular momenta. This is not difficult to see if one writes the explicit form of these terms and performs the corresponding averaging.

$$S_{0'0}$$

$$= -\frac{2\pi i}{\hbar} \delta(\omega - \omega') \left\{ \frac{1}{\hbar} \frac{V_{0'i} V_{i0}}{\left(\omega - \omega_{i0}^0 + \frac{i\Gamma_i^0}{2}\right)} + \frac{1}{\hbar} \frac{V_{0'k} V_{k0}}{\left(\omega - \omega_{k0}^0 + \frac{i\Gamma_k^0}{2}\right)} \right.$$

$$+ \frac{1}{\hbar^3} \frac{V_{0'i} V_{i0''} V_{0''k} V_{k0}}{\left(\omega - \omega_{i0}^0 + \dfrac{i\Gamma_i^0}{2}\right)\left(\omega - \omega'' + \dfrac{i\gamma}{2}\right)\left(\omega - \omega_{k0}^0 + \dfrac{i\Gamma_k^0}{2}\right)}$$

$$+ \frac{1}{\hbar^3} \frac{V_{0'k} V_{k0''} V_{0''i} V_{i0}}{\left(\omega - \omega_{k0}^0 + \dfrac{i\Gamma_k^0}{2}\right)\left(\omega - \omega'' + \dfrac{i\gamma}{2}\right)\left(\omega - \omega_{i0}^0 + \dfrac{i\Gamma_i^0}{2}\right)}$$

$$+ \frac{1}{\hbar^5} \frac{V_{0'i} V_{i0'''} V_{0'''k} V_{k0''} V_{0''i} V_{i0}}{\left(\omega - \omega_{i0}^0 + \dfrac{i\Gamma_i^0}{2}\right)^2 \left(\omega - \omega'' + \dfrac{i\gamma}{2}\right)\left(\omega - \omega''' + \dfrac{i\gamma}{2}\right)\left(\omega - \omega_{k0}^0 + \dfrac{i\Gamma_k^0}{2}\right)}$$

$$\left. + \frac{1}{\hbar^5} \frac{V_{0'k} V_{k0'''} V_{0'''i} V_{i0''} V_{0''k} V_{k0}}{\left(\omega - \omega_{k0}^0 + \dfrac{i\Gamma_k^0}{2}\right)^2 \left(\omega - \omega'' + \dfrac{i\gamma}{2}\right)\left(\omega - \omega''' + \dfrac{i\gamma}{2}\right)\left(\omega - \omega_{i0}^0 + \dfrac{i\Gamma_i^0}{2}\right)} + \cdots \right\}$$

$$= \frac{-2\pi i}{\hbar} \delta(\omega - \omega')\left\{ T(ii) + T(kk) + T(ik) + \cdots \right\} = \frac{-2\pi i}{\hbar} \delta(\omega - \omega') T_{0'0} \ .$$

$$(3.130)$$

The quantity $T_{0'0}$ is a matrix element known as the transition operator on the energy surface [3.2, 3]. V_{mn} is the matrix element projection of the dipole moment operator between the states $|m\rangle$ and $|n\rangle$ on the direction of polarization of the photon with frequencies ω, ω', ω'' etc., multiplied by the electric field strength, normalized to one photon per unit volume:

$$\mathscr{E} = \sqrt{2\pi\hbar\omega} \ . \qquad (3.131)$$

The matrix elements of the operator $\hat{S}$ are analogous to the quantities $C_{\mu'}$ in (3.79), which we also called the scattering matrix elements. One can prove that, for a particular problem using a particular approximation, both matrices will yield identical results.

In the interference of states with different angular momenta, all terms involving integration over intermediate states will be zero. We verify this for the example of the term $T(ki)$. The summation over the virtual states ($|0''\rangle$, $|0'''\rangle$ etc.) is equivalent to integrating over all frequencies (ω'', ω''' etc.) [3.2, 3, 5] and over all directions of propagation of a photon and all possible polarizations (we have condensed the last two quantities into the parameter Ω):

$$\sum_{0''} (\mathscr{E}'')^2 Y(\omega'') \rightarrow \int \frac{\hbar(\omega'')^3 \, d\omega'' \, d\Omega}{4\pi^2 c^3} Y(\omega'') \ . \qquad (3.132)$$

With this substitution and the equality[7] $V_{k0''} = \mathscr{E}''(d \cdot e^*)_{k0''}^0$ the term $T(ki)$ can

---

[7] The superscript "0" is simply a reminder that the matrix elements are taken between the initial functions.

be written as follows

$$T(ki) = \frac{1}{\hbar^2} \frac{V_{0'k} V_{i0}}{(\omega - \omega_{k0}^0 + i\Gamma_k^0/2)(\omega - \omega_{i0}^0 + i\Gamma_i^0/2)}$$

$$\times \int \frac{\hbar(\omega'')^3 (d \cdot e^*)_{k0''}^0 (d \cdot e)_{0''i}^0 d\omega'' d\Omega}{4\pi^2 c^3 (\omega - \omega'' + i\gamma/2)} . \tag{3.133}$$

The factor

$$\int (d \cdot e^*)_{k0''}^0 (d \cdot e)_{0''i}^0 d\Omega$$

will become zero if the states $|i\rangle_0$ and $|k\rangle_0$ differ in any of their angular momenta. One can prove this in analogy to (3.121–128) by averaging the interference terms to zero. The consideration of the interaction of states through the vacuum thus adds nothing new to the commonly used description of interference phenomena.

Let us return to the scattering matrix for atoms simultaneously excited to states with identical angular momenta. As already mentioned, for such states, the terms reflecting the vacuum interaction are nonzero. In order to give them a convenient form, we turn to the angular dependences of the states $|i\rangle_0$ and $|k\rangle_0$. Since these states differ only in the principal quantum numbers, their angular dependences are the same. From this it follows that the ratio of the projections of the dipole moments

$$\frac{\langle 0|d \cdot e|i\rangle_0}{\langle 0|d \cdot e|k\rangle_0} = \beta$$

does not depend on the polarization e. In our notation it reads

$$V_{i0''} = \beta V_{k0''} , \tag{3.134}$$

whence follows the equality

$$V_{0'k} V_{k0''} V_{0''i} V_{i0} = V_{0'k}(V_{i0''})^2 V_{k0} = (V_{0''k})^2 V_{0'i} V_{i0} . \tag{3.135}$$

These equalities prove that

$$T(ik) = T(ki) . \tag{3.136}$$

Let us now rewrite the expression for these terms making use of (3.135):

$$T(ik) = \frac{V_{0'k} V_{k0}}{\hbar^2 \left(\omega - \omega_{k0}^0 + \dfrac{i\Gamma_k^0}{2}\right)\left(\omega - \omega_{i0}^0 + \dfrac{i\Gamma_i^0}{2}\right)} \int \frac{\hbar(\omega'')^3 |(d \cdot e)_{0''i}^0|^2 d\omega'' d\Omega}{4\pi^2 c^3 \left(\omega - \omega'' + \dfrac{i\gamma}{2}\right)} . \tag{3.137}$$

After integrating over ω'' one can isolate the constant Γ_i^0. In accordance with the definition, this is equal to [3.2]

$$\Gamma_i^0 = \sum_0 \int \frac{\omega_{i0}^3 |(\boldsymbol{d} \cdot \boldsymbol{e})_{0i}^0|^2}{2\pi c^3} \, d\Omega \,, \tag{3.138}$$

where ω_{i0} is the frequency of the atomic transition.

Since there is only one final state in our model, only one term will remain under the summation sign in the expression for Γ_i^0. We ignore the difference between ω and ω_{i0}, i.e. we assume that the excitation is close to resonance. Some of the factors under the integral can then be collectively replaced by the factor Γ_i^0 and (3.137) is transformed to the form:

$$T(ki) = \frac{-i\Gamma_i^0 V_{0'k} V_{k0}}{2\hbar(\omega - \omega_{k0}^0 + i\Gamma_k^0/2)(\omega - \omega_{i0}^0 + i\Gamma_i^0/2)} = T(ik) \,. \tag{3.139}$$

Proceeding further in the same way, we obtain

$$T(kik) = \frac{-\Gamma_i^0 \Gamma_k^0 V_{0'k} V_{k0}}{4\hbar(\omega - \omega_{k0}^0 + i\Gamma_k^0/2)^2 (\omega - \omega_{i0}^0 + i\Gamma_i^0/2)} \,. \tag{3.140}$$

By exchanging the indices i and k we can find the equivalent expression for $T(iki)$.

For the sake of convenience in the subsequent summation, let us now rewrite the total sum in the form of four series:

$$
\begin{aligned}
S_{0'0} = -\frac{2\pi i}{\hbar^2} \delta(\omega - \omega') \Bigg\{ & V_{0'k} V_{k0} \\
\times \Bigg[& \frac{1}{\left(\omega - \omega_{k0}^0 + \dfrac{i\Gamma_k^0}{2}\right)} - \frac{\Gamma_i^0 \Gamma_k^0}{4\left(\omega - \omega_{k0}^0 + \dfrac{i\Gamma_k^0}{2}\right)^2 \left(\omega - \omega_{i0}^0 + \dfrac{i\Gamma_i^0}{2}\right)} \\
+ \cdots - \frac{i\Gamma_i^0}{2} & \Bigg(\frac{1}{\left(\omega - \omega_{k0}^0 + \dfrac{i\Gamma_k^0}{2}\right)\left(\omega - \omega_{i0}^0 + \dfrac{i\Gamma_i^0}{2}\right)} \\
& - \frac{\Gamma_i^0 \Gamma_k^0}{4\left(\omega - \omega_{k0}^0 + \dfrac{i\Gamma_k^0}{2}\right)^2 \left(\omega - \omega_{i0}^0 + \dfrac{i\Gamma_i^0}{2}\right)^2} + \cdots \Bigg) \Bigg] \\
& + \text{equivalent terms with the indices } i \text{ and } k \text{ exchanged} \Bigg\} \,.
\end{aligned}
\tag{3.141}
$$

It is easy to see that each series is a geometric progression. The summation results in

$$S_{0'0} = -\frac{2\pi i}{\hbar^2}\delta(\omega - \omega')$$

$$\times \frac{(\omega - \omega_{i0}^0)V_{0'k}V_{k0} + (\omega - \omega_{k0}^0)V_{0'i}V_{i0}}{\left(\omega - \omega_{i0} + \dfrac{i\Gamma_i^0}{2}\right)\left(\omega - \omega_{k0}^0 + \dfrac{i\Gamma_k^0}{2}\right) + \dfrac{\Gamma_i^0\Gamma_k^0}{4}} . \tag{3.142}$$

Let us transform the denominator to the form

$$\left(\omega - \omega_{i0} + \frac{i\Gamma_i}{2}\right)\left(\omega - \omega_{k0} + \frac{i\Gamma_k}{2}\right) ,$$

where $\omega_{i0} - i\Gamma_i/2$ and $\omega_{k0} - i\Gamma_k/2$ are solutions of the equation

$$(\omega - \omega_{i0}^0 + i\Gamma_i^0/2)(\omega - \omega_{k0}^0 + i\Gamma_k^0/2) + \Gamma_i^0\Gamma_k^0/4 = 0 .$$

They are equal to

$$\omega_{i0,k0} = (\omega_{i0}^0 + \omega_{k0}^0)/2 \pm \mathrm{Re}\{R/2\} , \tag{3.143}$$

$$\Gamma_{i,k} = (\Gamma_i^0 + \Gamma_k^0)/2 \mp \mathrm{Im}\{R\} , \tag{3.144}$$

where R, in turn, is given by

$$R = \left[(\omega_{i0}^0 - \omega_{k0}^0)^2 - \frac{(\Gamma_i^0 + \Gamma_k^0)^2}{4} - i(\Gamma_i^0 - \Gamma_k^0)(\omega_{i0}^0 - \omega_{k0}^0)\right]^{1/2} . \tag{3.145}$$

The new $\omega_{i0,k0}$ and $\Gamma_{i,k}$ are respectively the energy and damping constant of the atomic system with inclusion of the vacuum interaction.

In terms of the new notation

$$S_{0'0} = -\frac{2\pi i}{\hbar^2}\delta(\omega - \omega')$$

$$\times \frac{V_{0'k}V_{k0}(\omega - \omega_{i0}^0) + V_{0'i}V_{i0}(\omega - \omega_{k0}^0)}{(\omega - \omega_{i0} + i\Gamma_i/2)(\omega - \omega_{k0} + i\Gamma_k/2)} . \tag{3.146}$$

We note that v_{pq} are matrix elements of the dipole interaction operator between atomic states. They are defined without interaction of the states $|i\rangle_0$ and $|k\rangle_0$ with each other via the vacuum. The analogous matrix elements of the dipole interaction, but corresponding to the wavefunctions given in the best approximation (i.e., taking into account the mutual interaction of states $|i\rangle_0$ and $|k\rangle_0$ through the vacuum), are different from V_{pq}. We denote them by f for the absorption of a primary photon and by g for the emission of a photon. We note that

$$\Gamma_i^0 + \Gamma_k^0 = \Gamma_i + \Gamma_k$$

and that the $|d_{0q}|^2$ are proportional to Γ_q:

$$V_{0k} V_{k0} / V_{0i} V_{i0} = \Gamma_k^0 / \Gamma_i^0 \tag{3.147}$$

and, correspondingly,

$$f_{k0} g_{0k} / f_{i0} g_{0i} = \Gamma_k / \Gamma_i \ . \tag{3.148}$$

Moreover, one can show from (3.145) that

$$- \mathrm{Re}\{R\} \, \mathrm{Im}\{R\} = \tfrac{1}{2} (\Gamma_i^0 - \Gamma_k^0)(\omega_{i0}^0 - \omega_{k0}^0) \tag{3.149}$$

and from (3.143, 144):

$$- \mathrm{Re}\{R\} \, \mathrm{Im}\{R\} = \tfrac{1}{2} (\Gamma_i - \Gamma_k)(\omega_{i0} - \omega_{k0}) \ . \tag{3.150}$$

By subsequently using (3.147–150) one arrives at the equality

$$V_{0'k} V_{k0}(\omega - \omega_{i0}^0) + V_{0'i} V_{i0}(\omega - \omega_{k0}^0)$$

$$= f_{k0} g_{0'k}(\omega - \omega_{i0}) + f_{i0} g_{0'i}(\omega - \omega_k) \ .$$

The scattering matrix $S_{0'0}$ (3.146) can then be rewritten as:

$$S_{0'0} = - \frac{2\pi i}{\hbar} \delta(\omega - \omega')$$

$$\times \left[\frac{f_{k0} g_{0'k}}{\hbar\left(\omega - \omega_{k0} + \dfrac{i\Gamma_k}{2}\right)} - \frac{i\Gamma_i f_{k0} g_{0'k}}{2\hbar\left(\omega - \omega_{k0} + \dfrac{i\Gamma_k}{2}\right)\left(\omega - \omega_{i0} + \dfrac{i\Gamma_i}{2}\right)} \right.$$

$$\left. + \frac{f_{i0} g_{0'i}}{\hbar\left(\omega - \omega_{i0} + \dfrac{i\Gamma_i}{2}\right)} - \frac{i\Gamma_k f_{i0} g_{0'i}}{2\hbar\left(\omega - \omega_{i0} + \dfrac{i\Gamma_i}{2}\right)\left(\omega - \omega_{k0} + \dfrac{i\Gamma_k}{2}\right)} \right]$$

$$\equiv - \frac{2\pi i}{\hbar} \delta(\omega - \omega')[T_1 + T_2 + T_3 + T_4]$$

$$= - \frac{2\pi i}{\hbar} \delta(\omega - \omega')[T_1 + T_3 + 2T_2] \tag{3.151}$$

since $T_2 = T_4$. This expression can be compared with the Feynman diagram for a finite number of terms:

$$S_{0'0} = \underset{|0\rangle \quad |i\rangle \quad |0'\rangle}{\overset{\omega \qquad \omega'}{\diagram}} + \underset{|0\rangle \quad |k\rangle \quad |0'\rangle}{\overset{\omega \qquad \omega'}{\diagram}}$$

$$+ \underset{|0\rangle \quad |i\rangle \quad |0''\rangle \quad |k\rangle \quad |0'\rangle}{\overset{\omega \qquad \omega'' \qquad \omega'}{\diagram}} + \underset{|0\rangle \quad |k\rangle \quad |0''\rangle \quad |i\rangle \quad |0'\rangle}{\overset{\omega \qquad \omega'' \qquad \omega'}{\diagram}} \cdot$$

$$(3.152)$$

According to this diagram, the scattering matrix is determined by the following processes: by the scattering through the states $|i\rangle$ and $|k\rangle$ (determined with inclusion of the vacuum interaction) and by the scattering with an intermediate transition state from $|i\rangle$ to $|k\rangle$ and vice versa. The presence of the last two terms distinguish the present case of excitation of two states with the same angular momenta from the usual kind of interference of states. As will be seen below, these two terms suppress the interference term that arises when one neglects the vacuum interaction. Defining the scattering probability per unit time as $W_{0'0}$, we arrive at the expression [3.3]:

$$W_{0'0} = \frac{2\pi}{\hbar} |T_{0'0}|^2 \varrho_{\omega'} \quad (T_{0'0} = T_1 + T_3 + 2T_2),$$

where $\varrho_{\omega'} = \omega^2/(2\pi c)^3 \hbar$ is the possible number of final states. (The quantity $T_{0'0}$ has been discussed above; it is the transition matrix element on the energy surface and it is coupled with $S_{0'0}$ through the δ-function in the energy of the incident and scattered photons).

The intensity of the scattered light will also be proportional to the incident light power, which, as before, is defined by ϱ_ω (3.21), and to the number of scattering atoms N:

$$I \, d\Omega = N \int T_{0'0} \frac{\omega^2 \varrho_\omega \, d\omega \, d\Omega}{\hbar^2 (2\pi c)^3} \, . \qquad (3.153)$$

Entering the expression for the intensity is the quantity $|T_{0'0}|^2$:

$$|T_{0'0}|^2 = |T_1 + T_3 + 2T_2|^2 \, .$$

Let us consider one term of the intensity:

$$I(T_1 T_3^* + \text{c.c.}) = N \int \frac{\omega^2 \varrho_\omega}{\hbar^2 (2\pi c)^3} (T_1 T_3^* + \text{c.c.}) \, d\omega$$

$$= N \int \frac{\omega^2 \varrho_\omega}{\hbar^2 (2\pi c)^3} \frac{f_{k0} g_{0k} f_{i0}^* g_{0i}^*}{\hbar^2 \left(\omega - \omega_{k0} + \dfrac{i\Gamma_k}{2}\right)\left(\omega - \omega_{i0} - \dfrac{i\Gamma_i}{2}\right)} \, d\omega + \text{c.c.}$$

$$= N \frac{\omega^4 \varrho_\omega}{\hbar^2 c^3} \frac{(\boldsymbol{d} \cdot \boldsymbol{e}_\lambda)_{k0}(\boldsymbol{d} \cdot \boldsymbol{e}_\lambda)_{i0}^*(\boldsymbol{d} \cdot \boldsymbol{e}_r)_{0k}(\boldsymbol{d} \cdot \boldsymbol{e}_r)_{0i}^*(\Gamma_i + \Gamma_k)}{\omega_{ki}^2 + (\Gamma_i + \Gamma_k)^2/4} \, , \qquad (3.154)$$

where $(d \cdot e)_{pq}$ corresponds to the redefined wavefunctions. Here we make the same approximation as everywhere else in treating the density matrix and pump matrix, i.e. $\omega_{ik} \ll \omega_{k0,i0}$. One can thus assume that ω_{k0} and ω_{i0} in the numerator are equal. In both summands, we replace this quantity by ω.

Utilizing the previously introduced definition of the pump matrix (3.34) and observation matrix (3.66), we obtain

$$I(T_1 T_3^* + \text{c.c.}) = N \frac{\omega^4}{2\pi c^3} F_{ki} G_{ik} \frac{\Gamma_i + \Gamma_k}{\omega_{ki}^2 + (\Gamma_i + \Gamma_k)^2/4}$$

$$= K \Gamma_{ki} G_{ik} \frac{\Gamma_i + \Gamma_k}{\omega_{ki}^2 + (\Gamma_i + \Gamma_k)^2/4} . \tag{3.155}$$

In the same way, we find

$$I(T_1^2) = K \frac{F_{kk} G_{kk}}{\Gamma_k}, \qquad I(T_3^2) = K \frac{F_{ii} G_{ii}}{\Gamma_i} .$$

The sum

$$I(T_1^2) + I(T_3^2) + I(T_1 T_3^* + \text{c.c.})$$

$$= K \left[\frac{F_{kk} G_{kk}}{\Gamma_k} + \frac{F_{ii} G_{ii}}{\Gamma_i} + \frac{F_{ki} G_{ik}(\Gamma_i + \Gamma_k)}{\omega_{ki}^2 + (\Gamma_i + \Gamma_k)^2/4} \right] , \tag{3.156}$$

which neglects the last terms of the diagram (3.152), coincides with the description of the interference of states with different angular moments. However, unlike them, the matrix elements F and G in the case considered cannot be complex.

Let us find the last terms:

$$I((T_1 + T_3)2T_2^* + \text{c.c.}) = -K \frac{F_{ki} G_{ik}(\Gamma_i + \Gamma_k)}{\omega_{ik}^2 + (\Gamma_i + \Gamma_k)^2/4} . \tag{3.157}$$

The total intensity, which is the sum of (3.156) and (3.157)

$$I = K \left(\frac{F_{kk} G_{kk}}{\Gamma_k} + \frac{F_{ii} G_{ii}}{\Gamma_i} \right) ,$$

does not contain any terms that depend on the energy gap between the sublevels, and this means that the level crossing effect is absent.

We could restrict ourselves to this conclusion. However, we will prove by another approach that the free beats are absent for short pulse excitation. In order to make the calculation less awkward, let us introduce the concept of the amplitude of an excited state of an atomic system + field. It is represented by a diagram which differs from (3.129) in that it does not contain the last term

i.e. we have frozen the process at the stage at which a photon is absorbed by an atom and the real photon scattering has not yet taken place. Under continuous illumination at the time instant t, the Schrödinger picture yields

$$|\omega, t\rangle = a_i(t)|i\rangle_0 + a_k(t)|k\rangle_0$$

$$= \frac{e^{-i\omega t}}{\hbar}\left\{\frac{V_{i0}|i\rangle_0}{\left(\omega - \omega_{i0}^0 + \dfrac{i\Gamma_i^0}{2}\right)} + \frac{V_{k0}|k\rangle_0}{\left(\omega - \omega_{k0}^0 + \dfrac{i\Gamma_k^0}{2}\right)}\right.$$

$$+ \frac{V_{i0''}V_{0''k}V_{k0}|i\rangle_0}{\hbar^2\left(\omega - \omega_{k0}^0 + \dfrac{i\Gamma_k^0}{2}\right)\left(\omega - \omega'' + \dfrac{i\gamma}{2}\right)(\omega - \omega_{i0}^0 + i\Gamma_i^0/2)}$$

$$+ \frac{V_{k0''}V_{0''i}V_{i0}|k\rangle_0}{\hbar^2\left(\omega - \omega_{i0}^0 + \dfrac{i\Gamma_i^0}{2}\right)\left(\omega - \omega'' + \dfrac{i\gamma}{2}\right)\left(\omega - \omega_{k0}^0 + \dfrac{i\Gamma_k^0}{2}\right)}$$

$$+ \frac{V_{i0'''}V_{0'''k}V_{k0''}V_{0''i}V_{i0}|i\rangle_0}{\hbar^4\left(\omega - \omega_{i0}^0 + \dfrac{i\Gamma_i^0}{2}\right)^2\left(\omega - \omega'' + \dfrac{i\gamma}{2}\right)\left(\omega - \omega''' + \dfrac{i\gamma}{2}\right)\left(\omega - \omega_{k0}^0 + \dfrac{i\Gamma_k^0}{2}\right)}$$

$$+ \left.\frac{V_{k0'''}V_{0'''i}V_{i0''}V_{0''k}V_{k0}|k\rangle_0}{\hbar^4\left(\omega - \omega_{k0}^0 + \dfrac{i\Gamma_k^0}{2}\right)^2\left(\omega - \omega'' + \dfrac{i\gamma}{2}\right)\left(\omega - \omega''' + \dfrac{i\gamma}{2}\right)\left(\omega - \omega_{i0}^0 + \dfrac{i\Gamma_i^0}{2}\right)}\right.$$

$$\left. + \cdots\right\}. \tag{3.158}$$

For the steady-state case, this formalism leads once more to the above conclusion that interference is absent. Let us now assume pulse excitation in the form of a difference of two semi-infinite pulses, shifted a little relative to each other. This allows us to avoid difficulties linked with the transient process arising immediately after switching on the interaction. A semi-infinite pulse is an illumination with a broad spectrum, beginning at time $t = -\infty$ and terminating at some time t_1. When

$$|\omega, t\rangle = |\omega, t_1\rangle \exp\left[-\frac{i\hat{H}}{\hbar}(t - t_1)\right], \tag{3.159}$$

where $\hat{H}$ is the redefined Hamiltonian of a free atom (including the interaction through vacuum). From (3.158) and using (3.143–145) we find

$$a_i(\omega, t) = \frac{V_{io}(\omega - \omega_{ko}^0)\exp[-i\omega_{io}(t - t_1) - \Gamma_i(t - t_1)/2 - i\omega t_1]}{\hbar(\omega - \omega_{ko} + i\Gamma_k/2)(\omega - \omega_{io} + i\Gamma_i/2)} \tag{3.160}$$

and

$$a_k(\omega, t) = \frac{V_{ko}(\omega - \omega_{io}^0)\exp[-i\omega_{ko}(t - t_1) - \Gamma_k(t - t_1)/2 - i\omega t_1]}{\hbar(\omega - \omega_{ko} + i\Gamma_k/2)(\omega - \omega_{io} + i\Gamma_i/2)} .$$

The intensity is found as

$$I(e, e_\lambda) = K' \sum_{p,n} \int a_p(\omega, t) a_n^*(\omega, t) G_{pn}^0 \varrho_\omega \, d\omega .$$

It comprises four terms, two diagonals (with $|V_{io}|^2$ and $|V_{ko}|^2$) and two off-diagonals (with the product $V_{io}V_{ko}^*$ and its complex conjugate).

Let us consider one of the diagonal terms

$$I_{ii} = K' \int |a_i(\omega, t)|^2 G_{ii}^0 \varrho_\omega \, d\omega$$

$$= K' \int \frac{|V_{io}|^2(\omega - \omega_{ko}^0)^2 G_{ii}^0 \exp[-\Gamma_i(t - t_1)]\varrho_\omega \, d\omega}{\hbar^2\left(\omega - \omega_{io} + \dfrac{i\Gamma_i}{2}\right)\left(\omega - \omega_{ko} + \dfrac{i\Gamma_k}{2}\right)\left(\omega - \omega_{io} - \dfrac{i\Gamma_i}{2}\right)\left(\omega - \omega_{ko} - \dfrac{i\Gamma_k}{2}\right)} .$$

$$\tag{3.161}$$

Integrating over the residues (assuming, as previously, that the quantity ϱ_ω is constant in the region containing ω_{io} and ω_{ko}), we obtain

$$I_{ii} = K\left[\frac{(\omega_{io} - \omega_{ko}^0 - i\Gamma_i/2)^2}{(\omega_{ik} + i(\Gamma_k - \Gamma_i)/2)\Gamma_i(\omega_{ik} - i(\Gamma_k + \Gamma_i)/2)}\right.$$

$$\left. + \frac{(\omega_{ko} - \omega_{io}^0 - i\Gamma_k/2)^2}{(\omega_{ki} + i(\Gamma_i - \Gamma_k)/2)\Gamma_k(\omega_{ki} - i(\Gamma_i + \Gamma_k)/2)}\right] F_{ii}^0 G_{ii}^0 \exp[-\Gamma_i(t - t_1)] .$$

$$\tag{3.162}$$

Proceeding from (3.143–145), replacing Γ_i^0, Γ_k^0 and ω_{io}^0, ω_{ko}^0 by Γ_i, Γ_k, ω_{io}, ω_{ko} and using the relationship between V and Γ^0 [similar to (3.147)], some extremely cumbersome algebraic manipulation yields

$$I_{ii} = K \frac{F_{ii}^0 G_{ii}}{\Gamma_i} \exp[-\Gamma_i(t - t_1)] = K \frac{F_{ii}G_{ii}^0}{\Gamma_i} \exp[-\Gamma_i(t - t_1)] . \tag{3.163}$$

Here F_{ii}^0 and G_{ii}^0 are the pump and observation matrices for the primary wavefunction $|i\rangle_0$. Making use of the very same substitutions, we find the same expression for I_{kk} and we establish that both the off-diagonal terms $I_{ik} = I_{ki} = 0$. We thus have

$$I(t - t_1) = K \left[\frac{F_{ii}^0 G_{ii}}{\Gamma_i} \exp[-\Gamma_i(t - t_1)] + \frac{F_{kk}^0 G_{kk}}{\Gamma_k} \exp[-\Gamma_k(t - t_1)] \right] .$$

$$(3.164)$$

It is the fact that the off-diagonal terms (3.164) are zero that is responsible for the absence of beats; instead, the intensity decays as a sum of two exponents.

The above discussion refers to the case of a semi-infinite pulse. One can extend this to the picture of a very short exciting pulse without difficulty: we deduct from the intensity $I(t - t_1)$ the intensity from another semi-infinite pulse which terminates at time $t_2 < t_1$. This will correspond to excitation by a square pulse of duration $\tau = (t_1 - t_2)$:

$$I(t) = I(t - t_1) - I(t - t_2)$$

$$= K \left[\frac{F_{ii}^0 G_{ii}}{\Gamma_i} (1 - e^{-\Gamma_i \tau}) \exp[-\Gamma_i(t - t_1)] \right.$$

$$\left. + \frac{F_{kk}^0 G_{kk}}{\Gamma_k} (1 - e^{-\Gamma_k \tau}) \exp[-\Gamma_k(t - t_1)] \right] .$$

$$(3.165)$$

There are, of course, no beats.

In this way, the expression for the intensity retains only the diagonal terms. One can now confirm that, upon illumination by light of a sufficiently broad spectrum, no interference effects arise in the spontaneous radiation. At the same time, the density matrix formalism predicts the existence of such a term. However, the density matrix corresponds to a consideration of the first two terms only in the expression (3.92), i.e. one can say that this formalism corresponds to the first-order approximation of the scattering matrix and, in the case discussed in this section, this approximation turned out to be insufficient. We have thus demonstrated yet another constraint to the applicability of the density matrix.

One can in fact construct a density matrix formalism which takes into account the interaction of states through vacuum. Within this framework, the description of "interference" of states with identical quantum numbers coincides with the approach given in this section. For all other cases, it coincides with the normal density matrix formalism, which thus justifies the application of the latter for describing the interference phenomena.

3.10 Some Results from the Formalism of Irreducible Tensor Operators

The density matrix elements can be expanded in terms of the irreducible tensor operators of the rotational group

$$\sigma_{nk} = \sum_{\kappa,q} (-1)^q \varrho_q^\kappa (T_{-q}^\kappa)_{nk} \ , \tag{3.166}$$

where

$$(T_q^\kappa)_{nk} = (-1)^{J-K}[(2\kappa+1)/(2J+1)^{1/2}]\begin{pmatrix} J & \kappa & J \\ -n & q & k \end{pmatrix} \ . \tag{3.167}$$

Due to the properties of the 3j-symbols, we have the equality

$$-n + k + q = 0 \ ,$$

i.e. $q = n - k$. Therefore, in the sum over q in (3.166) there remains only one term.

The coefficients in this expansion ϱ_q^κ carry all the information that was contained in the density matrix components. In other words, all the parameters describing the interaction of an atomic system with light that are amenable to the density matrix formalism can also be expressed through these coefficients.

The coefficients ϱ_q^κ are transformed upon rotating the coordinate axes in the same way as the rotation group tensors. Thus they themselves are also tensors [3.20], known in nuclear spectroscopy as statistical tensors. In the following, this terminology will be used side by side with the optical terminology "polarization moments".

The formalism of irreducible tensor operators found its way into the area of optical investigation in about 1964 [3.21].

We cite here for reference some of the relations presented in [3.21]. One can easily achieve the inverse transformation of (3.168) by exploiting the properties of the 3j-symbols

$$\varrho_q^\kappa = (-1)^q [(2J+1)/(2\kappa+1)] \sum_{n,k} \sigma_{nk} (T_{-q}^\kappa)_{nk} \ . \tag{3.168}$$

The equation of motion for the polarization moments is deduced from the equation of motion of the density matrix (3.138)

$$i\hbar\dot{\sigma}_{nk} = H_{nn}\sigma_{nk} - \sigma_{nk}H_{kk} - i\hbar\Gamma\sigma_{nk} + i\hbar F_{nk}$$

by using the transformation (3.168). After making the substitution we get

$$i\hbar \sum_\kappa \dot{\varrho}_q^\kappa (T_{-q}^\kappa)_{nk} = (H_{nn} - H_{kk}) \sum_\kappa \varrho_q^\kappa (T_{-q}^\kappa)_{nk}$$

$$- i\hbar\Gamma \sum_\kappa \varrho_q^\kappa (T_{-q}^\kappa)_{nk} + i\hbar(-1)^q F_{nk} \ . \tag{3.169}$$

Let us multiply both sides of this equation by $(T^{\kappa'}_{-q'})_{nk}$ and sum over n and k. We assume that the difference $H_{nn} - H_{kk}$ does not depend on the index n, but only on the difference $n - k$. Then

$$\frac{1}{\hbar}(H_{nn} - H_{kk}) = q\Omega \ . \tag{3.170}$$

Let us define

$$(-1)^{q'}\frac{2J + 1}{2\kappa + 1}\sum_{n,k} F_{nk}(T^{\kappa'}_{-q'})_{nk} = \mathscr{F}^{\kappa'}_{q'} \ . \tag{3.171}$$

With the condition (3.170) and in order to simplify the sum over n and k, one can make use of the relationship [3.21]

$$\sum_{n,k}(T^\kappa_{-q})_{nk}(T^{\kappa'}_{-q'})_{nk} = \frac{2\kappa + 1}{2J + 1}\delta_{\kappa\kappa'}\delta_{qq'} \ . \tag{3.172}$$

In each of the sums over n and k only one term with $\kappa = \kappa'$ will remain and we obtain

$$\dot{\varrho}^\kappa_q = -(\Gamma - iq\Omega)\varrho^\kappa_q + \mathscr{F}^\kappa_q \ . \tag{3.173}$$

Because of the constraints imposed above, (3.173) is valid only when the levels are equidistant, which indeed they are when the degenerate states are perturbed by a magnetic field in the absence of other perturbations. Other fields, such as direct and alternating electromagnetic fields cause level-splitting that is not equidistant. The interaction of levels also leads to non-equidistant energies and gives rise to off-diagonal elements of the type H_{nk}. The statistical tensor formalism has explicit merit only in the special, but nevertheless frequently encountered, case of equidistant levels.

Upon optical excitation, the form of the pump tensor can be derived from (3.171) and (3.35)

$$F_{nn'} = F_0\langle n|\boldsymbol{d}\cdot\boldsymbol{e}_\lambda|\mu\rangle\langle n'|\boldsymbol{d}\cdot\boldsymbol{e}_\lambda|\mu\rangle^* \ . \tag{3.174}$$

It is known [3.15, 21] that a matrix element can be represented as a product of the reduced matrix element and the Klebsch–Gordon coefficients. Then

$$F_{nn'} = |\langle J_1\|d\|J_0\rangle|^2 C^{J_1 m}_{J_0 m_0 1 m-m_0} C^{J_1 m'}_{J_0 m_0 1 m'-m_0} a_{m-m_0} a^*_{m'-m_0} \ , \tag{3.175}$$

where a are the expansion coefficients of the polarization vector over circular unit vectors [they will be discussed in more detail below; see (2.187) and (2.122a)]. $\langle J_1\|d\|J_0\rangle$ is the reduced matrix element, C is the Klebsch–Gordon coefficient, in which the indices J and m indicate the angular momentum and the magnetic quantum number of the states $|n\rangle$, and J_0, m_0 are the corresponding quantities for $|\mu\rangle$. Substituting (3.174) and the expression for T^κ_{-q} (3.167) into the formula for the pump (3.171) and using the relationship between the

Klebsch–Gordon coefficients and the 6j-symbols (Sect. 5.5) [3.15], we find

$$\mathscr{F}_q^\kappa = (-1)^{J_1 + J_0} K' (2J_1 + 1)^{1/2} |\langle J_1 \| d \| J_0 \rangle|^2 \begin{Bmatrix} 1 & 1 & \kappa \\ J_1 & J_1 & J_0 \end{Bmatrix} \Phi_q^\kappa(e_\lambda) \ ,$$

(3.176)

where K' is a coefficient which depends on the energy characteristics of the exciting light, and

$$\Phi_q^\kappa(e_\lambda) = \sum_{i,k} (-1)^k a_i a_k^* \begin{pmatrix} 1 & 1 & \kappa \\ i & -k & q \end{pmatrix}; \quad i, k = 0, \pm 1 \ .$$

(3.177)

In the same way, transforming the expression for the observation matrix (3.66), we find an expression for the intensity of the spontaneous radiation for a transition whose final state possesses an angular momentum J_2:

$$I(e_r) = (-1)^{J_1 + J_2} d_{21}^2 K' \sum_\kappa (2\kappa + 1) \begin{Bmatrix} 1 & 1 & \kappa \\ J_1 & J_1 & J_2 \end{Bmatrix} \sum_q (-1)^q \varrho_q^\kappa \Phi_{-q}^\kappa(e_r) \ .$$

(3.178)

Here

$$d_{21} = \langle J_2 \| d \| J_1 \rangle \ ,$$

and the coefficient K' has the same meaning as before, but is now determined by the parameters of the system receiving the radiation.

The quantities ϱ_q^κ and Φ_q^κ possess the folowing properties

$$\varrho_{-q}^\kappa = (-1)^q (\varrho_q^\kappa)^*; \quad \Phi_{-q}^\kappa(e_r) = (-1)^q [\Phi_q^\kappa(e_r)]^* \ .$$

(3.179)

The components of the tensor Φ, for linearly polarized light whose polarization vector e is given by the angles θ' and ϕ', are

$$\Phi_0^0(e) = -1/\sqrt{3}, \qquad \Phi_q^1(e) = 0 \ .$$

$$\Phi_0^2(e) = (3\cos^2\theta' - 1)/\sqrt{30} \ ,$$

$$\Phi_{\pm 1}^2(e) = \sin\theta' \cos\theta' \, e^{\pm i\phi'}/\sqrt{5} \ ,$$

$$\Phi_{\pm 2}^2(e) = \sin^2\theta' \, e^{\pm 2i\phi'}/2\sqrt{5} \ .$$

(3.180)

For unpolarized light, the direction of propagation n is determined by the angles θ_n and ϕ_n:

$$\Phi_0^0(n) = -1/\sqrt{3}, \qquad \Phi_q^1(n) = 0 \ .$$

$$\Phi_0^2(n) = -(3\cos^2\theta_n - 1)/2\sqrt{30} \ ,$$

$$\Phi_{\pm 1}^2(n) = -\sin\theta_n \cos\theta_n \, e^{\pm i\phi_n}/2\sqrt{5} \ ,$$

$$\Phi_{\pm 2}^2(n) = -\sin^2\theta_n \, e^{\pm 2i\phi_n}/4\sqrt{5} \ .$$

(3.181)

For right-hand circularly polarized light, for which

$$a_- = (1 - \cos\theta_n)\,e^{i\phi_n}/2, \qquad a_0 = \sin\theta_n/\sqrt{2} \ ,$$

$$a_+ = (1 + \cos\theta_n)\,e^{-i\phi_n}/\sqrt{2} \ , \tag{3.182}$$

the components of Φ are given by

$$\Phi_0^0 = -1/\sqrt{3}, \qquad\qquad\qquad \Phi_0^1 = \cos\theta_n/\sqrt{6} \ ,$$

$$\Phi_0^2 = (1 - 3\cos^2\theta_n)/2\sqrt{30}, \qquad \Phi_{\pm1}^1 = \sin\theta_n/2\sqrt{3} \ ,$$

$$\Phi_{\pm1}^2 = -\sin\theta_n\cos\theta_n\,e^{\pm i\phi_n}/2\sqrt{5} \ , \tag{3.183}$$

$$\Phi_{\pm2}^2 = -\sin^2\theta_n\,e^{\pm 2i\phi_n}/4\sqrt{5} \ .$$

An important property of the tensors ϱ and Φ is that, upon rotating the coordinate axes, the new components with a given index κ are expressed through the old components (i.e. in terms of the unrotated coordinate system) with the very same index κ. The operation of rotation consists of the Wigner function D [3.15, 22], whose explicit forms are given in Table 3.1.

Let us discuss a few of the properties of the statistical tensors ϱ. We first consider the component with $\kappa = 0$. From the behaviour of the $3j$-symbols it follows, that the absolute value of the index q is smaller than or equal to κ: $|q| \le \kappa$. This means that when $\kappa = 0$, q can take only one value $q = 0$. And because

$$-n + k + q = 0 \ ,$$

we find that

$$n = k \ .$$

It is thus clear that in the expansion the components ϱ_0^0 [according to the formula (3.104)] enter only the diagonal terms σ_{nn} of the density matrix σ. The coefficient in front of σ_{nn} when $\kappa = 0$ (and therefore $q = 0$) will have a somewhat simpler form

$$\varrho_0^0 = (2J_1 + 1)^{-1/2}\sum_n (-1)^{J_1-n}\begin{pmatrix} J_1 & 0 & J_1 \\ -n & 0 & n \end{pmatrix}\sigma_{nn} \ ,$$

hence it is seen that

$$\varrho_0^0 = \sum_n (2J_1 + 1)^{-1/2}\,\sigma_{nn} = S_p(2J_1 + 1)^{-1/2}\sigma \ . \tag{3.184}$$

It is thus apparent that ϱ_0^0 is a scalar quantity and has the physical meaning of level population.

Also entering into the components of ϱ with $\kappa > 0$ are the off-diagonal terms of the density matrix σ. As before, the components with $q = 0$ consist of the diagonal terms σ_{nn}. The components with $q = \pm 1$ comprise terms with $n - k = \pm 1$ and $q = \pm 2$, and terms with $n - k = \pm 2$.

Upon optical excitation from a spherically symmetric state, the density matrix, due to the selection rule for dipole transitions will contain no off-diagonal terms with $n - k > 2$. Correspondingly, the matrix ϱ will have no terms with $\kappa > 2$. Even if these terms were to appear (which is not excluded in the case of stepwise excitation and for strong interaction with the optical field), it makes no difference since they will not be manifested in the spontaneous optical radiation.

The matrix ϱ thus contains 9 terms, one of which is the population term ϱ_0^0. The terms with $\kappa = 1$, of which there are three, are components of the orientation tensor; five of the components with $\kappa = 2$ constitute the tensor of alignment. The component ϱ_0^2 is called the longitudinal alignment; when all the other components, i.e. $\varrho_{\pm 1, \pm 2}^2$ are equal to zero, then this a uniaxial alignment and is fully characterized by the population of the Zeeman sublevels. The alignment tensor ϱ^2 can be reduced to the principal axes by performing a rotation of the coordinate system, which one can achieve using the rotation operator D^2. This will cause terms with $q = \pm 1$ to disappear, but the terms with $q = \pm 2$ may not vanish. The space orientation of the tensor is defined by the two principal axes. The alignment is thus described by a tensor in which the components with $q = \pm 2$ that do not vanish upon transforming to the principal axes are called biaxial.

Let us now treat a rather more complex case, corresponding to excitation not from a spherical state, but from a state possessing components of a statistical tensor of rank greater than zero, i.e. orientation, alignment and even higher order.

Let us make use of the pump matrix, given in the density matrix formalism (3.54) as

$$F_{mm'} = F_0' \sum_{\mu\mu'} (\boldsymbol{d} \cdot \boldsymbol{e}_\lambda)_{m\mu} (\boldsymbol{d} \cdot \boldsymbol{e}_\lambda)^*_{m'\mu'} \sigma_{\mu\mu'} . \tag{3.185}$$

We introduce the concept of polarization of the light density matrix f [3.23]. This matrix is characterized, in much the same way as the density matrix of atomic states, by the product of the coefficients of the components in the expansion of light polarization over circular components

$$\boldsymbol{e} = a_0 \boldsymbol{e}_0 + a_+ \boldsymbol{e}_+ + a_- \boldsymbol{e}_- .$$

The density matrix elements f have the form $a_0 a_+^*$, $a_+ a_-^*$ etc. Light of "pure" polarization, coinciding with one of the polarization bases is described only by a diagonal element of the type $a_0 a_0^*$, $a_+ a_+^*$ or $a_- a_-^*$. Light of the same polarization, but arbitrarily oriented in space is characterized by a matrix that is

Wigner's function D (Table 3.1), in which the values of the Euler angles are determined by the polarization direction.

The other interesting case is uniform illumination from all sides by completely unpolarized light. Its density matrix contains only diagonal elements and all are equal in magnitude:

$$|a_+|^2 = |a_-|^2 = |a_0|^2 \ .$$

Precisely such a matrix characterizes the vacuum fluctuation responsible for spontaneous emission.

Let us return to the formula (3.185). We write the product $(d \cdot e_\lambda)_{m\mu}$ in the form

$$\langle m|d|\mu\rangle e_p^* \cdot e_\lambda = d_{m\mu} e_p^* \cdot e_\lambda \ ,$$

where $p = m - \mu$. Now it is easy to see that

$$(d \cdot e_\lambda)_{m\mu} (d \cdot e_\lambda)_{m'\mu'}^* = d_{m\mu} d_{m'\mu'}^* a_{m-\mu} a_{m'-\mu'}^* = d_{m\mu} d_{m'\mu'}^* f_{mm'} \ , \tag{3.186}$$

where f is the light density matrix.

We now transfer all quantities in the expression (3.185) into the polarization moments

$$\mathscr{F}_q^\kappa = \sum_{m,\,m'} (-1)^{J_1 - m' + q} \sqrt{2J_1 + 1} \begin{pmatrix} J_1 & \kappa & J_1 \\ -m & -q & m' \end{pmatrix} F_{mm'} \ , \tag{3.187}$$

$$d_{m\mu} = (-1)^{J_1 - m} \begin{pmatrix} J_1 & 1 & J_0 \\ -m & q & \mu \end{pmatrix} (J_1 \| d \| J_0) \ , \tag{3.188}$$

$$\sigma_{\mu\mu'} = \sum_{\kappa_0 q_0} (-1)^{q_0 + J_0 - \mu'} \frac{2\kappa_0 + 1}{\sqrt{(2J_0 + 1)}} \begin{pmatrix} J_0 & \kappa_0 & J_0 \\ -\mu & -q_0 & \mu' \end{pmatrix} \varrho_{q_0}^{\kappa_0} \ ,$$

and finally

$$\mathscr{F}_q^\kappa = (-1)^{J_1 - J_0} (2J_0 + 1)^{-1/2} F_0 \sum_{\substack{\mu,\,\mu',\,\kappa_0,\,q_0 \\ m,\,m'}} (-1)^{q_2 + q_0} (2\kappa_0 + 1)$$

$$\times \begin{pmatrix} J_1 & \kappa & J_1 \\ -m & -q & m' \end{pmatrix} \begin{pmatrix} J_1 & 1 & J_0 \\ -m & q_1 & \mu \end{pmatrix} \begin{pmatrix} J_1 & 1 & J_0 \\ -m' & q_2 & \mu' \end{pmatrix}$$

$$\times \begin{pmatrix} J_0 & \kappa_0 & \bar{J}_0 \\ -\mu & -q_0 & \mu' \end{pmatrix} f_{m-\mu,\,m'-\mu'} \varrho_{q_0}^{\kappa_0} \ , \tag{3.189}$$

where

$$F_0 = |(J_1 \| d \| J_0)|^2 \sqrt{2J_1 + 1} \, F_0' \ .$$

From the properties of the Klebsch–Gordon coefficients and from the 3j-symbols [Ref. 3.15, p. 202–222], it is not difficult to derive the following relation

$$\sum_{\varepsilon,\,\beta,\,\phi,\,\gamma} \begin{pmatrix} f & j & c \\ \phi & i & -\gamma \end{pmatrix} \begin{pmatrix} f & d & e \\ \phi & \delta & -\varepsilon \end{pmatrix} \begin{pmatrix} b & a & c \\ \beta & \alpha & -\gamma \end{pmatrix} \begin{pmatrix} b & g & e \\ \beta & \eta & -\varepsilon \end{pmatrix}$$

$$= (-1)^{2g-j-f+2d+a-3b}$$

$$\times \sum_{k,\,\tau} (2k+1)(-1)^{2\tau} \begin{pmatrix} g & j & k \\ \eta & i & -\tau \end{pmatrix} \begin{pmatrix} d & a & k \\ \delta & \alpha & -\tau \end{pmatrix} \begin{Bmatrix} c & b & a \\ f & e & d \\ j & g & k \end{Bmatrix}. \tag{3.190}$$

In this sum, the summation indices are not independent:

$$\phi = \varepsilon - \delta \quad \text{and} \quad \gamma = \beta + \alpha = \phi + i \; .$$

This enables one to divide the sum over m, m', μ, μ' into a product of two sums

$$\sum_{\mu,\,\mu',\,m,\,m'} = \sum_{q_1,\,q_2} \sum_{\mu,\,\mu',\,\mu+q_1,\,\mu'+q_2} \; .$$

Transforming the second of these sums in accordance with the formula (3.190), we get

$$\mathcal{F}_q^\kappa = \frac{F_0}{\sqrt{2J_0+1}} \sum_{\kappa_0,\,q_0} (2\kappa_0+1)(-1)^{q_0+\kappa_0+1} \sum_{q_1,\,q_2} (-1)^{q_2} f_{q_1 q_2}$$

$$\times \sum_{\kappa',\,q'} (2\kappa'+1) \begin{pmatrix} \kappa_0 & \kappa & \kappa' \\ -q_0 & q & -q' \end{pmatrix} \begin{pmatrix} 1 & 1 & \kappa' \\ -q_1 & q_2 & -q' \end{pmatrix}$$

$$\times \begin{Bmatrix} J_1 & J_0 & 1 \\ J_1 & J_0 & 1 \\ \kappa & \kappa_0 & \kappa' \end{Bmatrix} \varrho_{q_0}^{\kappa_0} \; . \tag{3.191}$$

Let us isolate from (3.191) the exciting light tensor [it has in fact been determined previously (3.177)]:

$$\sum_{q_1,\,q_2} (-1)^{q_2} f_{q_1 q_2} \begin{pmatrix} 1 & 1 & \kappa' \\ -q_1 & q_2 & -q' \end{pmatrix}$$

$$= \sum_{q_1,\,q_2} (-1)^{q_2+\kappa'} f_{q_1,\,q_2} \begin{pmatrix} 1 & 1 & \kappa' \\ q_1 & -q_2 & q' \end{pmatrix} = (-1)^{\kappa'}\,\Phi_{q'}^{\kappa'} \; . \tag{3.192}$$

Making use of this definition, as well as the one following from the properties of

9j-symbols (the sum $\kappa + \kappa_0 + \kappa'$ must be even), we finally obtain

$$
\mathscr{F}_q^\kappa = (-1)^{1+\kappa} \frac{F_0}{\sqrt{2J_1+1}} \sum_{\kappa_0,q_0} (-1)^{q_0} (2\kappa_0 + 1)
$$

$$
\times \sum_{\kappa',q'} (2\kappa'+1) \begin{pmatrix} \kappa_0 & \kappa & \kappa' \\ -q_0 & q & -q' \end{pmatrix} \begin{Bmatrix} J_1 & J_0 & 1 \\ J_1 & J_0 & 1 \\ \kappa & \kappa_0 & \kappa' \end{Bmatrix} \varrho_{q_0}^{\kappa_0} \Phi_{q'}^{\kappa'} . \tag{3.193}
$$

Solving the equation (3.173) with such an expression for the pump one can obtain the form of the statistical tensor for the case where the initial state does not possess spherical symmetry. It is apparent that it can contain moments of higher order than the second; however, moments of higher order are not manifested in spontaneous radiation, nor in absorption, provided the absorption remains linear with the incident light intensity.

3.11 Radiation Polarization in the Statistical Tensor Formalism. Comparison of the Conclusions of Quantum Mechanical and Classical Approaches

Let us return to the special case, already discussed in Sect. 3.7, of emission from an atomic system from the transition $J = 1 \to J = 0$, excited from the state $J = 0$ by linearly polarized light e_λ, characterized by the angles θ' and ϕ'. The pump tensor in this case has the form:

$$
\mathscr{F}_0^0 = -\mathscr{F} \begin{Bmatrix} 1 & 1 & 0 \\ 1 & 1 & 0 \end{Bmatrix} \sqrt{\frac{1}{3}} .
$$

All the terms $\mathscr{F}_q^1 = 0$, because linearly polarized light contains no component with $\kappa = 1$. We rewrite the alignment tensor components as

$$
\mathscr{F}_q^2 = \mathscr{F} \begin{Bmatrix} 1 & 1 & 2 \\ 1 & 1 & 0 \end{Bmatrix} \times \begin{cases} \sin^2\theta' \exp(2i\phi')/2\sqrt{5} & \text{for } q = 2 , \\ \sin\theta' \cos\theta' \exp(i\phi')/\sqrt{5} & q = 1 , \\ (3\cos^2\theta' - 1)/\sqrt{30} & q = 0 , \\ \sin\theta' \cos\theta' \exp(-i\phi')/\sqrt{5} & q = -1 , \\ \sin^2\theta' \exp(-2i\phi')/2\sqrt{5} & q = -2 . \end{cases}
$$

Assuming a steady-state pump we have

$$
\varrho_q^\kappa = \frac{\mathscr{F}_q^\kappa}{\Gamma + iq\Omega} . \tag{3.194}
$$

The expression for the intensity of the light emitted in a random linear polarization e_r has the form of (3.178):

$$
I(e_r) = \mathscr{F} K' d_{01}^2 \left[\frac{1}{3\Gamma} \begin{Bmatrix} 1 & 1 & 0 \\ 1 & 1 & 0 \end{Bmatrix}^2 \right.
$$

$$
+ 5 \begin{Bmatrix} 1 & 1 & 2 \\ 1 & 1 & 0 \end{Bmatrix}^2 \left(\frac{(3\cos^2\theta'' - 1)(3\cos^2\theta' - 1)}{30\Gamma} \right.
$$

$$
+ \frac{2}{5} \sin\theta'' \sin\theta' \cos\theta'' \cos\theta' \, \frac{\Gamma\cos(\phi'' - \phi') + \Omega\sin(\phi'' - \phi')}{\Gamma^2 + \Omega^2}
$$

$$
\left. \left. + \frac{1}{10} \sin^2\theta'' \sin^2\theta' \, \frac{\Gamma\cos 2(\phi'' - \phi') + 2\Omega\sin 2(\phi'' - \phi')}{\Gamma^2 + 4\Omega^2} \right) \right] ,
$$

(3.195)

where the angles θ'' and ϕ'' determine the direction of the vector e_r.

The numerical values entering the 6j-symbols are equal:

$$
\begin{Bmatrix} 1 & 1 & 0 \\ 1 & 1 & 0 \end{Bmatrix}^2 = \begin{Bmatrix} 1 & 1 & 2 \\ 1 & 1 & 0 \end{Bmatrix}^2 = \frac{1}{9} .
$$

(3.196)

Let $\Omega = 0$, i.e. the levels are degenerate. If e_r is orthogonal to e_λ, then

$$
\sin\theta'' \sin\theta' \cos(\phi'' - \phi') + \cos\theta'' \cos\theta' = 0 .
$$

(3.197)

Taking into account this condition of orthogonality, we find from (3.196), after simple trigonometrical manipulation, that the radiation intensity in the polarization $e_r \perp e_\lambda$ is equal to zero for level degeneracy

$$
I(e_r) = 0 .
$$

This result was already obtained using the density matrix formalism (Sect. 3.8). When a magnetic field is applied (with the appearance of finite Ω), the terms with $q \neq 0$ decrease in magnitude, tending towards zero. These are the last two summands in (3.195). Simultaneously with their disappearance, the intensity $I(e_r)$ (3.195) becomes rather different from zero, i.e. when a magnetic field is applied, light with the polarization e_r will arise (an exception is the case θ'' or $\theta' = 0$, when the last two terms are simply equal to zero).

Let us now compare expression (3.195) with the expression (2.44) of Chap. 2. Let us take the 6j-symbols in the formula (3.195) out of the square brackets, which can be done thanks to (3.196), and let us transform the terms to another form that does not contain ϕ'' and ϕ':

$$
\frac{1}{3\Gamma} + \frac{1}{6\Gamma} (3\cos^2\theta'' - 1)(3\cos^2\theta' - 1)
$$

$$
= \frac{1}{\Gamma} \left(\cos^2\theta'' \cos^2\theta' + \frac{1}{2} \sin^2\theta'' \sin^2\theta' \right) .
$$

Substituting this into (3.195) we obtain

$$
I(e_r) = \mathscr{F}K'd_{10}^2\frac{1}{9}\left[\frac{1}{\Gamma}\left(\cos^2\theta''\cos^2\theta' + \frac{1}{2}\sin^2\theta''\sin^2\theta'\right)\right.
$$

$$
+ \frac{1}{2}\sin 2\theta''\sin 2\theta'\frac{\Gamma\cos(\phi''-\phi')+\Omega\sin(\phi''-\phi')}{\Gamma^2+\Omega^2}
$$

$$
\left.+ \frac{1}{2}\sin^2\theta''\sin^2\theta'\frac{\Gamma\cos 2(\phi''-\phi')+2\Omega\sin 2(\phi''-\phi')}{\Gamma^2+4\Omega^2}\right]. \qquad (3.198)
$$

The above expression differs from the classical formula for intensity (2.44), only by a multiplicative factor. One finds a similar correspondence for circularly polarized light. The description of interference phenomena in the transitions $J = 1 \rightarrow J = 0$ using the classical approach and using the density matrix formalism thus coincide with one another.

In the general case of arbitrary angular momenta J, the expression for the linearly polarized radiation intensity has the form

$$
I(e_r) = K'\mathscr{F}d_{12}^2\left(\frac{1}{3\Gamma}\begin{Bmatrix}1 & 1 & 1\\ J_1 & J_1 & J_0\end{Bmatrix}\begin{Bmatrix}1 & 1 & 0\\ J_1 & J_1 & J_2\end{Bmatrix}\right.
$$

$$
+ 5\begin{Bmatrix}1 & 1 & 2\\ J_1 & J_1 & J_0\end{Bmatrix}\begin{Bmatrix}1 & 1 & 2\\ J_1 & J_1 & J_2\end{Bmatrix}\left[\frac{1}{30\Gamma}(3\cos^2\theta''-1)(3\cos^2\theta'-1)\right.
$$

$$
+ \frac{1}{10}\sin 2\theta''\sin 2\theta'\frac{\Gamma\cos(\phi''-\phi')+\Omega\sin(\phi''-\phi')}{\Gamma^2+\Omega^2}
$$

$$
\left.\left.+ \frac{1}{10}\sin^2\theta''\sin^2\theta'\frac{\Gamma\cos 2(\phi''-\phi')+2\Omega\sin 2(\phi''-\phi')}{\Gamma^2+4\Omega^2}\right]. \right. \qquad (3.199)
$$

We introduce the abbreviations

$$
\begin{Bmatrix}1 & 1 & 0\\ J_1 & J_1 & J_0\end{Bmatrix}\begin{Bmatrix}1 & 1 & 0\\ J_1 & J_1 & J_2\end{Bmatrix} = A; \qquad \begin{Bmatrix}1 & 1 & 2\\ J_1 & J_1 & J_0\end{Bmatrix}\begin{Bmatrix}1 & 1 & 2\\ J_1 & J_1 & J_2\end{Bmatrix} = B
$$

(the quantities A and B are numbers that depend on the angular momenta of the initial and final states) and the whole term in the square brackets in (3.199) will be denoted α. Thus we can write (3.199) as

$$
I(e_r) = K'\mathscr{F}d_{12}^2\left(\frac{1}{3\Gamma}A + 5B_\alpha\right). \qquad (3.200)
$$

The quantity B can be either positive (this holds if $J_2 = J_1 \pm 0, 2$) or negative (in all other cases, i.e. $J_2 = J_1 \pm 1$). We multiply and divide the second term by $9B$

and rewrite (3.200) in the form

$$I(e_r)_{J_1 \to J_2} = K' \mathscr{F} d_{12}^2 \left[\frac{A-B}{3\Gamma} + 9B \left(\frac{1}{27\Gamma} + \frac{5}{9}\alpha \right) \right] . \tag{3.201}$$

Let us now compare this expression with (3.195), which can be expressed as

$$I(e_r)_{1 \to 0} = K' \mathscr{F} d_{10}^2 \left(\frac{1}{27\Gamma} + \frac{5}{9}\alpha \right) . \tag{3.202}$$

The first term in (3.201) is a spherically symmetric emission, since it has no dependence on angles. For $B > 0$ the remaining part differs only by the factor $9B$ from the emission from the transition $J = 1 \to J = 0$, (3.202) also excited by linearly polarized light. This second part coincides, to within a multiplicative factor, with the expression for the radiation of a classical dipole whose direction of oscillation is given by the polarization vector of the exciting light.

If however $B < 0$, then one can divide the total intensity into a spherically symmetric part and a residual anisotropic part, coinciding with the classical dipole emission excited not by linearly polarized light, but by unpolarized light whose direction of propagation n coincides with e_λ. An important distinction between the quantum mechanical description and the classical one is that only the former enables one to estimate the ratio of the spherically symmetric part to the residual anisotropic part of the radiation. In the classical approach the spherical part is simply absent.

In this way, one can represent the radiation from any transition as spherically symmetric radiation with the addition of the radiation from a classical dipole. The additive contribution decreases with increasing angular momentum quantum number of the transition approximately according to

$$\frac{\begin{Bmatrix} 1 & 1 & 2 \\ J_1 & J_1 & J_0 \end{Bmatrix} \begin{Bmatrix} 1 & 1 & 2 \\ J_1 & J_1 & J_2 \end{Bmatrix}}{\begin{Bmatrix} 1 & 1 & 0 \\ 1 & 1 & 0 \end{Bmatrix}^2} . \tag{3.203}$$

The above description of the fluorescence intensity is intended to provide a transparent comparison of quantum mechanical and classical approaches. A more common presentation is to divide the intensity into two other parts – the average intensity [the first term in (3.200)]

$$I = K' \mathscr{F} d_{12}^2 \frac{A}{3\Gamma} \tag{3.204}$$

and its interference part, consisting of the terms $\kappa > 0$. The latter is partially destroyed by fields and completely disappears upon integrating over all directions of the radiation (Sect. 3.9).

3.12   Biaxial Alignment

When exciting an atom by linearly polarized light or natural light of given propagation direction, an alignment arises in the excited state. An alignment is also induced when the light originates from all directions, provided that its intensity is different in different directions.

Let us consider some point in space through which light is passing in all directions. The vector $\mathscr{E}(t)$ of the electric field at this point varies in time in both magnitude and direction. The rate of change of the field strength is determined by the spectral characteristics of the light. The probability distribution of the light intensity can be described by the tensor with the components

$$I_{ik} = \overline{\mathscr{E}_i(t)\mathscr{E}_k^*(t)} \ , \tag{3.205}$$

where the bar indicates the time average, $\mathscr{E}_i(t)$ and $\mathscr{E}_k(t)$ are circular components of the electric field vector: $\mathscr{E}_i(t) = \mathscr{E}_i(t)e_i$. This tensor differs from the polarization tensor [3.24] and the coherent matrix [Ref. 3.25, Sect. 9.3] in that the latter describe the field states of a light wave in a given direction and are therefore of second order. The tensor I accounts for light passing through a point, whose field it describes over all directions. The instantaneous field vector can point in any direction in space and hence the tensor has order three. We note too that, if the polarization matrix describes the entire light beam throughout its spatial extent, then the tensor I may change drastically from one point in space to another.

One can go from this tensor representation to its expansion over the irreducible tensor operators [this transformation has already been treated above (3.177)]:

$$\Phi_q^\kappa = \sum_{i,k} (-1)^k I_{ik} \begin{pmatrix} 1 & 1 & \kappa \\ i & -k & q \end{pmatrix} \ . \tag{3.206}$$

The inverse transformation has the form

$$I_{ik} = (-1)^q \sum_\kappa (2\kappa + 1) \begin{pmatrix} 1 & 1 & \kappa \\ i & -k & q \end{pmatrix} \Phi_q^\kappa \ . \tag{3.207}$$

In this picture, the pump is characterized by the mean intensity Φ_0^0; the three quantities $\Phi_{+1}^1, \Phi_{-1}^1, \Phi_0^1$ define the direction and the degree of circular polarization; and the five quantities $\Phi_{\pm2}^2, \Phi_{\pm1}^2, \Phi_0^2$ characterize the magnitude and orientation of the alignment tensor.

The statistical state tensor excited by such light is calculated using (3.173). If the excited light propagating in all directions is unpolarized or linearly polarized, then the components with $\kappa = 1$ are absent in the tensor Φ and no orientation arises in the atom: the excited state is fully described by the population and alignment tensor.

In the classical model, emission from an atom excited by such a light field can be described by emission from three orthogonal linear dipoles, decoupled in phase (incoherent). Figure 3.3 illustrates this picture. The length of each arrow is proportional to the intensity of the light radiated by an ensemble of atoms with electric field vector directed along the dipole. Here there is no need to talk about the amplitude, since each of the dipoles is radiates in a given limited, but continuous frequency range. It is important to note that, in the most general case, the radiation intensity of all three dipoles is different. This case shows clearly the situation pertinent to biaxial alignment.

The radiation intensity of a superposition of dipoles, over which the emission of an ensemble of atoms is expanded, is found as a sum of the radiation intensities of the three dipoles. In the same way, one can obtain the direction and degree of polarization of the light emitted in any direction. However, the merit of this model lies not only in its ability to determine the radiation pattern diagram, but also in its simplicity and instructiveness for determining the intensity change of light of given polarization and propagation direction upon the application of a magnetic field. One can demonstrate this with a simple example. The intensity radiated by each dipole is zero along its axis and maximum in the direction perpendicular to the axis. Let us denote the maxima by I_x, I_y and I_z in a zero magnetic field for the three dipoles oriented along the x-, y- and z-axes, respectively. We apply a strong magnetic field along z. The dipoles along the x- and y-directions thus precess about the z-axis, such that in its lifetime every one of them is able to undergo a substantial number of rotations. The intensity in the xy-plane will thus be the same in all directions. One can write it in the form

$$I_{x,y}(\mathcal{H} = \infty) = (I_x + I_y)/2 \ . \tag{3.208}$$

The dipole directed along the z-axis is not perturbed by the magnetic field and its intensity remains as before

$$I_z(\mathcal{H} = \infty) = I_z \ . \tag{3.209}$$

The sign of the signal is easily determined. If the observation is carried out along

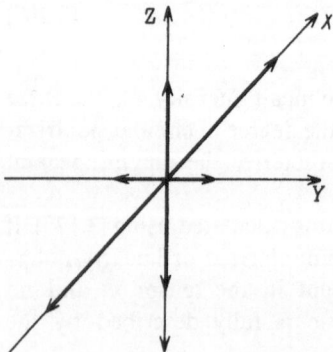

Fig. 3.3. The classical model of two-axis alignment

the y-axis, then the signal

$$S = I(\mathcal{H} = 0) - I(\mathcal{H} = \infty) = (I_x - I_y)/2 \tag{3.210}$$

is positive, i.e. the intensity decreases with the field, if $I_x > I_y$. Under the same conditions, the signal observed along x, $S = (I_y - I_x)/2$, is negative.

Let us now compare the model of the three dipoles with the statistical tensor model. In order to do this, we use the expression (3.178) to find $I_x - I_y$. Since for e_x we have $\theta = \pi/2$, and $\phi = 0$ and for e_y the angles are $\theta = \pi/2$, $\phi = \pi/2$, the non-zero components of the tensor Φ (3.180) will be:

$$\Phi_0^0 = -1/\sqrt{3}; \quad \Phi_0^2(e_x) = \Phi_0^2(e_y) = -1/\sqrt{30} \; ;$$

$$\Phi_{\pm 2}^2(e_x) = -\Phi_{\pm 2}^2(e_y) = 1/2\sqrt{5} \; ; \tag{3.180}$$

Substituting these expressions into (3.178) we find

$$I_x - I_y \propto \varrho_2^2; \quad (\varrho_2^2 = \varrho_{-2}^2 \text{ when } \Omega = 0) \; .$$

Hence, the statistical tensor component ϱ_2^2 is proportional to the difference in the dipoles' radiation intensity perpendicular to the axis of quantization.

In the three dipole model, the intensity of light in the direction perpendicular to the quantization axis z with polarization parallel to z is proportional to I_z. Expressed in terms of the statistical tensor components

$$I_z = K_1 \varrho_0^0 + K_2 \varrho_0^2 \; ,$$

where ϱ_0^0 is the state population, and $K_1 \varrho_0^6$ is the radiation intensity averaged over angles. In the model of three dipoles the latter can be given as

$$\bar{I} = \tfrac{1}{3}(I_x + I_y + I_z) \; .$$

Replacing $K_1 \varrho_0^0$ by $\bar{I}$, we find that the longitudinal component of the alignment is responsible for increasing the radiation intensity of the dipole directed along the axis of quantization to above the average radiation intensity of the ensemble over all directions

$$\varrho_0^2 \propto I_z - \bar{I} = I_z - \tfrac{1}{3}(I_x + I_y + I_z) \; .$$

Let us once more examine the formula (3.178) for the light intensity, assuming no precession of the components ϱ_q^κ, but only that the alignment is described by the two terms ϱ_0^2 and $\varrho_{\pm 2}^2$, which one is always able to achieve by choosing the coordinate axes properly. For the sake of convenience, let us rewrite (3.178) in the form

$$I = (-1)^{J_1 + J_2} K'(2J_1 + 1)^{-1/2} d_{12}^2 \left[\begin{Bmatrix} 1 & 1 & 0 \\ J_1 & J_1 & J_2 \end{Bmatrix} \varrho_0^0 \, \Phi_0^0 \right.$$

$$\left. + \begin{Bmatrix} 1 & 1 & 2 \\ J_1 & J_1 & J_2 \end{Bmatrix} 5(\varrho_0^2 \Phi_0^2 + \varrho_{-2}^2 \Phi_2^2 + \varrho_2^2 \Phi_{-2}^2) \right] . \tag{3.211}$$

The first term is the mean intensity. It is always positive, although its value can vary from one dipole transition to another. The sign of the interference term depends on J_2 and it reverses when J_2 changes by one (because the product $(-1)^{J_1 + J_2} \left\{ \begin{smallmatrix} 1 & 1 & 2 \\ J_1 & J_1 & J_2 \end{smallmatrix} \right\}$ is negative when $J_2 = J_1 \pm 1$ and positive when $J_2 = J_1$). The signal sign change on such "adjacent" transitions serves as a reliable indicator that the signals are evoked by the interference of states.

We will discuss one further characteristic of emission from an ensemble of atoms with biaxial alignment. Such an ensemble will emit light of different intensities for each of the three linear orthogonal polarizations and one can represent them in a descending or ascending order $I_z > I_y > I_x$. It is not difficult to show that, when observing radiation from the same initial level, but to the other lower level, this order will either be conserved or be reversed. We assume that the alignment tensor is reduced to the principal axes. It then possesses three components $\varrho_0^0, \varrho_0^2, \varrho_2^2 = \varrho_{-2}^2$ (provided the levels are degenerate; $\Omega = 0$). Let us write the intensities of the radiation polarized parallel to each of the principal axes

$$I_z = \bar{I} + 2Aa; \qquad I_y = \bar{I} - A(a + b); \qquad I_z = \bar{I} - A(a - b) , \qquad (3.212)$$

where we have introduced the following notation

$$\bar{I} = K' d_{12}^2 \left(\frac{-1}{\sqrt{3}} \right) (-1)^{J_1 + J_2} (2J_1 + 1)^{-1/2} \varrho_0^0 \left\{ \begin{matrix} 1 & 1 & 0 \\ J_1 & J_1 & J_2 \end{matrix} \right\} ,$$

$$A = K' d_{12}^2 (-1)^{J_1 + J_2} (2J_1 + 1)^{-1/2} \left\{ \begin{matrix} 1 & 1 & 2 \\ J_1 & J_1 & J_2 \end{matrix} \right\} ,$$

$$a = \varrho_0^2 \sqrt{\frac{5}{6}}; \qquad b = \varrho_2^2 \sqrt{5} .$$

The quantity $\bar{I}$ is the mean intensity. It depends on the transition probability, but is always positive. The second term in (3.212) varies in magnitude and in sign; nevertheless, it always remains smaller than $\bar{I}$ in absolute value. Let the coefficient A be changed only in magnitude upon moving to another spectral line (from the same radiating state). The ordering of the intensities in the series then remains the same. If, however the coefficient A changes sign, then the order of intensities also reverses: $I_z < I_y < I_x$.

3.13 Level Anti-crossings

The level anti-crossing phenomenon was in fact discovered accidentally. In an experimental investigation of the resonance fluorescence line, "incoherent" excitation was used by mistake but the signal, i.e. the intensity change in the vicinity of the crossing point still persisted. The authors of this discovery, *Eck* et al. [3.26], described and interpreted it and gave it the currently used name (a detailed treatment is given in [3.27–31]). The appearance of the signal can be described in terms of changes of the transition probabilities under the influence of a perturbation which mixes the initial states. The mixing in the vicinity of the initial crossing point is the highest, because in this region, the energy gap between the levels is the smallest. Crossings of the Zeeman components of the lithium fine structure $2\,^2P_{3/2}$ have been achieved by the application of a magnetic field. A second perturbation, which lifts the degeneracy, is the interaction between the electron orbitals and the nucleus of the atom, i.e. the perturbation evoking the hyperfine structure.

The behaviour of the atomic system, including the radiation characteristics, is fully described by the density matrix. The parts known as the level crossing signal and level anti-crossing signal are artificially isolated from the mathematical expression describing the characteristics of the radiation. As a result of this, there exists a certain amount of arbitrariness in their definitions. Although the concept of level crossing is quite well defined, at least when it exists in a clean form, the concept of level anti-crossings is highly indistinct. However, all the definitions of the latter converge to the fact that the phenomenon arises only when one perturbation, a smoothly varying one (thus it has to be an external one) induces level degeneracy, and the other, a constant one (an external or internal perturbation), lifts it.

Level anti-crossings may also refer to a scheme of energy levels that "cease" to cross upon the application of a perturbation that mixes the degenerate initial eigenstates.

The authors who discovered the phenomenon define pure the level anti-crossings as the variations of radiation intensity that arise upon exciting an atomic system with light of one of the polarizations σ_+, σ_- or π. This situation requires some explanation: Up to now we have considered atomic systems under conditions such that the atomic eigenfunctions do not change or are slightly changed when the levels are degenerate and the eigenpolarizations of the dipole transitions have been polarized σ_+, σ_- and π. If, in the expansion of the light over these polarizations, only one term differs from zero[8] then such light could not excite an atom to a coherent state. When the eigenfunctions of an excited atom are as we have just described them, excitation of the atom by light whose polarization matches one of the transition eigenpolarizations will not lead to any signal in the spontaneous radiation. When the states are mixed, the wavefunctions are changed and the signal is attributed to this. However, there is a weak point in this definition, because, when the characteristic wavefunctions are changed, so too are the characteristic polarizations of the transitions, and light of any polarization either populates both of the excited states coherently,

---

[8] *Cohen-Tannoudji* [3.23] defines as coherent the polarization of light, for which at least two terms of the expansion (3.185) are different from zero.

or populates neither of them. And two coherently populated states must give rise to the level crossing signal.

Let us proceed to the mathematical description of the level anti-crossing phenomenon. We write the Hamiltonian of the system

$$\hat{H} = \hat{H}_0 + \hat{H}_{in} + \hat{H}_M + \hat{V} = \hat{H}_0 + \hat{H}_p \ . \tag{3.213}$$

When $\hat{H}_p = 0$, the energies of all of the eigenstates are degenerate. The operator $\hat{H}_{in}$ (this can be the hyperfine interaction, fine splitting, etc.; in the present case its nature is not important) lifts the degeneracy. Let the operators $\hat{H}_{in}$ and $\hat{H}_M$ be diagonal. The operator $\hat{H}_M$ can also make some of these states degenerate with respect to the energy. For simplicity we assume that there are only two such degenerate levels; the presence of all other states is neglected. If the energies of these other states are far from the energy of the degenerate levels, then they will have no substantial influence on the pattern which we are going to derive now. These two eigenstates of the operator $\hat{H}_0 + \hat{H}_{in} + \hat{H}_M$ will be denoted by Ψ_1 and Ψ_2, and their energies by E_1 and E_2. An interesting case is when the latter part of the perturbation operator, $\hat{V}$, possesses off-diagonal matrix elements. Under the influence of the perturbation $\hat{V}$ new states Φ_1 and Φ_2 will be produced, whose energy levels will no longer cross, but will merely approach to within some distance of one another, depending on the off-diagonal matrix elements of the perturbation operator, and then diverge, having exchanged indices. At a point far away from the level crossing, the atoms response to this same action is insignificant, since the perturbation turns out to be small due to the high value of the denominator containing the difference of the level energies.

The wavefunctions in the vicinity of the crossing points are defined in terms of their expansion over the non-degenerate functions

$$\Phi_2 = C_2 \Psi_2 + C_1 \Psi_1 \ ,$$
$$\Phi_1 = C_1 \Psi_2 - C_2 \Psi_1 \ . \tag{3.214}$$

The coefficients are derived from the system of equations (Sect. 5.9)

$$C_1[(V_{11} + E_1) - \varepsilon] + C_2 V_{12} = 0 \ ,$$
$$C_1 V_{21} + C_2[(V_{22} + E_2) - \varepsilon] = 0 \ , \tag{3.215}$$
$$C_1^2 + C_2^2 = 1 \ ,$$

where $V_{ik} = \langle \Psi_i | \hat{V} | \Psi_k \rangle$ is the matrix element between the functions Ψ (i and k take the values 1 and 2) and ε are energies of the eigenfunctions of the total Hamiltonian. The energies ε are determined from the secular equation

$$\begin{vmatrix} V_{11} + E_1 - \varepsilon & V_{12} \\ V_{21} & V_{22} + E_2 - \varepsilon \end{vmatrix} = 0 \ , \tag{3.216}$$

which gives the following energy values for the new states

$$\varepsilon_{2,1} = \frac{V_{11} + V_{22} + E_1 + E_2}{2} \pm \frac{1}{2}\sqrt{(V_{22} - V_{11} + E_2 - E_1)^2 + 4V_{12}^2} \; .$$

$$(3.217)$$

Let us denote the difference between the energies of the eigenstates by $\hbar\delta$:

$$\hbar\delta = \varepsilon_2 - \varepsilon_1 \; .$$

Clearly

$$\hbar\delta_{21} = \sqrt{(V_{22} - V_{11} + E_2 - E_1)^2 + 4V_{12}^2} \; . \tag{3.218}$$

From the system of equations (3.215) we find the values of the coefficients

$$C_{2,1} = \frac{1}{\sqrt{2}}\left(1 \pm \frac{\omega_{21}}{\delta_{21}}\right)^{1/2} , \tag{3.219}$$

where

$$\omega_{21} = (V_{22} - V_{11} + E_2 - E_1)/\hbar \; .$$

Let us now discuss the emission of an atomic system and its dependence on the magnitude of the perturbation under excitation using a light of one of the three polarizations σ_+, σ_- or π. In the absence of perturbation, the polarized light will excite the atom only to a single eigenstate Ψ and there will be no change of the fluorescence in the vicinity of the crossing point. The signal arises only in the presence of the perturbation. The radiation intensity (3.67) is given by

$$I = K \sum_{i,k} \sigma'_{i,k} G'_{ki} \; .$$

The density matrix of the excited state for a constant time-independent excitation is given by

$$\sigma'_{ik} = \frac{F'_{ik}}{\Gamma + i\delta_{ik}} \; .$$

Here the pump matrix (3.35) is

$$F'_{ik} = K' \sum_{\mu} \langle \Phi_i | d \cdot e_\lambda | \mu \rangle \langle \Phi_k | d \cdot e_\lambda | \mu \rangle^* \; ,$$

where $|\mu\rangle$ is the lower state.

Let us express this through the pump matrix, which is defined by the function

$$F_{ik} = K' \sum_{\mu} \langle \Psi_i | \boldsymbol{d} \cdot \boldsymbol{e}_\lambda | \mu \rangle \langle \Psi_k | \boldsymbol{d} \cdot \boldsymbol{e}_\lambda | \mu \rangle^* \ .$$

The relation between the two pump matrices is

$$F'_{22} = C_2^2 F_{22} + C_1^2 F_{11} \ ,$$

$$F'_{11} = C_1^2 F_{22} + C_2^2 F_{11} \ , \tag{3.220}$$

$$F'_{12} = C_1 C_2 (F_{22} - F_{11}) \ .$$

The observation matrix G'_{Ki} in (3.67) can also be expressed in terms of the function Ψ. The transformation is the same as (3.220) and we refrain from repeating it here. Let us substitute into the expression for the intensity, the values F' and G' expressed in terms of F and G, and the coefficients C (3.219). We obtain

$$I = K \left[\frac{F_{11} G_{11}}{\Gamma} + \frac{F_{22} G_{22}}{\Gamma} \right.$$

$$\left. + (F_{22} - F_{11})(G_{11} - G_{22}) \frac{2|V_{12}|^2 / \hbar^2}{\Gamma(\Gamma^2 + 4|V_{12}|^2/\hbar^2 + \omega_{21}^2)} \right] \ . \tag{3.221}$$

We will discuss the dependence of the observed intensity on the convergence of the energy states Ψ. We recall that the energy difference between them is denoted by $\hbar\omega_{21}$. If the perturbation $\hat{V}$ is absent, i.e. $V_{12} = 0$, then the intensity will be given by the first two terms only and will not depend on ω_{21}. In the presence of $\hat{V}$ and in the region of small ω_{21} there will arise a change in the intensity – a maximum or a minimum. Moreover, the additive quantity will have Lorentzian form with the width $\sqrt{\Gamma^2 + 4|V_{12}|^2/\hbar^2}$ and an amplitude proportional to $|V_{12}|^2$. This variation in intensity has also been called the anti-crossing signal [3.26]. The analysis of formula (3.221) showed that this signal can also be produced under conditions in which the level crossing signal is absent. We note, however, that certain conditions must be satisfied in order to observe it, namely

$$F_{22} - F_{11} \neq 0; \qquad G_{11} - G_{22} \neq 0 \ .$$

These conditions imply that the excitation in the absence of the perturbation $\hat{V}$ has to support the population difference of the states Ψ_2 and Ψ_1 and that observation must be organized such that the population difference of the states Ψ_2 and Ψ_1 could have been observed.

The anti-crossing signal is also produced when states are populated coherently. The expression for it turns out to be rather awkward. We present its form without derivation for the case of excitation and observation of linearly polarized light. Furthermore, the vectors e_r and e_λ are either parallel or perpendicular to one another and at least one of them is perpendicular to the magnetic field (under these conditions, the pump matrix and the observation matrix do not contain imaginary terms)

$$
\begin{aligned}
I = K \Bigg\{ &\frac{F_{11}G_{11}}{\Gamma} + \frac{F_{22}G_{22}}{\Gamma} \\
&+ (F_{22} - F_{11})(G_{11} - G_{22}) \frac{2V_{12}^2/\hbar^2}{\Gamma(\Gamma^2 + 4|V_{12}|^2/\hbar^2 + \omega_{21}^2)} \\
&+ 2F_{12}G_{21} \frac{\Gamma^2 + 4|V_{12}|^2/\hbar^2}{\Gamma(\Gamma^2 + 4|V_{12}|^2/\hbar^2 + \omega_{21}^2)} \\
&+ [F_{12}(G_{22} - G_{11}) + G_{21}(F_{22} - F_{11})] \frac{2V_{12}\omega_{21}/\hbar}{\Gamma(\Gamma^2 + 4|V_{12}|^2/\hbar^2 + \omega_{21}^2)} \Bigg\} .
\end{aligned}
$$

(3.222)

It is seen that, upon coherent state population, the signal of the former type is augmented by other signals of two kinds: one of them is the level crossing signal, but broadened by the perturbation $\hat{V}$, the second has a dispersion form and will persist when the level crossing signal is absent, even if off-diagonal elements exist only for the pump matrix (coherent population of the states Ψ) or only for the observation matrix (e_r possesses coherency).

The level anti-crossing signal vanishes upon incoherent as well as coherent state population when $\hat{V} \to 0$. In the coherent case it converts to the normal level crossing signal.

Up to now we have not specified the form of the operator $\hat{V}$. This can be either an external or an internal perturbation. The anti-crossing will take place if this operator couples states, i.e. if $V_{12} \neq 0$. In particular, this can be the perpendicular component of a magnetic field that couples states with $\Delta m = \pm 1$ (Sect. 5.10) or an electric field that is perpendicular to the quantization axis, which couples the states with $\Delta m = \pm 2$.

Lehmann [3.28] treats the energy-level structure of an atom possessing a nuclear magnetic moment as level splitting of the fine structure in a magnetic field and anti-crossing, i.e. lifting of degeneracy in weak and zero magnetic fields under the influence of the nuclear perturbation. The level structure of cadmium 5^1P_1 is sketched in Fig. 3.4. The level splitting without the hyperfine interaction is indicated by the straight lines. The energy levels correspond to the eigenvalues of the operator

$$
\hat{H} = \hat{H}_0 + \mu_0 g_J(\hat{\boldsymbol{J}} \cdot \mathscr{H}) - \mu_0 g_I(\hat{\boldsymbol{I}} \cdot \mathscr{H}) .
$$

The levels with magnetic quantum number $m_I + m_J = 1/2$ cross in a field of 70 G, and those with magnetic quantum numbers $m_I + m_J = -1/2$ in a field $\mathscr{H} = -70$ G (the extension of the energy diagram into the region of "negative"

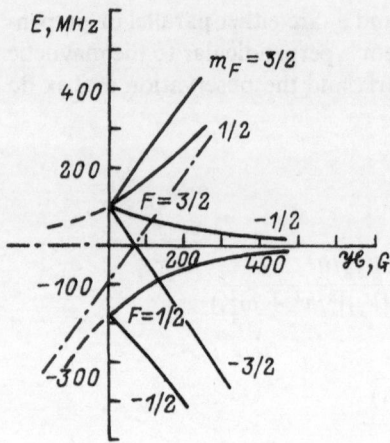

Fig. 3.4. The level structure of cadmium 5^1P_1

magnetic fields is shown by the broken lines). The hyperfine interaction lifts the degeneracy. The levels will no longer cross, but approach one another somewhere in the vicinity of the previous crossing point, reaching a minimum separation and once more diverging, having exchanged magnetic quantum numbers. The anti-crossing in the hfs is extremely well defined for ytterbium 1^1P_1 levels, and a signal of this type has been used in [3.30] in order to determine the hfs constant. *Series* [3.27] showed somewhat earlier that the magnetic resonance, in a coordinate system rotating at the Larmour frequency, is none other than the level anti-crossing. The rf perturbation field mixes the wavefunctions of the different Zeeman sublevels and changes their energy. If the Zeeman sublevels are degenerate (in a rotating coordinate system) in a given magnetic field and in the absence of an rf field, then the perturbing rf field (the rf in the laboratory coordinate system turns to be constant in the rotating frame of reference) will "push" the levels apart and they will not cross each other any more. If an observer had been placed in the rotating coordinate system, then he would have been able to register the anti-crossing signal in the fluorescence upon changing the magnetic field.

3.14 Interference Phenomena in Magnetic Resonance

Let us examine the behaviour of atoms that are placed in a static magnetic field with an alternating magnetic field superimposed perpendicularly to the former. Let us suppose that the Zeeman sublevels are populated differently. These are the prerequisites for observing magnetic resonance. Under the influence of the alternating field $\mathcal{H}_1$, transitions will take place between the Zeeman sublevels, and the rate of these transitions will attain a maximum at resonance, i.e. when the rf field frequencies equal the Zeeman splittings. The variation of populations of the sublevels is accompanied by a change of the polarization characteristics of the spontaneous emission and the coefficient of absorption. The magnetic resonances themselves are successfully described by the Bloch equations ([3.32] and Sect. 4.5, Eq. (4.1)). Here, however, we are interested in the relationship between magnetic resonance and interference phenomena.

We write the Hamiltonian of such a system

$$\hat{H} = \hat{H}_0 + \hat{H}_M + \hat{H}_{rf} \ .$$

We denote the eigenvalues of the Hamiltonian $\hat{H}_0 + \hat{H}_M$ by E_M, i.e.

$$(\hat{H}_0 + \hat{H}_M)\Psi = E_M\Psi \ .$$

The operator $\hat{H}_M$ has the form

$$\hat{H}_M = \mu_0 g(\hat{\boldsymbol{J}} \cdot \mathscr{H}) = \mu_0 g \hat{J}_z \mathscr{H} \tag{3.223}$$

and the interaction with the radio frequency field, rotating in the xy-plane is given by

$$\hat{H}_{rf} = \mu_0 g(\hat{\boldsymbol{J}} \cdot \mathscr{H}_1(t)) = \mu_0 g \mathscr{H}_1((\boldsymbol{e}_x \cos \Omega t + \boldsymbol{e}_y \sin \Omega t) \cdot \hat{\boldsymbol{J}}) \ .$$

We transfer to the coordinate system rotating about the z-axis with a frequency Ω. This enables us to reduce the problem of calculating the states to the well-known steady-state problem.

In the new coordinate system, the wavefunctions Ψ, the eigenfunctions of the operator $\hat{H}_0 + \hat{H}_M$ with eigenvalues E_M, are transformed to the new wavefunctions

$$\Psi' = \hat{S}\Psi = e^{i\Omega \hat{J}_z t}\Psi \ . \tag{3.224}$$

The eigenvalues of the operator $\hat{H}_0 + \hat{H}_M$ also change. From the Schrödinger equation it follows that

$$(\hat{H}_0 + \hat{H}_M)\Psi' = i\hbar \dot{\Psi}' = i\hbar(\dot{\hat{S}}\Psi + \hat{S}\dot{\Psi})$$

$$= -\hbar\Omega \hat{S}\hat{J}_z\Psi + \hat{S}E_M\Psi = (E_M - \hbar\Omega M)\Psi' \ ,$$

since

$$(\hat{H}_0 + \hat{H}_M)\Psi = i\hbar\dot{\Psi} = E_M\Psi \ .$$

Let the quantum number M take only two values M_1 and M_2, differing by one. In some magnetic field $\mathscr{H}'$, the corresponding states may become degenerate in energy. Actually, considering (3.223), we find

$$E_1 = E_{01} + \mu_0 g M_1 \mathscr{H}' - M_1 \hbar\Omega \ ,$$

$$E_2 = E_{02} + \mu_0 g M_2 \mathscr{H}' - M_2 \hbar\Omega \ ,$$

and their difference

$$E_2 - E_1 = E_{02} - E_{01} + \mu_0 g \mathcal{H}' - \hbar\Omega$$

may indeed become zero. In particular, if $E_{01} = E_{02}$, then the levels will become degenerate when

$$\hbar\Omega = \mu_0 g \mathcal{H}' \ .$$

Let us now consider the operator $\hat{H}_{rf}$. For convenience let us write it in the form

$$\hat{H}_{rf} = \tfrac{1}{2}\mu_0 g \mathcal{H}_1 (\hat{J}_+ e^{-i\Omega t} + \hat{J}_- e^{i\Omega t}) \ ,$$

where

$$\hat{J}_\pm = \hat{J}_x \pm i\hat{J}_y \ .$$

In the rotating coordinate system, the operator $\hat{H}_{rf}$ is transformed to $\hat{H}'_{rf}$:

$$\hat{H}'_{rf} = \hat{S}\hat{H}_{rf}\hat{S}*$$

$$= \tfrac{1}{2}\mu_0 g \mathcal{H}_1 (e^{-i\Omega t} e^{i\Omega \hat{J}_z t} \hat{J}_+ e^{-i\Omega \hat{J}_z t} + e^{i\Omega t} e^{i\Omega \hat{J}_z t} \hat{J}_- e^{-i\Omega \hat{J}_z t}) \ . \tag{3.225}$$

The operator $\exp(\pm i\Omega \hat{J}_z t)$ does not change the values of the magnetic quantum numbers M (Sect. 5.10) and they can be replaced by the eigenvalues. Since the operator J_+ increases the magnetic quantum number by unity:

$$\hat{J}_+ |M\rangle = C|M+1\rangle \ ,$$

where C is some coefficient, the expression

$$e^{i\Omega \hat{J}_z t} \hat{J}_+ e^{-i\Omega \hat{J}_z t}$$

can be replaced by

$$e^{i\Omega(M+1)t} \hat{J}_+ e^{-i\Omega M t} = e^{i\Omega t} \hat{J}_+ \ .$$

In the same manner, transforming the last term in (3.225), we obtain

$$\hat{H}'_{rf} = \tfrac{1}{2}\mu_0 g \mathcal{H}_1 (\hat{J}_+ + \hat{J}_-) \ .$$

The total Hamiltonian of the system can be given in the form

$$\hat{H}' = \hat{H}_0 + \hat{H}_M(\mathcal{H}') + \hat{H}_M(\Delta\mathcal{H}) + \tfrac{1}{2}\mu_0 g \mathcal{H}_1 (\hat{J}_+ + \hat{J}_-) \ , \tag{3.226}$$

where

$$\Delta\mathcal{H} = \mathcal{H} - \mathcal{H}' \ .$$

The eigenenergies of the Hamiltonian $\hat{H}_0 + \hat{H}_M(\mathscr{H}')$ are degenerate, and therefore the remaining terms in the Hamiltonian can be assumed as a perturbation

$$\hat{H}' = \hat{H}_0 + \hat{H}_M(\mathscr{H}') + \hat{V} \ .$$

One can proceed to solve this problem using the scheme of Sect. 3.13. The operator $\hat{V}$ gives rise to the presence of the off-diagonal terms. The eigenenergies of the Hamiltonian are given by

$$E_{2,1} = \frac{H'_{11} + H'_{22}}{2} \pm \frac{1}{2} \sqrt{(H'_{22} - H'_{11})^2 + 4|V_{12}|^2}$$

$$= \frac{\mu_0 g}{2} \left[(2M + 1)\varDelta\mathscr{H} \pm \sqrt{\varDelta\mathscr{H}^2 + \mathscr{H}_1^2 |\langle M|\hat{J}_\pm|M \pm 1\rangle|^2} \right] \ . \qquad (3.227)$$

Unlike the general case of anti-crossing, here there is no shifting of the energy gap minimum in relation to the point of degeneracy of the unperturbed states. This is due to the fact that the operator $\hat{V}$ has no diagonal elements.

If one remains in the rotating coordinate system, then the emission (absorption) will have characteristics identical to those of the ordinary anti-crossing case. If, however the observer is in a stationary coordinate system, he will detect modulation of the emission (absorption) at the frequency of the rf field.

In order to prove this, let us find the density matrix. It is coupled with the pump matrix. At the beginning of this section we mentioned the conditions under which one can observe magnetic resonance. They correspond to excitation with incoherent polarization and thus the pump matrix will contain only diagonal elements. When transforming to the rotating coordinate system its form will not be changed, and both π and $\sigma_\pm$ polarizations will be conserved upon changing from one coordinate system to the other.

In a rotating coordinate system the equation for the density matrix takes the form

$$\dot{\hat{\sigma}}' = \frac{-i}{\hbar}(\hat{H}\hat{\sigma}' - \hat{\sigma}'\hat{H}) - \Gamma\hat{\sigma}' + \hat{F} \ .$$

It contains the operator $\hat{H}$, which has the form of (3.226). This equation cannot be solved for a multilevel system. Thus, as before, we will restrict ourselves to the special case of two levels with M_1 and M_2. For this system, $V_{12} = V_{21}, \sigma_{21} = \sigma_{12}^*$, and thus

$$\dot{\sigma}'_{11} = \frac{-i}{\hbar} V_{12}(\sigma'_{21} - \sigma'_{12}) - \Gamma\sigma'_{11} + F_{11} \ ,$$

$$\dot{\sigma}'_{22} = \frac{i}{\hbar} V_{12}(\sigma'_{21} - \sigma'_{12}) - \Gamma\sigma'_{22} + F_{22} \ , \qquad (3.228)$$

$$\dot{\sigma}'_{12} = \frac{-i}{\hbar} V_{12}(\sigma'_{22} - \sigma'_{11}) + i\sigma'_{12}(\omega_{21} - \Omega) - \Gamma\sigma'_{12} \ .$$

In a steady state all the terms on the left-hand side are zero. What then remains are algebraic equations, whose solutions will have the following form

$$\sigma'_{11} = \frac{F_{11}}{\Gamma} + \frac{2V_{12}^2(F_{22} - F_{11})}{\hbar^2\Gamma(\Gamma^2 + (\omega_{21} - \Omega)^2 + 4V_{12}^2/\hbar^2)} \ ,$$

$$\sigma'_{22} = \frac{F_{22}}{\Gamma} + \frac{2V_{12}^2(F_{11} - F_{22})}{\hbar^2\Gamma(\Gamma^2 + (\omega_{21} - \Omega)^2 + 4V_{12}^2/\hbar^2)} \ , \qquad (3.229)$$

$$\sigma'_{12} = \frac{V_{12}(F_{22} - F_{11})(\omega_{21} - \Omega - i\Gamma)}{\hbar\Gamma(\Gamma^2 + (\omega_{21} - \Omega)^2 + 4V_{12}^2/\hbar^2)} \ .$$

Let us turn once again to the stationary reference system. The expressions for the diagonal elements σ' will not change, but the off-diagonal elements, in accordance with (3.226) will acquire an oscillatory factor:

$$\sigma_{ik} = \sigma'_{ik} \exp[-i\Omega(M_i - M_k)t] \ . \qquad (3.230)$$

The observed intensity is

$$I = K \sum_{i,k} \sigma_{ik} G_{ki} \ .$$

The elements of the observation matrix G_{ki} are determined by the conditions of observation: direction and polarization. When observing coherent polarizations, the matrix G will contain only diagonal terms, which have no time dependence. The dependence of I on the frequency of the rf field and its strength $\mathcal{H}_1$ is what constitutes the usual magnetic resonance signal:

$$I = K \sum_{i=1,2} \sigma_{ii} G_{ii}$$

$$= \frac{K}{\Gamma}(F_{11}G_{11} + F_{22}G_{22}) + \frac{K}{\Gamma}\frac{2V_{12}^2(F_{11} - F_{22})(G_{22} - G_{11})}{\hbar^2(\Gamma^2 + (\omega_{21} - \Omega)^2 + 4V_{12}^2/\hbar^2)} \ . \qquad (3.231)$$

Upon observing the coherent polarization, for example the linear polarization at some specified angle to the z-axis, the observation matrix will contain off-diagonal terms and oscillations will arise in the intensity at frequencies Ω (if one assumes that $M_1 - M_2 = \pm 1$):

$$I = K\left[\sum_{i=1,2} \sigma_{ii} G_{ii} + 2\mathrm{Re}\{\sigma_{12} G_{21} e^{-i\Omega t}\}\right]$$

$$= K\left[\sum_{i=1,2} \sigma_{ii} G_{ii} + \mu_0 g \mathcal{H}_1 \langle M_1|\hat{J}_+ + \hat{J}_-|M_2\rangle \right.$$

$$\left. \times \frac{(F_{11} - F_{22})|G_{21}|\sqrt{\Gamma^2 + (\omega_{21} - \Omega)^2} \cos(\Omega t - \phi)}{\hbar\Gamma(\Gamma^2 + (\omega_{21} - \Omega)^2 + \mu_0^2 g^2 \mathcal{H}_1^2|\langle M_1|\hat{J}_+ + \hat{J}_-|M_2\rangle|^2/\hbar^2)}\right] \ , \qquad (3.232)$$

where

$$\tan \phi = \frac{\Gamma \operatorname{Re}\{G_{21}\} + (\omega_{21} - \Omega)\operatorname{Im}\{G_{21}\}}{(\omega_{21} - \Omega)\operatorname{Re}\{G_{21}\} - \Gamma \operatorname{Im}\{G_{21}\}} .$$

The model of a two-level scheme described well the behaviour of the atomic state with $J = 1/2$ or $F = 1/2$, which split in a magnetic field into two magnetic sublevels, from which follow the π and σ transitions, however the σ transitions with M_1 and M_2 have different polarizations (Fig. 3.5). If the lower level has a total angular momentum equal to 1/2, then the transition probabilities π and σ will be identical. The observation of radiation from the π transition is described by a diagonal matrix in which $G_{11} = G_{22}$ and, as a consequence, in the expression for the intensity there remain only terms that do not depend on the perturbing rf field

$$I = K \frac{G_{11}(F_{11} + F_{22})}{\Gamma}$$

i.e. in the π polarization magnetic resonance is not observed and in the polarization σ_+ ($G_{22} = 0$) the intensity depends on $\omega_{21} - \Omega$, i.e. magnetic resonance is seen. This is also true for observations of the polarization, except that now the sign of the magnetic resonance is reversed.

Let us now examine the light whose polarization e_r must be observed in order to see beats. For this purpose it is necessary that the off-diagonal matrix element G_{21} be different from zero. Let us express this in the following way:

$$G_{21} = \langle \tfrac{1}{2}|d \cdot e_r|\tfrac{1}{2}\rangle \langle -\tfrac{1}{2}|d \cdot e_r|\tfrac{1}{2}\rangle^* + \langle \tfrac{1}{2}|d \cdot e_r|-\tfrac{1}{2}\rangle \langle -\tfrac{1}{2}|d \cdot e_r|-\tfrac{1}{2}\rangle^* .$$

The matrix elements of the σ transitions are of the same magnitude, which we denote as d_σ, but have different signs. Let us denote the numerical value of the matrix element of the π transition through d_π. Then

$$G_{21} = e_r \cdot e_0 d_\pi d_\sigma (e_r \cdot e_+ - e_r \cdot e_-) .$$

We expand e_r over the coordinates

$$e_r = a_z e_z + a_x e_x + a_y e_y ,$$

where the coefficients a may be complex. We then obtain

$$G_{21} = -2 d_\pi d_\sigma a_z (a_x + i a_y) . \tag{3.233}$$

Thus, in accordance with (3.233) beats are seen if the polarization of the observed light contains components both along the constant magnetic field $a_z e_z$, and along the alternating field $a_x e_x$. A phase shift between these components leads only to a phase shift of the beats in the emitted light.

The two-level scheme also provided a good description of the magnetic resonance in an intermediate field for any level structures, provided that only a single pair of levels with $\Delta M = 1$ is close to the resonance. One can then assume that the remaining levels are not perturbed by the rf field. As a rule, the probabilities π and σ of transitions from these levels are different and magnetic resonance must be observable in radiation of any polarization; the sign and magnitude, however, of the observed resonance depend on the choice of the polarization. Of course, one can always search for a polarization for which the ordinary magnetic resonance signal is absent, $G_{11} - G_{22} = 0$. However, this polarization will contain π as well as σ components and will possess an interference term, namely an intensity modulation at a frequency Ω on a constant background:

$$\frac{K}{\Gamma} (F_{11} G_{11} + F_{22} G_{22}) \ .$$

3.15 Application of Interference Signals

The Hanle and level-crossing effects, quantum beats, resonance beats and parametric resonance are all attributed to interference phenomena. Anti-crossing and magnetic resonance are also accompanied by interference phenomena. Their magnitude and profile of these signals depend on the atomic structure, on the perturbation applied to it, and on the conditions of the experiment. The observation of interference phenomena enables one to determine a number of atomic constants, constants of interaction, and/or the size of the external perturbations. In all the expressions describing the various interference phenomena, there necessarily enter two quantities: the separation between the interfering sublevels ω and the relaxation constant Γ. In experiments to observe interference phenomena, one changes the quantity ω, usually by applying external fields, most frequently magnetic but occasionally also electric fields. The relationship between ω and the field value depends on the structure of the atom.

The simplest and the first discovered interference effect is the Hanle effect. It is observed in weak magnetic fields for isolated and unperturbed states. For these states the relationship between the magnetic field and ω is linear and is identical for every pair of the adjacent magnetic sublevels. In this situation the intensity variation has the form

$$I = I_0 + \frac{K'}{\Gamma^2 + 4(\mu_0 g \mathcal{H})^2/\hbar^2} = I + I' \ .$$

When $I' = I'_{max}/2$ the magnetic field is $\mathcal{H} = \hbar\Gamma/2\mu_0 g$, and from this one can immediately determine the relaxation constant Γ. For a free atom Γ corresponds to the inverse of the radiation lifetime and for an atom in a medium it depends on collisions and on diffusion of the radiation. Therefore, by investigating the dependence of the signal width on external conditions one obtains information about the interaction of atoms. In principle it is quite simple to extract the Landé factor from the beats observed in the emission of such simple states. The signal of free beats is described by two factors: the time dependent damping and the oscillating term. The frequency of the latter in a known magnetic field immediately gives the Landé factor g.

A number of atoms possess hyperfine structure. For such atoms, the dependence of ω on $\mathscr{H}$ becomes nonlinear; the levels with different total momenta can have different g factors, and moreover, each magnetic sublevel itself depends on the magnetic field as a result of interactions. These interactions depend on the hfs constants: the constant A of the magnetic dipole interaction of the electron orbitals with the nucleus and the constant B of the electrical quadrupole interaction. If the structure is such that one observes isolated crossing signals, then based on the magnetic field corresponding to the levels' degeneracy with a known g factor, it is possible to calculate these constants. If the structure is not resolved, i.e. if the Hanle signal and the crossing signals overlap, then information can nonetheless be extracted. For this purpose one usually calculates the signals for a number of values of the parameters, varying these until agreement is achieved between the computed and experimentally observed dependences.

An electric field also changes the energy structure of an atom, and this variation is determined by the Stark constant β. The application of an electric field can lead to level degeneracy and the appearance of a level-crossing signal, which immediately enables one to calculate the constant β. An alternative method of determining β is the measurement of crossing signals in a magnetic field with a superposed electric field parallel to the magnetic field. In this method, a lower strength electric field is sufficient and this simplifies the experiment.

Finally, if the atomic constants and constants of interaction are known, then based on the interference signals one can derive information about the external perturbations. The final chapter of this book (Chap. 5) describes the dependence of ω on external fields and on the structure of atoms, as well as the matrix elements of the dipole transitions required in order to calculate the interference signals. The next chapter is devoted to a description of experimental work on the interference of states. In an experiment, it is impossible to realize the "pure" conditions for which the description given in the present and second chapter is exact. The experimentally observed signal may be distorted by radiation diffusion, by collisions and most frequently by cascaded transitions. The emission process occurring in the radiating volume actually induces the coherent population of levels. The next chapter provides insight into all the additional details concerned with arranging the observation of interference signals and with the extraction information from these signals.

4. Experimental Observation of Interference Signals

4.1 Basic Experimental Scheme

Any experiment on interference phenomena in atoms and molecules requires a coherent ensemble of particles and a radiation analyzer. The coherency, as a rule, is induced by an external source, whose nature can be extremely diverse: however, it must necessarily be anisotropic, i.e. its interaction with the ensemble of particles in different directions is different. The techniques used to induce coherency are: resonant excitation by light; excitation by collisions with a flux of direct particles and, conversely, acquisition of coherency by a flux of particles upon its passage through a thin foil (beam-foil technique); interaction with laser radiation; interaction with an rf field via the magnetic resonance transitions in quantum particles; and optical pumping in the sense that *Kastler* [4.1, 2] attached to this terminology, proposing optical illumination as a means to change the characteristics of ensembles of atoms in the lower state in order to observe magnetic resonance in this state. The production of coherency in an ensemble without external interactions will be discussed in Sect. 4.3.

4.2 Ensembles of Particles

In order to prepare a coherent ensemble of particles, the first prerequisite is of course an appropriate ensemble of particles. It is not possible in all cases to induce observable coherency.

State coherency was observed for the first time in the year 1923 [4.3] (although at that time the terminology "coherent state" did not yet exist) in the form of resonance fluorescence in vapours. Thus for historical reasons, atomic vapours, usually enclosed in a cell, were the first suitable media discovered. The vapour pressure must be small. *Wood* [4.4] writes that the emission of light by the "resonating" gas is not observed. This was because the investigation was not carried out at sufficiently low pressure. A proper choice of the vapour density is essential in order to observe the interference phenomena. On one hand, the lower the density, the smaller the number of radiating atoms; but on the other hand, an increase in the vapour density enhances the radiation diffusion and the rate of collisions, leading to a decrease and distortion of the interference signals.

Moreover, when exciting by light, a high vapour density can disturb the light "entering" the cell, the excitation is concentrated in the vicinity of the entrance window. The optimal density depends mainly on the properties of the optical transition for which the resonance fluorescence is excited.

When measuring the lifetime of the excited neon states [4.5], the pressure has been varied over a wide range. Neon has two resonance levels $^3P_1$ and $^1P_1$, the former being excited by the line at 74.3 nm and the latter by that at 73.6 nm. The transition probabilities of these lines are very different, and thus, at a relatively high pressure, the Hanle effect was clearly observed on the line at 74.4 nm, but on the 73.6 nm line at the same pressure it was distorted to such an extent by diffusion that measurement was not possible.

Historically, the first cells used for the observation of fluorescence were of a design known as a "Wood's horn": this is a cylindrical cell with a plane parallel window on one end, which is curved and pulled into a cone at the other end. The role of the pulled cone is to minimize the scattered light. With time this design evolved and was transformed in some cases into a highly complex construction.

There are only a few elements for which it is possible to prepare a vapour cell without any special tricks. Most materials are more or less non-volatile and in order to induce coherency in them one uses atomic and molecular beams. The optimal particle density in the beam is determined by the same conditions as in a cell, but with additional constraints imposed by the beam technique itself.

Studies by astrophysicists have shown that the sun's prominences provide a suitable medium for the formation of an alignment. However, the coherency [4.6] can also be produced in special cases at atmospheric pressure. Since the time of publication of Refs. [4.7, 8], a gas discharge has been employed in order to observe interference signals from a number of elements.

4.3 Techniques for Inducing Coherence

In most studies the coherent ensemble is achieved as a result of external forces.

The symmetry of the external forces together with the quantum characteristics of the particles determines the moments that can be induced in the ensemble [4.9–12]. It is obvious that an isotropic field acting on an isotropic ensemble cannot induce moments higher than zero, i.e. population.

The first moment, orientation, is induced only when the system (the ensemble and the factor acting on it) has properties that physically differ upon rotating to the right and left, i.e. a "screw" must exist in the system. A trivial example is the illumination of an isotropic ensemble by circularly polarized light. A less trivial example is the presence of orientation in the initial state of an ensemble and its transfer to other states even by an isotropic action, for instance upon spontaneous decay. A rather special case is the presence of alignment in the initial ensemble with an axially symmetric external action, but where the symmetry axes are inclined to each other. Rotations to the right and left will be different since in one case the fastest transition will be from the alignment axis to the axis of the action, whereas in the other case the reverse will be true. It is obvious that if the angle of "inclination" is zero or $\pi/2$, then no orientation will arise, since under these conditions right and left rotations are indistinguishable.

Let us now suppose that the initial ensemble is isotropic and the action axially symmetric. Under these circumstances only moments of even order can be induced, and within the linear approximation only zero and second order moments, i.e. population and alignment. In particular, under the influence of collisions with a beam of electrons or ions, or in the beam–foil technique, only population and alignment are produced [4.13,14].

Of the various external actions we first consider optical excitation, and, first of all, that by light from a resonance lamp. The light polarization is produced by a polarizer and must be circular in order to induce orientation in the excited state, or linear if one is interested in alignment. In the latter case, it is not obligatory to use a polarizer, since a directed ray of natural light already possesses axial symmetry and also induces alignment. However, the magnitude of alignment produced by unpolarized light is much smaller.

Laser Illumination. The production of coherency by laser illumination is akin to excitation by means of a resonance lamp. The frequency of the laser radiation must again be in resonance with the atomic or molecular transition and the same requirements are imposed with regard to the directions of polarization of the excited and observed light. In Sect. 3.2 we discussed the excitation of an ensemble by monochromatic polarized light. This may serve as a basis for formulating the description of excitation by laser radiation. In particular, it follows from Sect. 3.2 that the profiles of all the interference signals are the same as those induced by light from a resonance lamp, provided the absorption Doppler linewidth is broad (compared to the radiative Doppler linewidth) and the laser radiation intensity is low.

However, there are also some differences. They are attributed to the spectral distribution of the radiation and to the spectral density. The resonance lamp utilized in experiments with resonance fluorescence produces a line with a finite Doppler width. In the spectral range corresponding to the natural linewidth, the light flux (except in rare cases) is not large, and therefore nonlinear effects are absent. But because the whole ensemble, i.e. particles of all velocities, takes part in the fluorescence, the intensity is quite sufficient for measurement. However, laser radiation is either close to monochromaticity or when oscillating on a few modes, its spectrum resembles a "comb" with strong sharp maxima. Therefore, only a fraction of atoms will be excited, but since this fraction will interact with rather higher density radiation, this will frequently lead to non-linear effects, that superimpose on the interference.

These difficulties are of different degree depending on the exact nature of the experiment, but they are not insurmountable.

The most significant advantage of a laser lies in the fact that it is possible to tune it to any wavelength and excite atoms (and molecules) from any energy level, not necessarily from the ground state. This is not only because one can tune the laser radiation, but also a result of its high intensity, which is particularly important for exciting from levels that are already excited by some means, since the density of atoms (molecules) in the excited states is usually low.

A very important application of laser radiation is for exciting molecules, even from the lower states, since the laser energy is concentrated in a single frequency, which can be made resonant with one of the transitions in the molecule. However, the utilization of this molecule's emission is not acceptable, because the radiated energy is distributed over a large number of transitions, thereby inducing resonance transitions between a large number of levels, and moreover, the excitation of each of these transitions is weak.

Electron Impact. Electron impact can be used to excite and produce coherency over a large number of levels, including higher levels. However, electron-impact excitation has low selectivity and the emission process involves particles in a whole series of states. As a result, it requires the use of monochromators or interference filters in the observation channel. Simultaneous excitation of a series (group) of levels has yet another negative consequence – these are cascaded transitions. As is known, when determining the lifetime from the intensity time decay [4.15] using electron pluse excitation, cascaded transitions are the main source of error. They also influence the Hanle effect, since, upon spontaneous emission, the coherency of the emitting level is transferred to the lower, final state of the atom or molecule.

A significant advantage of electron impact for inducing coherency is its simple time variation.

Another drawback of excitation by an electron beam is the influence of a magnetic field on the trajectory of the electron beam [4.16]. As a result of this influence, not only does the magnitude of the alignment depend on the magnetic field, but also its direction; as a consequence of this the form of the Hanle signal becomes complicated.

Experimentally, excitation by electron impact is most frequently achieved by allowing a vapour to interact with an electron beam produced by a specially designed electron gun. However, the group in Grenoble [4.9–11, 17] has used for this purpose a high frequency capacitor discharge; in it electrons move substantially in the direction of the electric field. This direction also defines the axis of symmetry of the coherent ensemble produced upon excitation.

Ion Impact. In some studies [4.18, 19] ion impact has been used to induce coherency. Inasmuch as an ion is a heavy particle, magnetic fields only slightly distort its trajectory, and the Hanle signal, in this sense at least, remains "pure".

Beam–foil. The beam–foil technique was developed in order to measure the lifetime of atoms. It is based on the formation of excited atoms upon neutralization of the ion as it passes through a thin carbon or metallic film. The excited atoms proceed on a trajectory in the direction of motion of the ion that produced them and their velocity is known. With time, the atoms' fluorescence and the radiation intensity decreases. This is reflected in an intensity change with distance from the film. This intensity variation enables one to establish the law of decay of the excited state in time and thus to determine the lifetime.

It turns out that, upon neutralization of the ion, an atom in a superposition of states is produced. The symmetry of the neutralization process, i.e. the beam impact on the film, has one axis of infinite order. This permits the occurrence of an alignment; in fact, it permits only a uniaxial alignment. The formation of alignment upon such an impact offers an opportunity to simultaneously observe and utilize the interference phenomena in a single technique.

The most interesting feature of coherency formation by the beam–foil technique is the possibility of observing beats using a constant (time-independent) excitation. Similarly, in the observation channel, no time resolution is required: the intensity variation along the beam replaces both the former and latter time variations.

4.4   Observation of Interference Phenomena

Let us assume that a coherent ensemble has been prepared. It is now necessary to observe the coherency. We first examine the observation of the interference phenomena of excited states. In a number of cases this is done simply by measuring the degree of polarization of the light emitted by the ensemble. Various schemes can be used to measure the light polarization. One possibility is to measure the light intensity transmitted through a polarizer, whose axis can be rotated by the experimenter. One records the intensity upon gradually rotating the linear polarizer. In another case, if the symmetry axis of the coherent production process is well defined, then one can simply measure the intensity in two positions of the polarizer – parallel and perpendicular to the above-mentioned symmetry axis.

The rotating polarizer method, in which a phase detector is used to measure the difference between the maximum and minimum intensity of the transmitted radiation through a polarization analyzer, is another widely used technique.

Another technique of observation and investigation of coherency is a rather sensitive one which destroys this coherency by using an external magnetic (in rare cases electric) field. Thus one measures the dependence of the degree of polarization or the radiation intensity (or light absorption) on the field strength. When observing the level crossing, in contrast to the Hanle effect, the presence of an external field is a must, since the coherency between the "crossing" levels is not otherwise manifested. In these experiments the application of a field gives rise to the possibility of producing coherency, whereas its variation enables one to ascertain the positions of level crossings.

In a number of cases there is another reason why the application of a magnetic field is desirable, namely an insufficient coherency, as a result of which the degree of polarization is close to zero. However, the intensity change upon the application of a magnetic field can be measured far more accurately than the degree of polarization. This is related to purely technical factors: in particular, when the polarizer is rotating, the transmitted light intensity changes, not only because of the inclination of the polarizer axis, but also as a consequence of variation of light loss, which inevitably accompanies this because the polarizers are not perfect. Actually this method, i.e. the application of external fields, has wide possibilities and is more informative than the first method of direct measurement of the polarization. We note that the historical Hanle experiment [4.3] recorded precisely the influence of a magnetic field on the intensity of resonance fluorescence.

Interference of Non-degenerate States. It has already been mentioned above that the variation of coherency is manifested not only in the degree of polarization of the spontaneous radiation, but also in the variation of its intensity in a selected direction. The observation is carried out either directly or through a polarizer. This is possible due to the rigorous and simple relationship between the polarization and the radiation pattern (Sects. 2.1, 3.7)

Inasmuch as all interference between non-degenerate states is manifested in the time evolution of the radiation parameters, it is usually observed in the radiation intensity or in the magnitude of the absorption coefficient, the primary concern of an observer is the time resolution. The salient technical features of the experiment depend on this. The intensity measurement itself has the same standard character as intensity measurements in all other branches of optical spectroscopy As a rule, one also encounters the same difficulties – either measurement of low intensities or measurement of small intensity variations on the background of a relatively high intensity.

4.5 Hanle Effect in Atoms in the Ground State

Up to now we have primarily discussed the interference between excited states of an atom. It was already pointed above that it is also possible to induce coherency in the lower state of an atom. For this purpose one can use the techniques known as optical pumping or optical orientation of atoms. The same mechanism can also be used to produce alignment. However, since the majority of atoms that are easily accessible to the polarized pump cannot in the lower state possess moments higher than the first, more attention is paid to the optical orientation of atoms [4.1, 20]. We discuss this vast topic only briefly insofar as it is relevant to the subject of interference of states.

In thermal equilibrium, an ensemble of atoms with a non-zero angular momentum is isotropic. The purpose of an optical pump is to produce an anisotropic distribution of angular momenta. For unexcited atoms this is usually much more difficult to achieve than for excited atoms. The reason is that in thermal equilibrium at moderate temperatures, there are no excited atoms at all. Excited atoms are produced only under the action of a controlled excitation agent, which can transfer anisotropy to the atoms (e.g. a ray of light or a beam of particles). For such excitation the concentration of excited atoms increases linearly with excitation intensity (at least up to very high values, determined by the relationship of the excitation rate to the rate of spontaneous decay). However, the degree of anisotropy of the excited state depends not on the intensity of the excitation, but on its type – on the degree of polarization of the light, on the particle energy, etc. With this, since fluorescence is the source of information about the state of the excited atoms, then in a mixture of excited and unexcited atoms we record only the excited atoms, no matter how small a minority they represent.

The situation turns out to be different when one prepares an anisotropic ensemble of unexcited atoms. An optical investigation of the anisotropy of unexcited atoms yields the polarization diagram of absorption. In this, the "oriented" atoms, i.e. those influenced by the external factor, and the remaining atoms, which are an isotropic ensemble, participate equally. Thus the degree of orientation of the whole ensemble obviously depends on the competition between the process of orientation and the rate of the relaxation process which destroys the polarization moments. Therefore, in contrast to the case of orientation of excited atoms, the degree of orientation of an ensemble of unexcited atoms depends on the intensity of the orienting agent.

To orient atoms in the ground state it is thus not sufficient to have a process which merely orients the atoms, it is also necessary for this process to be powerful enough. This situation severely restricts the number of elements for which it is possible to realize orientation of the atoms.

The lower level anisotropy can arise as a result of spontaneous decay of the excited anisotropic state. This is attributed to the isotropic vacuum fluctuation, which stimulates spontaneous decay. Therefore, in principle, it is possible to produce an anisotropic ground state by the same techniques used to induce anisotropy in the excited state. In practice, however, one is forced to use resonance excitation by light with circular polarization.

Optical orientation of the ground state is possible also in the case where the excited state is perfectly isotropic (as a result of intensive depolarizing collisions or simply because the excited state has no angular moment at all). In this case the flux of atoms from the excited state to the ground state is isotropic. But nevertheless an anisotropy of the ground state may arise if the excitation process is anisotropic, i.e. when the probability of the excitation depends on the orientation of the atom. Both variants of the orientation process are schematically shown in Fig. 4.1 where the magnetic structure of transitions $^2S_{1/2}-^2P_{1/2, 3/2}$ of the principle doublet of an alkaline atom is presented. For the sake of simplicity the nuclear spin is ignored. The figures near the arrows denote the relative probabilites of σ_+ transitions which can be excited by circularly polarized light directed along the magnetic field. If the single D_1 line ($^2S_{1/2}-^2P_{1/2}$) is used, then the pumping leads universally to the depletion of the ground sublevels $^2S_{1/2}$,

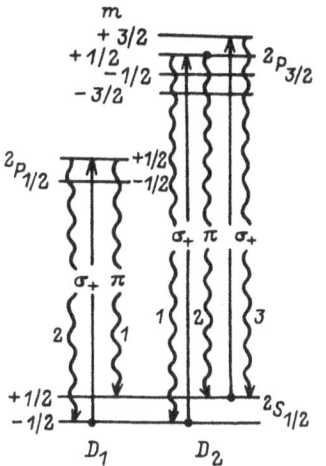

Fig. 4.1. The structure of the resonant transition lines D_1 and D_2 of the alkaline metals (without nuclear moments). The relative probability of the transition is indicated by the numerals 1, 2, 3

$m = -1/2$. Indeed, according to the selection rules, the D_1, σ_+ light can excite an atom from the ground sublevel $m = -1/2$ to the excited state $m = +1/2$. If the excited atom is not subjected to depolarizing collisions it will return as a result of spontaneous decay to the initial sublevel or (with half the probability) will make a transition to the sublevel $^2S_{1/2}$, $m = +1/2$, which cannot be excited by D_1, σ_+ light. If the excited atom is depolarized by collisions it decays to either of the ground sublevels with equal probability. In both cases the multiple excitation events lead to the depletion of the same sublevels though with different rates.

A different picture arises when using the line D_2 for pumping. When depolarization of the excited state is absent, the optical pump obeys the law of conservation of angular momentum: circularly polarized radiation brings with each photon a unit projection of momentum to the atomic system, which is partially conserved after spontaneous emission. In the absence of relaxation of the angular momentum in the lower state, the pump must end up by accumulating all the atoms into the sublevel $S_{1/2}$, $m = 1/2$. If the momentum in the excited state is destroyed by collisions within the lifetime of the state, then the process is determined by the probabilities of exciting the atom to states with different orientations of the momentum, since its return from the excited state to the various sublevels is now equiprobable. The sublevels with lower excitation probability, i.e. $S_{1/2}$, $m = -1/2$ will be preferentially populated. In this way, the relaxation of momentum in the excited state leads to a change of sign of the orientation of atoms.

The picture presented here is described in a coordinate system in which the resulting orientation turns out to be longitudinal. Its quantitative evaluation, for which one can make use of the distribution of population, can be obtained by solving the kinetic differential equations in which the probabilities of the dipole transitions are used, and on which the model description of the relaxation process of angular momenta of the ground and excited states is based.

The Hanle effect may be observed for oriented atoms in the presence of a magnetic field perpendicular to the direction of orientation. The description of the Hanle effect is simplest in a coordinate system with the axis of quantization (z-axis) along the magnetic field. One can transfer to these coordinates after calculating the orientation parameters (density matrix of the lower state) by rotating the coordinate system using the Wigner function D (Table 3.1).

In essence the effect is completely analogous to the Hanle effect in the excited state, differing only in the constants that characterize the level width. For the excited state the minimum width is determined by the natural lifetime τ. In the case of the Hanle effect in the ground state, the role of τ is played by the relaxation time of the orientation under the action of a set of processes which act on the angular momentum of the atom. The most important of them is thermal relaxation due to collisions between atoms and with the tube wall, and the absorption of the orienting light. For alkali metal atoms oriented in the $S_{1/2}$ state, the time τ may be as long as 1 s. Provided all other conditions remain constant, the width of the Hanle signal in the lower state may be many orders narrower than in the excited states.

The effects in the lower state are most frequently described by the Bloch equations. Historically this is related to the fact that the optical polarized pump was originally proposed [4.1] in order to transfer the techniques of magnetic resonance from condensed matter, where magnetic resonance was described by the Bloch equations, to atomic vapours. Therefore, we briefly discuss this approach. The equation has the following form

$$\frac{dM}{dt} = \gamma M \times \mathscr{H} + \frac{M - M_0}{T} \, ,$$

where T is the effective relaxation time, γ the hydromagnetic ratio, and M_0 has the sense of a magnetic moment, which is established under the action of the light pump in the absence of a magnetic field. This vector is directed along the light ray. Let the pump light by directed along the x-axis, and the magnetic field $\mathscr{H}$ along the z-aixs. From the above equation, the stationary solutions for the magnetization along the three axes follow as

$$M_x = \frac{M_0}{1 + \Omega^2 \tau^2}, \qquad M_y = \frac{M_0 \Omega \tau}{1 + \Omega^2 \tau^2}, \qquad M_z = 0 \, , \tag{4.1}$$

where $\Omega = \gamma \mathscr{H}$.

We point out the difference between the expression (4.1) and the previously derived expression (2.44), which describes the Hanle effect in the excited state. They differ by a factor 2 at frequency Ω. This is due to the fact that, traditionally, with excited atoms one investigates the alignment, which is described by the interference of magnetic sublevels of momentum components differing by two units. In the ground state one studies the orientation, which corresponds to the interference between adjacent magnetic sublevels.

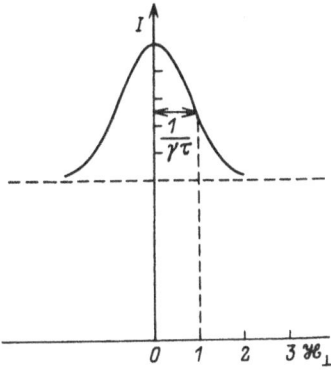

Fig. 4.2. The characteristic dependence of the light intensity transmitted through a cell containing oriented atoms from the transverse magnetic field

In the simplest case of a two level system, the Bloch equations describe the optical pumping exactly. In this case the absorption of circularly polarized light is linearly related to the projection of the moment M on the direction of the light ray. For a multilevel system, i.e. a system with an angular momentum $J \geq 1/2$ this relation is approximately valid.

The characteristic dependence of the light intensity, I, transmitted through a cell containing an atomic vapour, as a function of the transverse magnetic field is shown in Fig. 4.2.

The Hanle effect for unexcited atoms is interesting mainly because of its application for measuring very weak magnetic fields. The basis for this implementation is the dependence depicted on Fig. 4.2. It is seen from the figure and from the expression (4.1) that the magnetic field region in which the light intensity changes significantly is determined by the characteristic width $\mathcal{H} \sim (\gamma\tau)^{-1}$. The higher the relaxation time, the lower is the range of measurement of magnetic fields and the steeper the light intensity dependence on the field, i.e., the higher the sensitivity to a change in field strength. When utilizing this technique of field measurement in practice, one introduces an auxiliary alternating field $\mathcal{H}_1 \cos \Omega' t$ parallel to the unknown field $\mathcal{H}_0$. This causes the light transmitted through the cell to acquire a modulation at the frequency Ω':

$$I(\Omega') = K' J_0\left(\frac{\omega'}{\Omega'}\right) J_1\left(\frac{\omega'}{\Omega'}\right) \frac{\omega_{12}^0 \sin \Omega' t}{1 + (\omega_{12}^0 \tau)^2} , \tag{4.2}$$

where $\omega' = \gamma \mathcal{H}_1$ and J_0 and J_1 are the zero and first order Bessel functions. The signal at frequency Ω' does not depend on the relation of Ω' to τ – it is an inertialess coherent response to the parametric modulation of the system's energy. This is a special case of parametric resonance $p = \mp 1, l = 0$ (3.110).

Figure 4.3 presents the experimental dependence [4.21] of the amplitude of the optical signal at frequency Ω' on the strength of the unknown field $\mathcal{H}_0$. In this work a sensitivity of about 10^{-9} Oe was demonstrated for a measuring time of 1 s.

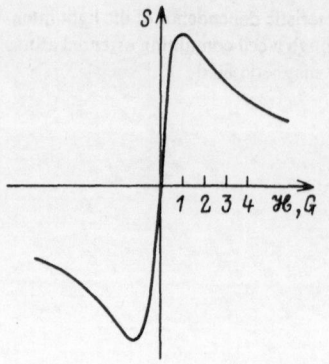

Fig. 4.3. The experimental Hanle signal taken from [4.21]. The magnetic field is given in Gauss multiplied by 10^{-6}

An illustrative example of the application of a Hanle magnetometer is given in [4.22]. It was used to measure the magnetic field of a bulb of gaseous He, whose nuclei were polarized by using the techniques of optical orientation. The experimental scheme is shown in Fig. 4.4a. The Hanle magnetometer bulb with rubidium vapour was placed near to the helium bulb inside a magnetic shield. Initially the helium oriented in a direction perpendicular to the pump ray of the magnetometer. Then a weak magnetic field was applied to the helium bulb, which led to the precession of the helium nuclei at a frequency of 6×10^{-3} Hz.

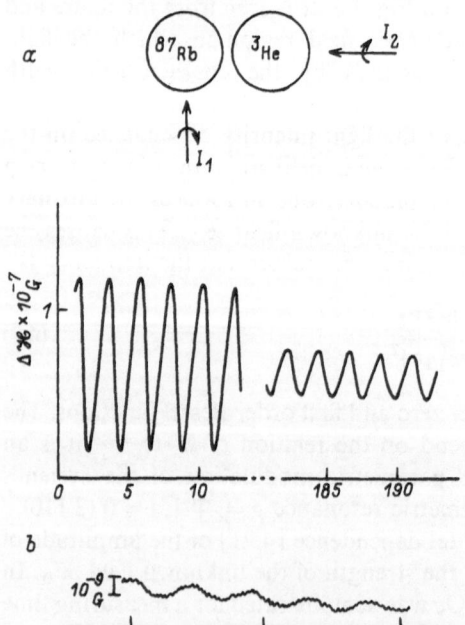

Fig. 4.4. (a) Experimental scheme for recording the magnetization of ^{3}He. I_1 and I_2 are the directions of the rubidium and helium light beams. **(b)** Extracts of the records of the modulation of the helium magnetization in a magnetic field of 10^{-6} G

The Hanle magnetometer registered this modulation. Figure 4.4b presents parts of the record of this modulation with large interruption time. As can be seen, the coherent precession of the helium nuclei was maintained even after 11 h.

The Hanle effect enables one to measure fields in the vicinity of zero strength. Fields that are stronger than $(\gamma\tau)^{-1}$ can be measured with the same sensitivity by the magnetic resonance technique [4.23]. This can be considered as a generalized Hanle effect in a coordinate system rotating relative to the laboratory at the frequency of the alternating field that induces the resonance.

In summarizing we add that, upon introducing accessory modulating fields, the Hanle-magnetometer is able to register all three components of a weak magnetic field. Such a magnetometer is described, for instance, in Ref. [4.24].

4.6 Manifestation of the Interference of States in Collisions

It has long been known that when atoms are excited by electron impact they emit partially polarized light. Since the interference of states has become used as an investigative technique in areas of spectroscopy, electron beam excitation has been widely used to produce coherent ensembles of excited particles.

As has been emphasized, collisions with directed beams also lead to excitation in aligned states. Although ion beam excitation is quite rare compared to the use of electrons, one can employ it for producing alignment [4.19, 18]. Coherence can be induced not only by collisions of atoms with other particles, but also with solids. For example, in the beam–foil technique an ion undergoing collisions in the foil does not simply transform into an excited atom, it can also acquire an alignment.

The collisions of heavy atoms with charged particles of high energy can lead to the excitation of an electron from the inner shells of the atom. Via this process the ensemble of atoms can also acquire an alignment.

Collisions between two atoms can also result in alignment of states [4.25]. The basic difference between atom–atom collisions and the collisions with electrons or ions is the smallness of the velocities and energies. It is easy to control the velocities of charged particles by using fields, and in experiments with electron and ion beams the energy is quite sufficient to excite atoms. However, atom–atom collisions, due to the small energy involved, do not usually result in transitions to other electron states; the atom remains in its initial state or makes a transition to the energetically closest component of the hyperfine or fine structures. Such collisions, if they are directed or simply anisotropic, can induce alignment. The experimental observation of such collisions is quite difficult, and the experiments described in the literature give only an indirect verification of the alignment effect by collisions. Thus for example, in [4.26] the inversion of the Hanle effect on a few lines of the spontaneous radiation of the nitrogen oxide molecules for increasing power of the laser excitation, observed by *Weber* [4.27], is attributed to the self-alignment of the molecular beam by collisions of the molecules in the beam source [4.26]. The self-alignment of atoms in an atomic beam, observed by *Toschek* [4.28], is explained by the author as being a result of collisions at the exit channel of the beam source and by the anisotropy of the scattering cross-section of the "aligned" atoms.

In Refs. [4.29, 30] the influence of anisotropic collisions on the absorption of light in neon was reported. The effect was observed in the magnetic field dependence of the absorption of polarized light, resonant with the dipole transition of the neon. Despite the fact that a number of factors influence the absorption as a function of magnetic field, the authors succeeded in showing by quantitative measurements the difference in the isotropic and anisotropic relaxation constants of the polarization moments. In this way they demonstrated the formation of alignment upon anisotropic collisions, because the anisotropy of collisions (the authors call it the "wind effect") is related to the thermal motion of atoms in a plasma (Sect. 4.12), and as a result of such collisions a hidden alignment (Sect. 4.11) is produced.

Also associated with the formation of alignment by means of collisions are some new features of measurements of the cross-sections of collisions. Indeed, the production of alignment is related to differently probable collisional transitions between different pairs of Zeeman sublevels of a state or of its neighbouring states. Therefore, the measurement of the transition cross-section between the fine or hyperfine sublevels gives a result which depends on the experiment conditions. Thus, for instance, Refs. [4.28, 31] are devoted specially to detecting such a difference; they investigate the cross-section of the collisional transition between the $P_{3/2}$ and $P_{1/2}$ states of gallium upon collision with inert gases.

The interaction of atoms in collisions can mix up the polarization moments. In the year 1964, *Fano* [4.12] studied the problem of such intertwinement under the action of external fields. He showed that uniform fields couple polarization moments of the same parity, and that the gradient of the electric field – inclined to the axis of the statistical tensor (Sect. 3.10) – couples polarization moments of different parities. In particular, an ensemble of particles initially aligned in a nonuniform electric field, whose gradient does not coincide with any of the axes of alignment ($\kappa = 2$), can acquire the orientation ($\kappa = 1$). Later on the effects of nonuniform fields and anisotropic collisions were calculated theoretically [4.32] and observed experimentally [4.33].

Under the influence of collisions, the polarization moments of atoms can easily be destroyed, and such a process is actually more common than the above-mentioned ones. In the year 1965, *Dyankov* and *Perel* investigated the decay of polarization moments under isotropic collisions [4.34]. In this work it was shown that each of the polarization moments decays independently of the others; consequently, each is characterized by its own lifetime. For anisotropic collisions the rate of decay depends not only on the index κ, but also on q [4.34]. The index κ is the range of the statistical tensor and the index q indicates its orientation in space (Sect. 3.10).

Finally, for collisions between atoms, it is possible for the colliding particles to exchange their polarization moments; see, for example, [4.35].

4.7 Quantum Beats upon Pulse Excitation

The beats upon pulse excitation are attractive primarily because of their simplicity. The luminescence oscillation is not provoked by any external action and it illustrates the oscillation characteristics of the radiation intensity emitted by atoms at the characteristic frequencies. The beat pattern produced by the interference of only two excited states is particularly transparent. Like the phenomenon itself, the experimental technique is also simple: after a short-pulse excitation of the atomic vapour, one needs to oscillograph the kinetics of damping of the luminescence. Despite the apparent simplicity and attractiveness of such an experiment, it was realized only after the observation of interference of non-degenerate states in other more indirect manifestations. This is related to the considerable difficulties in obtaining a powerful and short excitation pulse in the epoch before the advent of tunable lasers. We describe briefly one of the first works on the observation of quantum beats [4.36].

This work investigates the kinetics of luminescence of cadmium vapour from the transition $5\,^3P_1 - 5\,^1S_0$, 3261 Å. The triplet state $5\,^3P_1$ was split in a magnetic field into three sublevels, of which only two, $m = \pm 1$, were excited by light polarized perpendicular to the magnetic field. In this way a simple scheme consisting of two excited levels and one lower state was realized. The selected cadmium level $5\,^3P_1$ is characterized by a long lifetime, 2.4×10^{-6} s. This enabled beats to be observed at the relatively low frequency of 1 MHz, since pronounced beats appear when the splitting of the levels significantly exceeds their width. The cadmium vapour was excited by a light pulse of duration 10^{-7} s from a cadmium lamp, formed by an electrooptic shutter. The luminescence intensity was registered in the direction transverse to the exciting ray by a photomultiplier. For every excitation pulse its photocathode emitted approximately 100 photoelectrons, far too few to allow direct oscillographing. (It must be emphasized that even this yield of photoelectrons is a big achievement: in an analogous study [4.37], which was conducted in search of beats on the 2537 Å line of mercury, an average of one photoelectron was registered for every 50 excitation pulses!).

In order to register beats under these conditions, a very simple and at the same effective method of storage was used. An electrooptic shutter was switched on and off in a stroboscopic regime 50 times per second. Every luminescence pulse was displayed synchronously on the screen of an oscilloscope. A set of oscillograms from the screen were photographed on a single photographic film. With this the noise overshoots on the oscillograms could be gradually averaged, but the regular features of the process were accumulated and displayed. Figure 4.5 depicts two strongly contrasting prints from the negatives, each of which

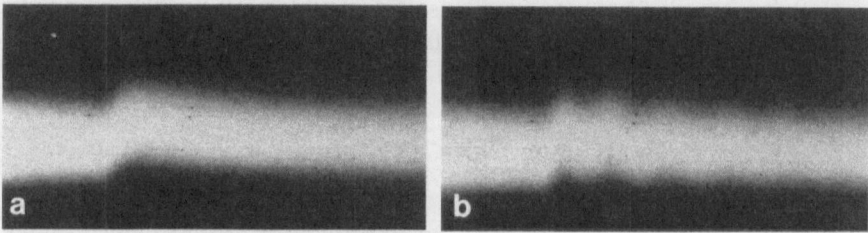

Fig. 4.5. Prints from the negatives produced by accumulating oscillograms of the cadmium fluorescence damping: (**a**) zero magnetic field; (**b**) magnetic field of 10^{-6} G

accumulated approximately 10 thousand oscillograms. The first is a control and corresponds to the case of level degeneracy where beats are absent. The broad white strip shows the averaged noise overshoots. Its envelope gives the kinetics of the process – the fast rise at the moment of the excitation and the smooth decay of the spontaneous radiation intensity. The second synthetic oscillogram illustrates the beats with a frequency of 1 MHz. The original of this negative was subjected to microphotometery, and the results are presented in Fig. 4.6. Although theoretically the oscillations of the luminescence intensity in this system should have minimum values of zero, in reality many factors contribute to make the modulation depth rather less. Among these are the finite apertures for the light beams involved in observation and excitation, the finite duration of the excitation pulse and the admixture of the odd cadmium isotopes with other beat frequencies.

Investigations of beats have been developed in three different directions. As the first of these one can point to the development of experiments on pulse excitation of atoms by an electron beam. The development in this direction is based on the steady progress of the technique of fast registration of light pulses and on the adoption of high-speed multichannel information-gathering systems. The technique of electron pulse excitation has been used to conduct investigations of beats in a system of sublevels of magnetic and fine splitting of a number of excited states of helium atoms [4.38].

The second very special development direction of the free beats technique has concerned the technique of excitation of atoms and ions known as "ray–foil" (beam–foil technique). A fast ion beam with a velocity greater than 10^8 cm/s pierces a thin (10^{-5} cm) carbon foil. As a result of charge recombination, a beam of fast atoms excited to a wide variety of states is produced behind the foil. This

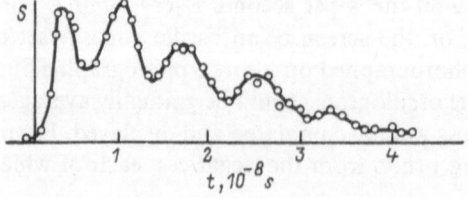

Fig. 4.6. The microphotogram of the negative presented in Fig. 4.5b

leads to the formation of coherent superpositions of states, attributed to the components of hyperfine and fine structures, and for hydrogen atoms and hydrogen-like ions including states with different orbital moments. The appearance of beats is linked to the alignment of the orbital momentum of the atom along the beam direction. As a result of subsequent spin–orbit interaction, the spin and orbital components will undergo precession about the direction of the resultant momentum. Moreover, for all atoms there exists a common preferred initial direction of the orbital angular momentum. This will ensure that the atomic beats in every cross-section of the beam are in phase. For the hydrogen atom an interesting and plausible explanation can be given to the beats observed at the frequency of the fine splitting $2\,S_{1/2}$–$2\,P_{1/2}$. Upon passing through the carbon foil the proton will acquire an electron shell which lags behind it at the moment of its emergence from the foil. The electron tail will then 'overtake' the nucleus and continue to oscillate longitudinally at the frequency of interference of the levels $2\,S_{1/2}$ and $2\,P_{3/2}$. (The pattern is, of course, additionally complicated due to the presence of hyperfine structures.)

Due to the rapid motion of the excited and spontaneously emitting atoms, there will be a natural sweep of the fluorescence intensity with time – the beat pattern will be stationary in space displaying a spatially periodic variation of the brightness of the luminescent beam. Since charge recombination occurs extremely fast (10^{-13} s), the time resolution of this method is determined by the spatial resolution of the optical system that accepts the radiation from the selected cross-section of the beam. In practice, the resolution reaches 10^{-10} s. An example may be found in Ref. [4.39].

The advent of lasers greatly enhanced the possibilities to investigate and exploit the phenomenon of quantum beats. This is related to two properties of lasers – the extremely high brightness of their radiation and the possibility of generating extremely short pulses. The first property alone, the high brightness, allowed modifications of the experiments on the spectroscopy of beats in fast atomic and ionic beams [4.40]. In these experiments the excitation of the beam was carried out not in the process of charge recombination, but by the laser radiation intersecting the beam of particles. Such optical excitation is preferable in a number of cases due to its high selectivity – only the transition in interest is excited. The brightness, even of relatively low power cw lasers, is sufficient for a resolved optical transition to be saturated during the flight time of the particles across the small intersection region of the beams. The high velocity of the particles will make the Doppler shift of the resonance frequencies significant, permitting the effective frequency of the radiation to be tuned by changing the angle of intersection of the atomic beam and the light beam. The analysis of beats is carried out as before by recording the fluorescence of the beam as a function of distance from the point of excitation to the place of registration.

Figure 4.7 shows the beat pattern and its Fourier spectra, adapted from the work [4.40] for excitation of the resonance transition of barium ions $6\,^2S_{1/2}$–$6\,P_{3/2}$ by an argon laser. Along the ordinate axis are plotted the number of registered photon pulses. The Fourier spectra are given on a linear scale.

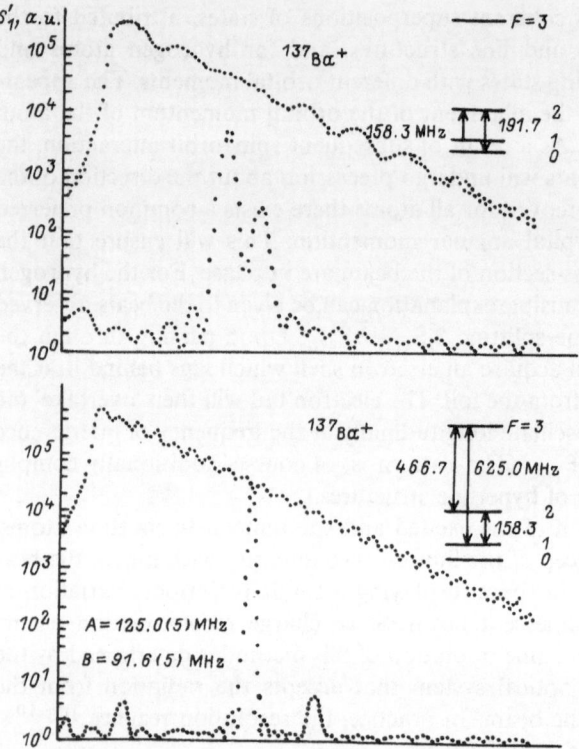

Fig. 4.7. The barium ion beat pattern and its Fourier spectrum taken from [4.40]

Finally, the third and most promising direction is related to the utilization of pulsed frequency tunable dye lasers for excitation. For pumping one can use the radiation of the second or third harmonics of a pulsed neodymium laser (1.06 μm), of excimer lasers, or of nitrogen electrodischarge lasers. All these lasers generate pulses of the order of 1 ns duration. This in turn determines the pulse width of the dye laser. If required, much shorter pulses can be produced.

Laser-excited beats were first demonstrated in [4.41]. The beats were observed in the magnetic structure of the state 6s 6p $^3P_1$ of ytterbium. They could be observed directly on an oscilloscope without using storage. This fact alone demonstrates the new experimental level achieved with the application of lasers.

The utilization of lasers enables one to significantly extend the list of systems accessible to investigations. With the help of stepwise excitation one can access states that are not coupled to the lower state by a radiative transition. Since the laser excitation can easily saturate the transition, the population of the excited state turns out to be comparable to the population of the lower state. An effective example of the implementation of this technique is provided by the study [4.42], in which the beat frequencies induced by the interference components of the fine structures of the level $n\,^2D_{3/2,\,5/2}$ of sodium for n from 9 to 16

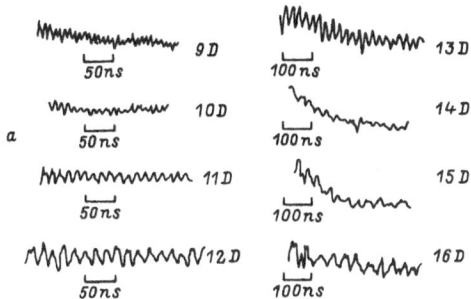

Fig. 4.8. (a) Beats in the radiation of the sodium nD levels under coherent excitation. (b) Fourier spectra of two of these signals

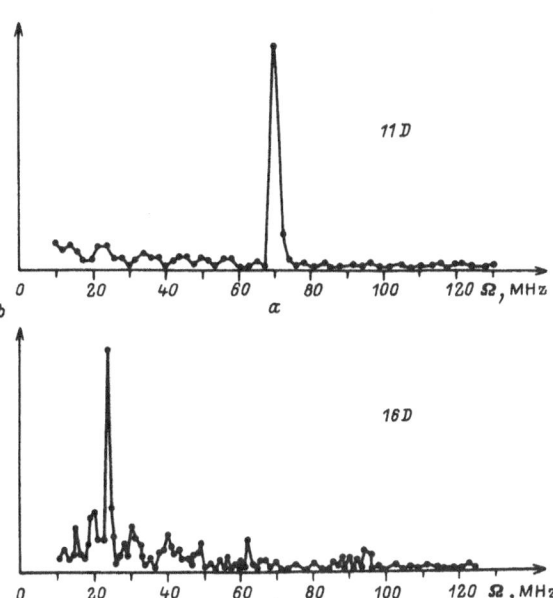

were measured systematically. The beats were observed in the spontaneous radiation from the transition to the 3P level. Figure 4.8 shows the form of observed signals, recorded by an averaging technique over a large number of exciting pulses. Figure 4.8b presents the Fourier spectra of a few of these signals. They allow one to determine the frequency of the beats precisely. In these experiments the time resolution is limited by the response speed of the photomultiplier and was characterized by a cutoff frequency of 150 MHz.

Potentially a very high time resolution can be achieved by recording beats in absorption. Such a technique was demonstrated for the first time in [4.43], where the interference of sublevels of the hfs states of sodium $^3P_{1/2}$ was observed. The superposition of sublevels of the hfs was produced by the exciting laser pulse, and the beats were registered with the help of a pulse from the second laser, exciting the atom from the state $3P_{1/2}$ to the state $20\,^2S_{1/2}$, whose population was controlled by the ionization of the atoms in an external electric field. Both of the dye lasers were simultaneously excited by a pulse from a common pump laser. Between the pulses of the two lasers a regulated delay was introduced with the help of an optical delay line of variable length. By gradually increasing the delay it was possible to trace the whole beat pattern. The probability of excitation of the atoms by the second pulse beats at the frequency of the hyperfine splitting, 192 MHz. In this case the resolution limit of the method was determined by the pulse width of the lasers and did not depend on the response of the detection apparatus. The drawback of this variation of the method is its low measurement productivity: for each excitation pulse, only one small element of the beat pattern can be registered. It is obvious that this shortcoming can be overcome by using, in place of a single controlled pulse, a series of pulses, generated for example by mode-locked lasers.

As has been emphasised, the technique of free beats is particularly attractive due to the fact that the observation of the beats is performed after the termination of the excitation process, which ensures the absence of state perturbation regardless of the excitation power. Here, however, it is necessary to make two assumptions. First, the above is true only when observing beats in the spontaneous radiation. When registering beats by absorption one must take into account the perturbing action of the controlling radiation. Second, with powerful laser excitation the concentration of excited atoms can be very high, so that the collective effects of the atoms' interaction through the common field of the spontaneous radiation could turn out to be substantial. In fact, such collective effects can be useful, since they increase the intensity of the radiation and make it directional. This situation will be of particular importance in the long wavelength region, where photomultipliers fail to function. Reference [4.44] provides an example of beat observation under the influence of collective effects. This study investigated the effects of superradiation, photon echo and nutation in atomic and molecular vapours.

4.8 Coherent Resonances

Following pulse excitation beats appear as a transient process. When periodic excitations act on the system it is possible for stationary coherent states to arise. The most familiar case of stationary coherency is the resonance beat, which arises upon harmonic modulation of the intensity of the process used to introduce coherency into the system.

Resonance Beats. The most illustrative resonance beat is in luminescence. Let there exist atoms with split excited states populated by light of coherent polarization, i.e. allowing the excitation of atoms from a common lower state to two (or more) upper levels. The intensity of the exciting light is modulated. The phenomenon consists in the resonant growth of the modulation depth of the spontaneous radiation (as well as in the coherent polarization) as the frequency of modulation of the light approaches the frequency of splitting of the upper level. That this is a nontrivial phenomenon is manifest in the fact that the modulation depth of the spontaneous radiation at resonance is independent of the lifetime of the excited state – the modulation occurs at an arbitrarily high frequency. Its lack of inertia can be explained by the fact that it is due to the coherency of the sublevels of the excited state and not to the modulation of their population.

Resonance beats in the luminescence of cadmium vapour from the transition $5\,^3P_1 \to 5\,^1S_0$ were demonstrated almost simultaneously in two quite similar studies [4.45, 46]. The luminescence of the cadmium vapour was excited by radiation from a high frequency cadmium lamp, modulated in intensity via a modulation of the power source. The modulation frequency was fixed and was chosen to be much higher than the width of the $5\,^3P_1$ level, so as to emphasize the resonance character of the effect.

When modulating the lamp through its power source the modulation depth of the light decreases sharply with increasing frequency. This is related to the inertness of the spontaneous radiation of the atoms, which continue emitting light after the termination of the excitation, and also to the inertness of the gas discharge – the temperature of the electrons cannot follow the fast variation of the discharge power source. As a consequence, at a frequency of almost 1 MHz the modulation depth of the resonance line at 3261 nm amounted to a few percent. Nevertheless, the utilization of phase-sensitive detection in these unfavourable conditions enabled the resonance beats to be reliably registered. We present here results from [4.46]. In this work the cadmium vapour was excited by light polarized at an angle of 45° with respect to the direction of the magnetic field, so that a superposition of all three sublevels $m = 0, \pm 1$ of the state $5\,^3P_1$ appeared. Because these levels are equally spaced, the system possesses two resonance frequencies of beats, corresponding to interference between the sublevels $m = \pm 1$ and to the pairwise interference of the sublevels $m = 1, m = 0$ and $m = -1, m = 0$. The magnetic field strength was varied at a fixed modulation frequency of 1 MHz. The luminescence light was registered in a polarization inclined at 45° to the magnetic field. The signal from the photodetector passed to the phase-sensitive detector. Figure 4.9 presents the measured dependence of the signal on the magnetic field together with the calculated curve. In accordance with the calculation, the resonances appear at field strengths of 0.42 and 0.48 Oe. As is seen from the figure, the modulation of the luminescence in these two resonances takes place with a phase shift of 90°. The resonance shape, of course, depends on the choice of the reference voltage of the phase-sensitive detector.

Prior to their observation in luminescence, resonance beats were observed in absorption experiments on optical orientation of atoms by modulated light [4.47]. The authors of this work used the phenomenological Bloch equations to describe the motion of spins in a magnetic field upon transverse pumping by modulated light. The effect was detected by recording the resonance decrease of the absorption of the pump light when the modulation frequency of the light

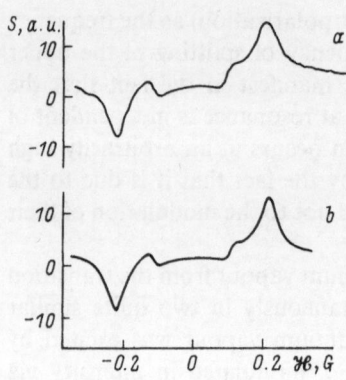

Fig. 4.9. Beat signal in the cadmium fluorescence: (a) experimental curve; (b) calculated curve

coincided with the precession frequency of the spins. That this effect is a manifestation of resonance beats was only realized later, after the demonstration of resonance in luminescence.

The essence of the effect is easy to understand qualitatively. It is based on the process of optical orientation of atoms, which can be treated as the transfer of angular momentum from circularly polarized radiation to the atoms. The orientation of atoms along the light ray is accompanied by a variation in their optical absorption (usually the absorption decreases). In the absence of a magnetic field or for a field oriented along the light ray, the degree of orientation is limited by the relaxation process only, which competes with the orienting action of the light. A transverse magnetic field changes the pattern dramatically. An atom oriented by a light ray perpendicular to the magnetic field turns out to be in a superposition of states with definite energy and definite projection of momentum on the magnetic field axis. The superposition state is stationary, which finds its classical interpretation in the precession of the atomic angular momentum about the direction of the field. Since the phases of this precession are uniformly distributed, the atomic system will, on average, be deprived of any orientation[1]. Nevertheless, if the orienting ray is modulated in intensity at the precession frequency of the atoms, then an in-phase precessing group of atoms occurs, which interact with the light in approximately the same way as those permanently oriented along the ray, since a periodically switched light always illuminates the atoms at one and the same precession phase, specifically that along the direction of the ray. The outcome of this is a resonant change in the average absorption of light by an atomic vapour as a function of magnetic field.

In essence, the phenomenon described here is beat resonance in a nonlinear modification. The nonlinearity is related to the fact that the light ray changes the ground state of the atom, which, in linear theory, is always described by the diagonal density matrix.

---

[1] This conclusion is valid if the precession frequency is much higher than the rate of optical orientation (see Sect. 4.4).

The advent of lasers extended the scope of beat resonances in two directions. First, the laser radiation is easy to modulate, even at very high frequencies, by making use of ultrasound, electrooptical, or mechanical modulators, since laser radiation requires only a tiny modulator aperture. An example of the application of modulated laser excitation for spectroscopic purposes is presented in [4.48], which describes experiments conducted to determine the g-factors of the molecular levels of selenium by the beat resonance technique.

A second important laser application in beat spectroscopy is connected with the implementation of the multimode regime of operation. The intermodal beats replace the modulation of intensity and have been successfully used in experiments on beat resonances at extremely high frequencies [4.49].

The wide applications of lasers in spectroscopic studies and, in particular, in the investigation of the induced two-quantum processes such as resonance Raman scattering, have led to the phenomenon of beat resonance being "discovered" for a second time in new manifestations. Beat resonance went by the name of "mode-crossing" in the work [4.50], in which the absorption (or amplification) of multimode laser radiation was investigated in a three-level system. Transitions between the states 0–1 and 0–2 were induced, and the frequency separation between the levels 1 and 2 was equal to an integral multiple of the intermodal interval. The radiation power, P, absorbed or emitted by the medium under the action of the laser radiation is given in third-order perturbation theory by the expression:

$$P \propto I_1 I_2 (N_{12} - N_0)[\Gamma_{12}^2 + (\omega_{12} - \Delta)^2]^{-1} \; ,$$

where I_1 and I_2 are the intensities of a two mode laser, N_{12} is the sum of the population of levels 1 and 2, N_0 is the population of the level 0, Γ_{12} is half the sum of the homogeneous width of the levels 1 and 2, Δ is the intermodal interval, and ω_{12} is the frequency interval between the levels 1 and 2.

If the transitions 0–1 and 0–2 were to possess only homogeneous broadening, then the phenomenon would be almost trivial. In practice, however, these transitions are inhomogeneously broadened such that the structure 1–2 is completely unresolved and the absorption coefficient of each of the modes, taken separately, does not depend on the splitting of the states 1–2. Only the interference of states 1 and 2 leads to the resonant dependence of the intensity.

Finally, beat resonance was "discovered" yet again in investigations of the resonance absorption of laser radiation in sodium vapour and in connection with this became the subject of a number of theoretical studies [4.51] under the name of "population trapping". Experimentally, the effect was manifest in the intensity decrease of the resonance luminescence of cadmium vapour excited by a two mode laser, when the frequency of the hyperfine splitting (1772 MHz) of the lower state of sodium coincides with the intermodal spacing. In essence, this is an exact analogy of the experiment on optical orientation of atoms by modulated light in a transverse field. Here, however, we are concerned not with the magnetic splitting of levels, but with the hyperfine splitting. The decrease of luminescence intensity resulting from the decrease of light absorption in resonance was treated in terms of "trapping" of the radiation by atoms in the sublevels of the lower state. Different versions of this experiment, in which the continuum states played the role of the excited state, have also been performed. The phenomenon was thus manifest in the decrease of the photoionization occurring when the frequency difference of the two ionizing spectral lines matched the spacing between the levels of the system, at least one of which was populated. In these studies the emphasis was placed on the relation between the populations of the two lower and one upper states, whereas in earlier works on beats it was the occurrence of a coherent superposition of states that was of prime interest.

The fact that a variety of different observations all lead back to the phenomenon of beat resonance reflects the presence of two equally important, but substantially different methodological approaches to the understanding of the phenomenon.

The first, dynamical approach is based on the representation of beat resonance as the result of an ordered summation of a set of elementary processes of free beats with respect to time. This approach, of course, follows from the picture of atomic excitation by "white" light or by some other exclusively nonresonant excitation, such as electron impact. When an atom is excited by light of linewidth $\Delta\omega$, the process of excitation will take place within a time interval of the order of $\Delta\omega^{-1}$, corresponding to 10^{-10} s for the narrowest (non-laser) spectral lines. Therefore this process is pulse type, and it is followed by the free evolution of the state with its characteristic damping and beats. A modulation of the light leads to the ordered (phased) summation of the elementary beats into a macroscopic and stationary process.

The second approach is stationary in its basic nature. It treats the interaction of two coherent frequencies ω_1 and ω_2 with a three-level system, in which the coherent fields coupling the states 0–1 and 0–2 establish coherency of the states 1 and 2.

Basically both approaches lead to the same result, since a modulated "white" light has a spectral representation which is a superposition of independent harmonics, each of which prossesses two lateral coherent satellites shifted to lower and higher frequencies by the frequency of modulation [4.52]. On the other hand, from this very same picture we can arrive at a description of multimode laser excitation which includes the Doppler broadening of the absorption lines of atoms.

4.9 Other Resonances

The stationary coherency of non-degenerate states can be introduced into the system not only by modulating the intensity of the excitation, but also by other periodic actions. By periodically modulating the interval between the energy sublevels in the presence of coherent excitation, one may induce parametric resonance. The theory of this is presented in Chap. 3. It is also possible to modulate the direction of the vector characterizing the anisotropy of the exciting process, for example, the direction of the polarization plane of the light. The resonance that results has been called phase resonance [4.53]. Finally, the width of the interfering levels can be modulated, which leads to the relaxation of the coherent resonance [4.54]. All these resonances are rather more complex than the beat resonance. Essentially, this is because they are resonances with an infinite set of maxima of coherency, whereas beat resonance corresponds to only a single resonance condition. Complex coherent resonances, have been demonstrated on model systems but did not receive widespread attention from spectroscopists. Parametric resonance, however, drew considerable attention, owing to the simplicity of its realization.

In almost all works on parametric resonance, the modulation of the interval between the levels was produced by an alternating magnetic field. The first experimental demonstration of the effect was undertaken in the work [4.55] on cadmium vapour. Resonance luminescence from the transition $5\,^1S_0-5\,^3P_1$ was excited by light from a cadmium lamp on the line 3261 Å, polarized perpendicular to the magnetic field and modulated at a frequency $\Omega' = 1$ MHz. The modulation of the luminescence was registered at the frequency multiples $\Omega_{p-l} = (p-l)\Omega'$ as a function of the average value of the magnetic field. The luminescence attains maximum amplitude of modulation at the frequency of modulation of the field [$p - l = 1$; see (3.110)]. Depending on the average distance between the sublevels $\bar\omega_{12}$, one can observe a large number of resonances, which occur when $\bar\omega_{12}$ is equal to a whole number of quanta of the alternating field: $\omega_{12}^0 = l\Omega'$. The amplitudes of the resonances do not depend on the modulation depth of the field, which is characterized by the modulation index ω'/Ω', where ω' is the oscillation amplitude of the frequency interval ω_{12} between the interfering levels: $\omega_{12} = \omega_{12}^0 + \omega' \cos \Omega't$. The first resonance maximum on the first harmonic ($p = 0, l = 1$) will be attained when the modulation index $\omega'/\Omega' = 1.8$: when $\omega'/\Omega' = 1.8$, the function $J_0(\omega'/\Omega')J_1(\omega'/\Omega')$ attains its maximum; see (3.110). Thus, in order to achieve the maximum effect it is necessary to move the interfering levels periodically, changing the separation between them at the frequency of their average separation; moreover, the oscillation amplitude of the levels must exceed their average separation. It is this last condition in particular which severely limits the application of parametric resonance, making it a convenient means for investigating very small splittings only, since the required power of the alternating field increases quadratically with the average splitting $\tilde\omega_{12}$. At a frequency of the order of 1 MHz, this power is already sufficient to excite and sustain the gas discharge. For large field amplitude the requirements on its uniformity and on its coaxial alignment with the direct field are stringent. If the field is not perfectly coaxial, then the pattern will become highly complex, since the transverse components of the alternating field are able to induce transitions between states whose momentum projections differ by one; in other words, the parametric resonance will be mixed with ordinary magnetic resonance.

Parametric resonance of atoms in the ground state has practical applications in the measurement of very weak fields (Sect. 4.5). It enables one to measure level splittings ω_{12}^0 smaller than the width of the levels. Under these conditions the maximum amplitude of the absorption modulation will be attained in the vicinity of zero-field resonance, when $\omega_{12}^0 = 1/\tau$; see (4.2). The modulation depth of the absorption is then independent of the frequency Ω' of the modulating field, which is selected purely for reasons of technical convenience. Hence, strictly speaking, in this special case, one should call the phenomenon not parametric resonance, but parametric modulation of absorption.

To conclude our review of coherent resonances (for more detail see [4.56]), let us compare the methodological importance of these phenomena with the most closely related technique of double radio-optical resonance (DRR). We note that the essence of DRR lies in inducing magnetic resonance in the system of sublevels in question by applying an alternating field and recording the resonance through the optical channel. DRR enables one to measure, with unlimited instrumental resolution, level splittings within the bandwidth $0\text{--}10^{10}$ Hz. This method is applicable primarily to long-lived states, since, as the state width increases, the power of the alternating field required to saturate the transition rises quadratically. In order to saturate the magnetic dipole transition of linewidth 10 MHz one requires an alternating field strength of the order of 10 Oe. To maintain such a field strength requires all the more energy, the higher the frequency of the field. In the microwave range, an oscillating power of the order of hundreds of Watts is required. Therefore the DRR technique is chiefly applied to the investigation of ground state structures, for which narrow linewidths are characteristic.

An important merit of coherent resonances is the absence of additional perturbation of the state in question, which is unavoidable in the DRR technique. The perturbing action of the alternating field leads to a broadening of the transitions under investigation, and in multilevel structures to a shifting of the resonances as a function of the applied field power. In contrast to this, the width of a coherent resonance is determined only by the width of the states under investigation.

We note one further difference between the DRR and beat techniques. The DRR method can be used to investigate magnetic or electric dipole transitions that obey specific selection rules for one-photon transitions; in particular, the selection rule for the magnetic quantum number is $|\Delta m| = 0, 1$. The techniques based on interference of states have the more relaxed rule $\Delta m = 0, \pm 1, \pm 2$.

Thus we see that the beat techniques, although they provide the very same information as DRR, are more universal and do not perturb the state in question. Nevertheless, for investigating sublevels of the ground state, DRR is quite a feasible technique and is frequently used, especially for problems of frequency standardization and magnetometry.

4.10   Self-Alignment of Atomic States in a Plasma

The phenomena of self-alignment, hidden alignment (Sect. 4.11) and self-orientation (Sect. 4.12) occurring in a plasma occupy a special position among experiments to observe interference phenomena. This is because, in these cases, the coherency in the ensemble is produced as a result of internal process in the plasma itself.

Gas discharge plasmas were first investigated long ago. It was discovered early on that their emission is slightly polarized and that it is difficult to obtain from them totally unpolarized light. However, the causes of this polarization of the light remained unknown up to the end of the 1960s. At the time when investigators' attention was taken by the alignment of atoms excited by a laser radiation, the polarization of spontaneous radiation was "rediscovered" and again became the subject of serious study. The dependence of the degree of polarization and of the spectral line intensity observed in a given direction on the direction and on the magnitude of an external magnetic field left no doubt that the polarization of plasma emission is induced by the alignment of the emitting atomic states. The alignment in a plasma is produced without any external action and because of this it is called self-alignment.

The main reason for the appearance of the alignment is the diffusion of radiation. The radiation is absorbed not only by the ground state, but also by the excited states, provided the concentration of atoms in these states is sufficient. For different atomic transitions the conditions for the occurrence of diffusion of radiation are different: the "degree" of diffusion is proportional to the product of the transition probability and the population of the absorbing state.

What is the consequence of diffusion? Primarily, it leads to an increase of the photon lifetime in the volume of the plasma. This time can become arbitrarily long. Moreover, the diffusion of radiation definitely leads to the alignment of atomic states.

Actually, the light flux in the plasma is anisotropic. In a long cylindrical discharge tube, the light flux along the axis of the cylinder is greater than that in the radial direction. At the same time, at any point of the discharge not lying on its axis, the flux of light in the radial direction is greater than that in the tangential direction. In order to produce an alignment by means of light excitation, the latter need not necessarily be polarized or be a beam of natural light. Any arbitrary light can be decomposed into beams of different directions. Each of these beams will induce an alignment of its own direction and sign. If the intensities of these beams are not equal in different directions, the integral alignment will not be equal to zero.

It is not difficult to guess that the alignment tensor will have the same symmetry as the angular intensity distribution of the light, and since the anisotropy of the light in a cylindrical discharge tube is described in the general case by a biaxial tensor, the alignment will, in general, also be biaxial. If one chooses the coordinate axis in such a way that the z-axis is parallel to the tube axis, with the x-axis along a radius, then the alignment tensor will contain only the terms ϱ_0^2 and $\varrho_{\pm 2}^2$.

Exactly this form of alignment was observed experimentally in the direct current discharge in a mixture of neon and helium [4.7]. The experiment was conducted in the following way: in a discharge tube, a direct current discharge was sustained in pure neon or in a mixture of neon and helium at a pressure of about 1 torr. The radiation was observed through the wall of the discharge tube. An optical device enabled different sections of the discharge tube to be projected onto the slit of a monochromator. A magnetic field was applied to the observed part of the tube in one of three directions: along the discharge tube axis, along the direction of observation or perpendicular to both these directions. In order to modulate the static field, an alternating field of audio-frequency was superimposed coaxially with the static magnetic field. The alternating part of the photomultiplier photocurrent was amplified by a phase-sensitive detector, controlled by the same audio-frequency signal generator that modulated the magnetic field. The registered quantity is effectively $dI/d\mathscr{H}$, the derivative of intensity with respect to the magnetic field.

The recorded signals are depicted on Fig. 4.10. It is apparent that the sign of the signal in different parts of the discharge tube in the axial field corresponds to the pattern presented in Fig. 4.11. This is confirmed by the signals registered when the magnetic field was applied perpendicular to the axis.

Figure 4.10b shows the signal observed from a part of the discharge tube located approximately half a radius from the edge. The light from this section was observed at an angle with respect to the principal axes of the alignment. We note that when observing the Hanle signal at an angle to the initial axis of the alignment, the Lorentzian lineshape will be mixed with a dispersion profile. If the direction of observation forms an angle α with the axis of alignment in the wall region of the discharge closer to the observer, then the corresponding angle at the far wall will be $-\alpha$. Hence, due to the symmetry of the alignment distribution, the same holds true for every elemental volume located along the line of observation. The dispersion signals from the nearest and farthest regions of the wall are equal in magnitude and opposite in sign and would have cancelled one another were it not for the absorption of light. Because of absorption, only a part of the light emitted by the distant region in the direction of the observer will reach the detector, and the dispersion contributions to the signal will not be compensated. Therefore the curve c in Fig. 4.10 is asymmetric.

The same work [4.7] also describes an experiment which proves that light reabsorption participates in the production of an alignment. The idea is encompassed in the fact that one can change the relative intensity of the light flux without changing anything in the discharge itself. On both sides of the discharge tube were placed prisms oriented to produce total internal reflection (Fig. 4.12). Thus all visible light was reflected back into the tube. The light flux in the x-direction was thereby increased, whereas the light flux in other directions remained as before. In an axial magnetic field the signal from the edge image of

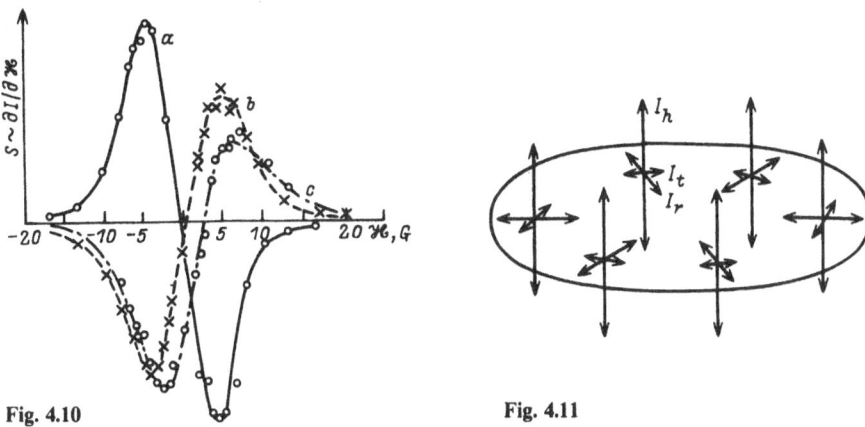

Fig. 4.10

Fig. 4.11

Fig. 4.10. The Hanle signals from different parts of the discharge tube: (a) from the centre of the tube; (b) from the edge of the tube; (c) from the part between (a) and (b)

Fig. 4.11. The alignment distribution over a cross secton of the discharge tube

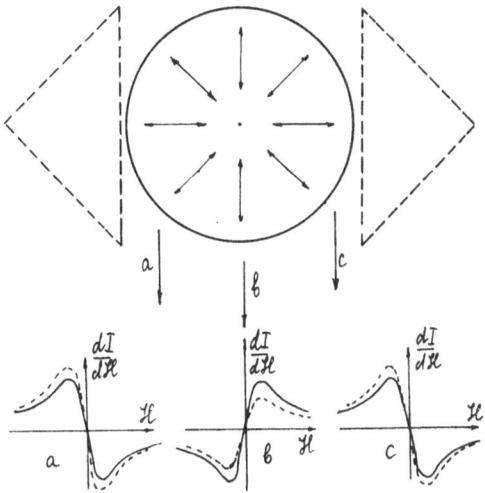

Fig. 4.12. The signal change due to additional illumination. The dotted line corresponds to the signal obtained in an experiment with prisms of total internal reflection

the tube must, according to the model, increase and decrease at the centre upon "switching" the prisms. In the experiment the magnitude of the signal changed in the expected direction, thus confirming the above discussed model of its formation.

The authors of [4.8] also came to the conclusion that the alignment of the lower excited states of neon in their experiment is caused by the reabsorption of light. Their conclusion is based on the coincidence of the computed and the experimentally measured relative intensities of the Hanle signals of 26 transitions.

The magnitude of the self-alignment signals, i.e. the intensity variation of the observed radiation in the spectral lines upon applying a magnetic field, is not large. They are of the order of 1% or less. There are various reasons for this small value of the signals: first, the excitation by reabsorption of light is not the main, but only an additional mechanism of excitation of atoms; second, the reabsorption of different lines will lead to alignment of different signs, and finally, at high vapour density, when the absorption is high, and when the excitation due to radiation diffusion can compete with the "primary" excitation of the same states, the anisotropy of the light will decrease due to the very same absorption. It is clear that the magnitude of the self-alignment signals can vary over a wide range under different conditions and on different lines.

Nevertheless, the present-day technique enables one to measure confidently signals whose values are 10^{-3}–10^{-4} of the line intensity. The high sensitivity is facilitated by the fact that the signal is clearly manifest in the intensity difference of the orthogonal and linearly polarized light components. By measuring this difference the noise in the radiation will be substantially reduced. Moreover, in such measurements one only measures the intensity variation induced by the magnetic field variation; the integral line intensity is not important.

One can cite a number of works that have usefully exploited this technique; see the review [4.57]. The error in determining the linewidth does not, as a rule, exceed 10%. As is most experiments, it is rather difficult to estimate the systematic error. One of the sources of systematic error is the influence of the magnetic field on the discharge: the discharge will be compressed at the wall and its configuration changed, leading to a change in the radiation intensity. Therefore, the alignment signal will be observed against a background that itself varies slightly in a magnetic field.

In an electric discharge, in addition to the radiation, there is another source of anisotropy of the excitation of atomic states – the electron impact. The observation of alignment in the plasma of a high frequency discharge of the capacitor type was reported in 1958 in the publication [4.17]. Under the influence of the electric field, the electrons oscillate along the direction joining the two electrodes. The states excited by electron impact acquire an alignment along this direction.

The authors of [4.58] also relate the self-alignment observed in a hollow cathode discharge to electron impact. In a direct current discharge at low pressure an alignment produced by an electron impact has also been observed. It has been observed for highly excited states of inert gases [4.59] and cadmium ions [4.60]. The picture of the alignment mechanism here is more complex, than in the hf discharge. The potential gradient along the discharge axis is small, and the drift velocity of the electrons is small compared to the thermal velocity. However, the gradient of the field between the tube axis and the wall-layer region is large. This gradient will drive the electrons into motion along the tube radius. Thus, it is these electrons that are considered to yield the directed flux which causes the alignment, and which had been ascribed to electron impact [4.7, 60, 61]. It has been possible to observe this mechanism due to the fact that the electron motion turns out to be inclined to the wall, because of the drift along the tube axis, and the axis of the alignment produced by the electron impact does not coincide with the tube radius. As a result of this the alignment signal is slightly asymmetric – the Lorentzian type signal is mixed with a dispersion lineshape. At the same time, however, the alignment signals related to the diffusion of radiation remain strictly symmetric.

Moreover, it has been demonstrated in the experiment, that the strongest dipole is the one directed along the tube radius. This is not true in the case of optical excitation, where a radially directed dipole is always of intermediate strength compared to the axial and the tangential ones.

Example of the Combination of Externally Induced Alignment with Self-Alignment. As has already been mentioned, the classic technique for producing an alignment is the excitation of vapour by light from the corresponding spectral lamp. In all such experiments there has to be significant absorption of the resonance light in the vapour: if there is no absorption, there can be no fluorescence. However, if there is absorption, there must be also self-alignment, usually it is extremely small, to be noticed. An interesting combination of both types of alignment is described in the work [4.62]. The experiment was intended to measure the lifetime of the resonance state Ne^3P_1.

The experimental scheme was similar to the classical type, with the gas illuminated by a directed beam of light. The resonance fluorescence was observed in a perpendicular direction and the magnetic field was directed along the third perpendicular direction. Since the resonance transitions lie in the vacuum ultraviolet region, polarizers were not used in the experiment, and a multichannel plate was used as the input window. This transmits the light fairly well and supports the pressure difference between the resonance cell and the rest of the system. The latter contained the specially designed spectral lamp and the vacuum monochromator. The monochromator selected from the whole spectrum only the two adjacent resonance lines $\lambda = 74.4$ nm and $\lambda = 73.6$ nm. The photodetector was also designed specially and was placed at the side wall inside the resonance cell. It was anticipated that the Hanle signal would consist of two signals of Lorentzian type but of different width. However, in the experiment a complex signal profile was registered. It is depicted in Fig. 4.13.

One can explain this complex lineshape in the following way: one of the lines $\lambda = 74.4$ nm is characterized by a low transition probability, hence it is absorbed

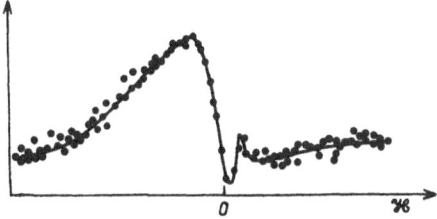

Fig. 4.13. The intensity of the neon resonance line as a function of magnetic field

insignificantly in the light source and the emission on this transition reaches the resonance cell in the form of a narrow spectral line. This radiation is slightly absorbed in the cell and the resonantly scattered light falls on the detector without noticeable absorption. The dependence of the intensity of this scattered light on the magnetic field is a narrow contour with a minimum at zero magnetic field (Fig. 4.13). The second line, whose oscillator strength is 15 times greater than that of the first is emitted from the source broadened and apparently self-reversed. At the cell input the line is still more self-broadened due to the strong absorption of the line centre. In the working region of the cell, excitation is produced by the line-wings. So what is happening with the scattered line? In the direction perpendicular to the beam of the exciting light, the spectral distribution of the scattered light will have a Doppler profile. Because of the strong absorption, a significant portion of this light will not reach the detector. However, upon diffusing to the chamber wall this light will produce secondary alignment (self-alignment), whose axis is perpendicular to the primary axis. It is this that gives rise to the alignment signal with a maximum at zero magnetic field and a width 15 times larger than the signal on the first of the two considered lines; this is the symmetric part of the signal presented in Fig. 4.13. The antisymmetric part arises for the following reasons: Since the aperture of the detector in the experiment was large, about 90°, light fell on it at oblique angles, and if the scattered light is not directed at a rightangle with respect to the exciting light, then its spectrum will "remember" the spectral composition of the exciting light, tending to reproduce it as the angle approaches zero. Thus, in the scattered light the intensity of the frequencies far from the centre is larger than in the Doppler contour, and it easily penetrates through the thickness of the vapour. The light incident at oblique angles gives rise to the antisymmetric admixture to the signal. In ordinary experimental conditions, the antisymmetric parts compensate one another, but in the experiment considered here this did not happen because of the absorption, which causes the light intensity to decrease as it passes through the cell. In such a way the curve in Fig. 4.13 can be decomposed into the narrow signal of the ordinary alignment (weak), the broad signal of the self-alignment (strong) and the antisymmetrical residue signal of the ordinary alignment (also strong). The cited example shows the intricate and complex relationship betwen the absorption and alignment signal in real physical objects.

Self-Alignment of Metastable States. Figure 4.14 is a sketch of the lower energy levels of cadmium. The transition between the ground state $^1S_0$ and the metastable levels $^3P_2$ and $^3P_0$ is forbidden and they cannot be excited by optical means; nevertheless, an alignment on the level $^3P_2$ has been observed [4.63]. The experimental scheme is quite simple: light from a cadmium spectral lamp passed through a modulator and through the discharge tube. The line $\lambda = 508.5$ nm ($6\,^3S_1$–$5\,^3P_2$) was selected by a monochromator. A polarizer placed in front of the monochromator slit allows the selection of light that is linearly polarized either along the tube axis or perpendicular to it. The phase detector enables one to suppress the unmodulated light from the discharge tube.

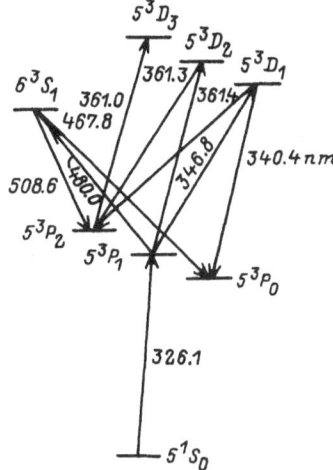

Fig. 4.14. The energy levels of cadmium

The dependence of the transmitted light intensity on the magnetic field applied to the discharge tube along the direction of propagation of the light was investigated. The photocurrent changed only due to the variation of absorption. From the signal analysis it follows that the state $5\,^3P_2$ is aligned. The signal is narrow because the state is a long-lived one, despite the destruction of the polarization moments by collisions.

The spontaneous radiation of the state 6^3S_1 is partially absorbed by the atoms in the state $5\,^3P_2$. Thus, it carries a memory, or record, of the alignment of the absorbing metastable state, and its intensity dependence as a function of the magnetic field has the form of double Lorentzian contours. The broader contours are related to the Hanle effect of the state 6^3S_1; the second profile is narrow and it is attributed to the alignment of the lower $5\,^3P_2$ state. In the literature two mechanisms have been proposed for the production of alignment of a metastable state: optical pumping and cascaded transfer of alignment.

4.11 Hidden Alignment

Signals similar to alignment signals arise even on lines emitted by states with $J = 0$, which by their nature cannot possess any polarization moments higher than zero [4.7]. Similar signals are observed on the π components of other lines, which are described only by the diagonal elements of the density matrix (within the limit of linear approximation with respect to intensity), which do not depend on the magnetic field. The similarity between these signals and the interference signals compels one to assume that the population changes in a magnetic field (and only this can describe the appearance of the signals) are somehow related to the alignment.

We now describe a model of signal formation of this type. It is based on the phenomenon known as "hidden alignment" [4.64, 65]. This consists in the alignment of the electron orbitals in each of the atoms along the direction of their motion in a gas (of course, since one is dealing with individual atoms, we can only discuss probabilistic characteristics of these properties).

The mechanism of formation of hidden alignment can be described as follows: If the intensity of the illuminating light, which possesses a finite spectral width, has a uniform angular distribution, the probability that a moving atom absorbs a quantum of light incidént on it from the side is higher than in the case of light "catching up" with the atom or incident in the opposite direction to be the atom's motion, because the intensity of the light of the resonance transition corresponds to the centre of the Doppler line (Fig. 4.15). The probability of interaction of an atom with light propagating in the direction of motion of the atom (or in the opposite direction) is determined by the light intensity at a frequency shifted with respect to the line centre v_0 by a quantity $v_0 v/c$. It is predominantly the excitation of atoms by light incident at a right angle to the atoms motion that will induce an alignment of the excited states of the ensemble of atoms moving in a given direction. The alignment axis coincides with the direction of motion. Such type alignment has been called hidden alignment, since it vanishes when averaging over all velocity directions of atomic motion. However, this does not mean that hidden alignment has no influence on the characteristics of emission and absorption of a gas.

In particular, the hidden alignment of an ensemble influences the spectral content of its emission (and the absorption line profile). The mechanism of this influence is this: because of the alignment, each of the subensembles with a given velocity directions emits predominantly perpendicular to its motion; therefore, the intensity at the line centre is higher than in the absence of alignment, and the intensity in the line-wings is lower. Consequently, the spectral profile of the radiation from an ensemble with hidden alignment is narrower than the iso-tropic one. Of course, it is hard to imagine that one could notice the distortion of the lineshape due to hidden alignment by simply measuring the line profile. However, since a magnetic field partially destroys the hidden alignment, the

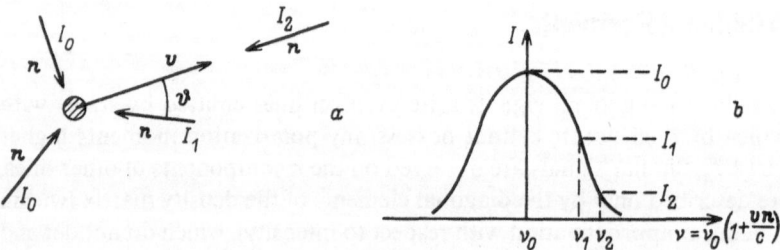

Fig. 4.15a, b. The relation between the direction of an atom's velocity and the direction of exciting light. $I_0 = I_\theta = \pi/2$, $I_1 = I_0 \exp\left(\dfrac{-v^2}{a^2} \cos^2 \theta\right)$, $I_2 = I_\theta = 0$, $\pi = I_0 \exp\left(\dfrac{-v^2}{a^2}\right)$; $v_1 = v_0[1 + (v/c)\cos\theta]$, $v_2 = v_0(1 + v/c)$

influence of the latter on the spectral profile can still be observed. Such an experiment is described in [4.66]. This work investigated not the spontaneous emission line, but the absorption line. A helium-neon laser, whose frequency could be tuned within the range of the emission line $\lambda = 632.8$ nm, was used as a light source. The medium of interest, the direct current neon discharge was placed in a magnetic field perpendicular to the illuminating ray. The experiment demonstrated that the magnetic field decreases the absorption at the line centre, but increases it at the boundaries, i.e. makes the line broader. This means that hidden alignment decreases the linewidth. We note that, due to the properties of the alignment, the other line of absorption from the very same level may show the opposite effect: hidden alignment can also broaden the spectral line.

The spectral line narrowing is accompanied by an increase in the average coefficient of absorption for the reemitted light, i.e. by an increase in the average number of photon-reabsorption events. Thus the hidden alignment is accompanied by an increase in the population of the excited states reacting to the diffusion of radiation.

The higher excited states of atoms in a discharge are populated by optical and collisional processes from the ground state and lower excited states. The intensity of the spectral lines emitted by the higher excited states depends on their population. Therefore, the variation of population in the "lower" excited states due to the action of the magnetic field, will produce an intensity variation in a number of spectral lines from the higher excited states – the latter will also depend on the magnetic field. This dependence is unrelated to the alignment of the emitting state; it is manifest in the lines emitted by the spherically symmetric states, i.e. by the states with $J = 0$. We recall that it was the observation of exactly such signals that led to the discovery of hidden alignment.

The calculations [4.64] illustrate that the hidden alignment signals have a more complex profile than the ordinary Hanle signal:

$$I \propto \left(\frac{1}{\Gamma^2 + \Omega^2} + \frac{1}{\Gamma^2 + 4\Omega^2} \right) . \tag{4.3}$$

In an experiment it is rather difficult to distinguish signals of such a profile from true Lorentzian signals. However, the width of the hidden alignment signal is larger than the macroscopic alignment width of the same state. A comparison of the widths of the signals carried out in the work [4.5] verified the formula (4.3). From the signal of the neon line $\lambda = 607.4$ nm [transition $p_3(^3P_0)-S_4(^3P_1)$], under the assumption that it is related to the alignment of the state $^3P_1$ and described by the formula (4.3), the lifetime of the state $^3P_1$ was determined to be: $\tau = 22.5 \pm 1.8$ ns. On the other hand, the Hanle signal originating from the level $^3P_1$ was measured on the line 73.6 nm by the classical scheme. In this scheme the signal is described by a true Lorentzian lineshape (3.92):

$$I \propto \frac{1}{\Gamma^2 + 4\Omega^2} .$$

The quantity τ calculated by the above expression is in good agreement with that calculated from the hidden alignment signals: $\tau = 23 \pm 2$ ns.

In a real experiment the mechanisms of formation of the macroscopic and hidden alignment work concurrently. Moreover, it is hard to imagine the hidden without the macroscopic alignment, since, in a finite volume containing luminous vapour, the macroscopic alignment will always be present, and in an infinite volume the spectral lines merge into a continuous spectrum; under these conditions hidden alignment is not produced.

The technique of measuring the relaxation time and the radiation decay constant by exploiting the hidden alignment has two interesting features. The first of them is that the hidden alignment signals are observed on spectral lines emitted not by the states in question, but rather by higher states. In systems such as inert gases, this permits one to extend the measurement from the vacuum ultraviolet to the visible region thereby simplifying the technical demands. The second feature is related to the fact that the coherence time, which is manifest in the hidden alignment signals, depends on the gas pressure only as a result of collisional process and does not depend on radiation diffusion (Sect. 4.15). Actually, hidden alignment arises afresh with every act of light reabsorption, and there will be no transfer of memory about the emitting state: hidden alignment is produced by isotropic illumination of the type that can arise upon the destruction of the aligned states in the process of spontaneous decay.

The diffusion of radiation is not the only mechanism of formation of hidden alignment. The works [4.25, 26, 28] illustrate that in a gas hidden alignment of the lower state is produced as a result of collisions.

4.12   Self-Orientation

Another manifestation of hidden alignment is the fact that a weak magnetic field induces orientation in a plasma, i.e. a macroscopic magnetic moment, which disappears again when the magnetic field is increased [4.67]. The mechanism of this phenomenon involves, in addition to the hidden alignment, the anisotropic collisions of atoms in the plasma. The latter can be characterized as follows: Let us hypothetically isolate a single atom out of the whole ensemble and denote its thermal velocity by v_a. Let us choose the coordinate system in such a way that v_a is in the z-direction (furthermore, let the x-axis be perpendicular to the magnetic field, which is so far equal to zero). Let us determine the probability that the atom in question collides with other atoms and let us find the angular distribution for such collisions (Fig. 4.16). If we assume that the cross-section for collisions does not depend on the relative velocity of the colliding particles, then the probability of collision will be given by the velocity distribution of the atoms in a coordinate system which moves together with the atom in question, i.e. with the velocity v_a. The probability of collision is proportional to the integral taken over the product of the number of atoms possessing a given relative velocity

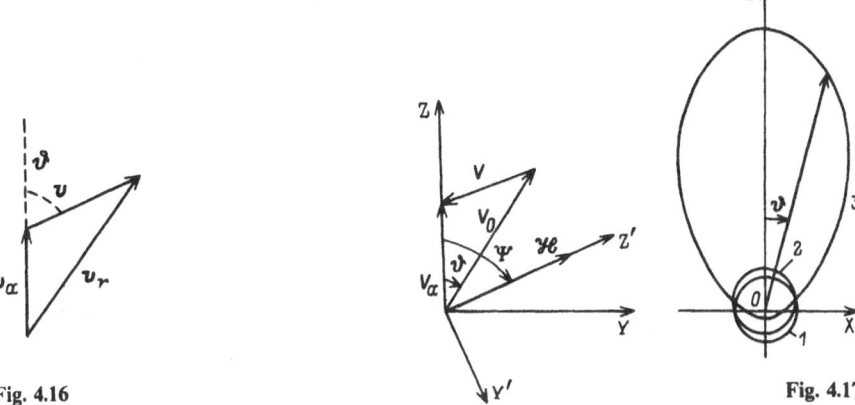

Fig. 4.16

Fig. 4.17

Fig. 4.16. The axis of atomic collision. v_a is the velocity of the atom of interest, v_r the velocity of striking atom, and v the axis of collision

Fig. 4.17. The probability distributions of collisions of atoms. Angular distributions of the probability of collisions of atoms with the velocity v_a. Here θ is the angle between v_a and the axis of the collisions: (1) $v_a = 0$; (2) $v_a = 0.1 v_{prob}$; (3) $v_a = v_{prob}$ (v_{prob} is the most probable velocity of thermal motion)

with the magnitude of the velocity. The angular distribution of the probability of collisions in a plane passing through $0z$ is presented in Fig. 4.17 in polar coordinates. The distance between the origin and any arbitrary point on the curve is proportional to the probability of a collision with a final trajectory at an angle θ to the axis. Figure 4.17 shows that the anisotropy of collisions increases very rapidly with the velocity of the atom. It is characterized by the second moment in the expansion of the distribution of collision probability (in a moving coordinate system) over the spherical functions $Y_q^\kappa : C_0^2$. Its velocity dependence is depicted in Fig. 4.18. Also presented here is the dependence of the total number of collisions on velocity.

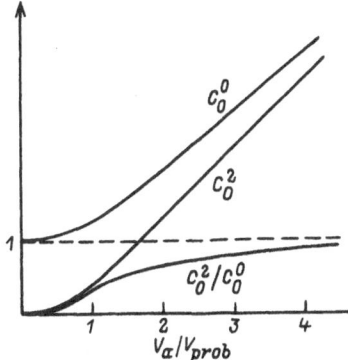

Fig. 4.18. The atomic velocity dependence of the moments C_0^0 and C_0^2

The axis of the anisotropy tensor coincides with the velocity vector v_a, as does the axis of hidden alignment of the atoms. In a weak magnetic field (in which the Larmour frequency is comparable with the level width) the alignment axis is inclined with respect to the axis of anisotropy of collision and precesses about it. An orientation arises under such conditions. The orientation vector is perpendicular to the axes of both tensors (alignment and orientation) and its direction will thus depend on the velocity vector v_a and on the direction of the magnetic field. However, after averaging over all v_a the orientation vector will coincide with the direction of the magnetic field.

This orientation will have an influence on the optical properties of a plasma: It will acquire dichroism and rotation of the polarization plane. In the experiment, the laser radiation passed through a system consisting of the discharge tube, placed between crossed polarizers, and of the photodetector. Figure 4.19 shows the dependence of the light intensity transmitted through such a system on the magnetic field applied along the laser beam [4.68]. An experiment has been conducted on a direct current plasma in neon (tube diameter 4 mm, current 56 mA and pressure about 1.5×10^{-1} torr). The laser radiation frequency coincided with the transition in neon ($\lambda = 607.4$ nm) whose lower absorbing level is $^3P_1$. The probing radiation from a dye laser was tuned in a controlled manner within the range of this absorbtion line. The laser light was modulated, enabling its registration to be separated from the spontaneous emission of the discharge itself. In the absence of a magnetic field light should not pass through the cell placed between the crossed polarizers. The slight transmission when $\mathscr{H} = 0$ is related to mjsalignment and imperfection of the polarizers.

The decaying of the wings of the curve in Fig. 4.19 is a result of the line splitting in a magnetic field. The different frequency distributions of the coefficients of refraction and absorption for the right- and left-circularly polarized light will lead to a rotation of the polarization plane and dichroism, respectively. This is the ordinary Faraday effect. Hence, the behaviour of the curve in weak

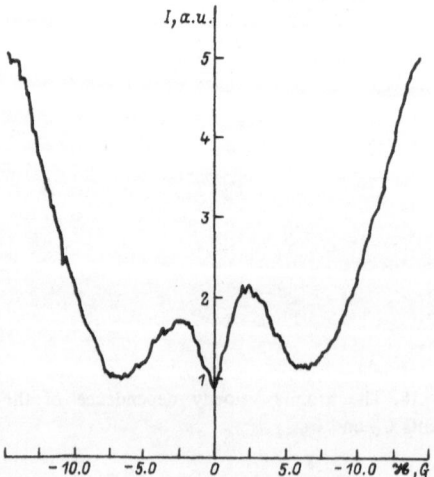

Fig. 4.19. The magnetic-field dependence of the light intensity transmitted through a neon discharge ($\lambda = 607.4$ nm)

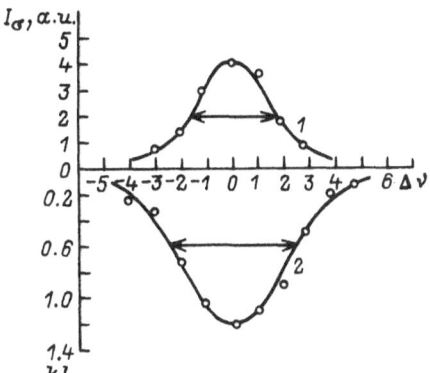

Fig. 4.20. The dependences of dichroism (1) and of absorption coefficient (2) on the distance from the centre of the spectral line

magnetic fields is related to the orientation arising from the hidden alignment. Its form is determined by the influence of the magnetic field on the hidden alignment tensor. Like any other continuously produced alignment, its axis rotates in a magnetic field and its magnitude decreases. For the initial increase in field, the most significant role is played by the increase of the angle between the axes of collision and alignment, and the orientation increases. Furthermore, the role of variation of the angle will turn out to be insignificant; the destruction of the alignment by the magnetic field is far more important. As a result, the orientation in fields of the order of

$$\mathscr{H} \approx \frac{\Gamma}{2\mu_0 g}$$

will reach a maximum and subsequently decreases.

Figure 4.20 [4.69] presents results of the other experiment, which measures the difference in the coefficients of absorption of the right- and left-circularly polarized waves for the same transition in neon. The sign is the same for all frequencies and the width of the spectral distribution is slightly smaller than the Doppler width of the absorption line.

4.13 Interference of Atomic States in Astrophysics

The observed self-alignment in the plasma of the outer layers of the sun can be attributed to the phenomena treated in the previous three sections.

As far back as 1900 at the time of the sun's eclipse on the 28th of May, Robert Wood found that the green line of the coronal emission ($\lambda = 530.3$ nm) is slightly polarized (*Hyder* refers to this work of Wood in his publication [4.70]). The polarization of the spectral lines of the gaseous layer of the sun has attracted much attention since an attempt was made to analyse the performance of a solar magnetometer by measuring very weak fields [4.71].

The principle of operation of a magnetometer [4.72] is as follows: In a longitudinal magnetic field the components of the Zeeman structure polarized in the right and left sense, will shift on the frequency scale in the opposite direction to the line centre. This means that the right and left edges of the Doppler line will become enriched, one in the right- and the other in the left-circularly polarized light. Two slits of the spectrograph will cut off the edges of the Doppler broadened spectral line. The degree of circular polarization is then determined in each of these fluxes. The difference in these quantities is proportional to the magnetic field.

In Ref. [4.72] a technique is proposed to measure the transverse component of a field by making use of the degree of linear polarization. In a transverse field, owing to the very same Faraday effect, the edges of the line are polarized linearly, but the polarization at each edge is identical. The line centre, in contrast, is enriched in another linear polarization. Concerning the measurement of weak magnetic fields by this method, *Hyder* in his paper [4.71] wrote: "*Severnyi* [4.72] has used the 5250 Å line (which has no net linear polarization) and two polarizable lines of different strengths to study the rotation of the direction of the transverse magnetic field with depth in the solar atmosphere. ... I shall show that Severnyi reported rotations could be due to resonance polarization effects in the lines that he selected . . . ."

Actually, in the sun's atmosphere, particularly in the corona and prominences, the same alignment occurs as in a discharge tube: the excited states are aligned under the influence of the anisotropic illumination. The alignment is destroyed by collisions and by the "transverse" component of the magnetic field. The latter only partially destroys the polarization and upon this changes its direction. The degree of destruction of the alignment is determined by the relationship between the radiation decay time, the cross-section of the collisions, and the direction and strength of the magnetic field. In the subsequent works of the astrophysicists mentioned above, the "disturbing" influence of the polarization of spectral lines has already been used for diagnosing the magnetic fields [4.73-75]. In particular, the method established itself successfully in investigations of the prominences, where the magnetic fields are of the order 1–15 G. This is exactly the region, where the Hanle effect is strongest on most of the atomic lines. This problem is rather more complex than in the terrestrial environment, because the human observer cannot alter any of the parameters of the "experiment", not even the direction of observation. On the other hand, the radiation of the sun, its corona and prominences contains a far richer spectrum than laboratory sources of light, and one can select spectral lines whose polarization characteristics are sensitive to the expected strength of the magnetic fields. Here one has to take into consideration the influence of all processes on the atomic emission.

The astrophysics literature relating to the effect discussed here is very rich in theoretical works [4.74-79]. A number of these discuss the importance of the Hanle effect, further developing its quantum mechanical theory. Various quantities are calculated, taking into account alignment formation and destruction by

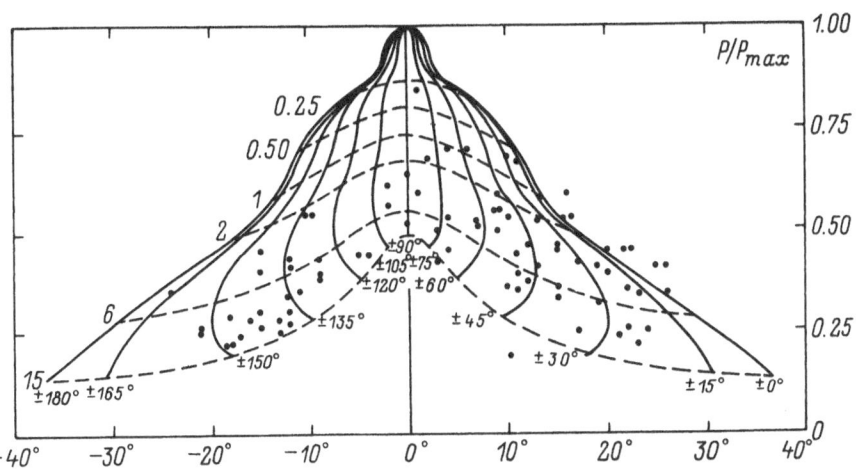

Fig. 4.21. The polarization diagram for the case of horizontal magnetic fields ($\varphi = \pi/2$). The curves result from the computation of p/p_{max} and φ for various values of $\mathcal{H}$ and θ. Full lines correspond to constant values of θ while dotted lines give the variation of p/p_{max} as a function of ϕ for $\mathcal{H}$ constant and expressed in Gauss. The points refer to the average values of p/p_{max} and ϕ observed on 82 different prominences. It is clearly seen that the intensity of magnetic field in quiescent prominences in most cases lies between 1 and 15 G, which is exactly the range of intensities where the Hanle effect has a clearly visible influence

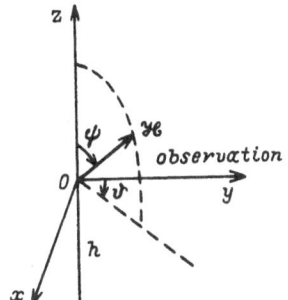

Fig. 4.22. The reference system

all possible processes. We present here two plots which characterize the style of these astrophysical studies.

Figure 4.21, adapted from Ref. [4.74], is calculated for the case in which the magnetic field is horizontal, i.e., parallel to the surface of the sun (Fig. 4.22), and the helium line is observed from the level $3d\,^3D$. Plotted along the ordinate is the ratio of the actual degree of polarization of the observed radiation to the highest possible maximum value. The latter depends on the height above the "surface" of the sun, since the radiation anisotropy increases with height. The diagram is drawn for one specific height. The abscissa corresponds to the angle ϕ, i.e. the direction of the polarization relative to the tangent to the solar disc. The solid lines connect points with identical values of θ, where θ is the angle between the

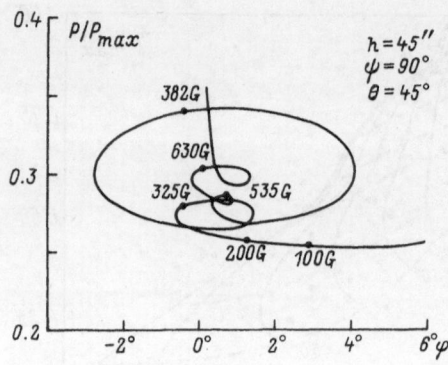

Fig. 4.23. An example of level-crossing loops of the diagram in the particular case of a horizontal magnetic field ($\psi = 90°$) with an angle $\theta = 45°$ between the field and the line of sight. The corresponding level-crossings, which obey $|\Delta m| = 2$ occur between sublevels stemming from $3d^3 D_1$ and $3d^3 D_{2,3}$

magnetic field and the line of observation, but of different magnetic field strengths. The broken lines connect points of equal field strength, but with different angles θ. This diagram enables one to relate any combination of the pair of measured quantities – the degree of polarization and direction of polarization – to the pair of unknown quantities – the strength of the magnetic field and its direction. Points on the diagram are measured quantities; the data presented here represent a large series of measurements in 82 prominences.

The same work [4.74], as well as [4.75] treat the influence of crossings of the Zeeman fine-structure sublevels of helium in nonzero magnetic fields. An example of this type of calculation is illustrated on Fig. 4.23, which is taken from Ref. [4.75]. The plot is sketched for the same spectral line and in the same coordinates, but only one line is shown on it. It is drawn for a horizontal magnetic field ($\psi = 90°$) and an angle between the field and the line of observation of $\theta = 45°$. The field strengths in Gauss are indicated on the curve. The intricate pattern of the curve is related to the crossing effect of the Zeeman sublevels $3d^3 D_1$ and $3d^3 D_{2,3}$ of the helium states. Reference [4.75] also takes into consideration the anti-crossing of levels, by which one understands the transition from the Zeeman effect to the Pashen-Back effect. Such an interpretation is analogous to the previously mentioned (Sect. 3.13) interpretation of the behaviour of the hyperfine structures of the Zeeman components as anti-crossing in a zero magnetic field: corresponding to the magnetic field variation, there is also a variation of the wavefunctions, which in turn changes the relative transition probabilities. This leads to a change in the degree of polarization of the radiation, which is very important in astrophysics research.

4.14 Cascaded Transitions

A cascaded transition refers to the sequential transition of an atom from an initial state to an intermediate state and thence to the final state. Cascaded transitions can transfer coherence from one state to the next. This phenomenon is also used to induce coherency in the state to be studied, but quite often it is merely a nuisance.

A description of the first experiments on stepwise excitation is given in the work of *Mitchell* and *Zemanski* [4.80] where the phenomenon is given the name "stepwise emission". Stepwise excitation was observed in the vapour of mercury and cadmium. In both cases the level $^3P_1$ (Fig. 4.24) was the intermediate level. The process turns out to be intensified by the addition of impurities of foreign gases: upon collisions with these impurity atoms the mercury or cadmium atoms can make transitions to the long-lived state $^3P_0$ (the double arrow in Fig. 4.24) thereby increasing the concentration of atoms that can absorb visible light.

The polarization of the radiation emitted upon stepwise excitation was observed in the year 1929 by *Hanle* and *Richter* [4.81]. *Mitchell* and *Zemanski* described this discovery in their monograph [4.80] as follows: ". . . we passed light from a water cooled mercury arc through a crystal of icelandic spar and focused two images in the form of beams into the resonance tube, which contained mercury vapour and approximately 2 mmHg of nitrogen. They recognized that the two rays seen as fluorescence were differently tinged. The ray with electric field vector perpendicular to the x-axis was blue-green, whereas the second ray, polarized parallel to the y-axis, was yellow-green. It was immediately understood that this phenomenon depends on the polarization". The same work includes a description of fluorescence polarization in the case of exitation through the level $^3P_0$. It is obvious how the alignment of the level $7\,^3S_1$ in Hanle and Richter's experiment influenced the visually observed colour of the ray. It is easiest to treat this case in terms of the polarization of the Zeeman transitions. Let us choose the z-axis – i.e. the axis of quantization – to be along the direction of linear polarization of the exciting ray. In such a coordinate system only the sublevel $m = 0$ of the state $^3S_1$ (solid arrow on Fig. 4.24) will be populated. The emission from the transition $^3S_1$–$^3P_1$, $\lambda = 435.8$ nm, is possible only for the sublevel $\mu = \pm 1$, since the transition $J = 1, m = 0 \rightarrow J = 1, \mu = 0$ is forbidden.

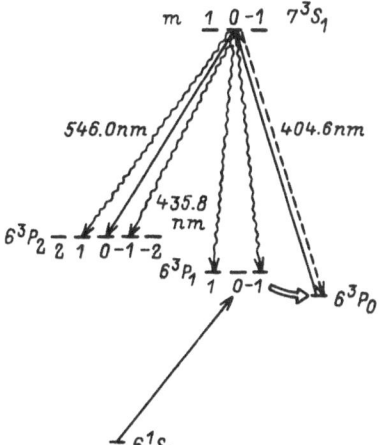

Fig. 4.24. The energy levels of mercury

The transitions for $\mu = \pm 1$ occur with the emission of σ-polarized light (wavy lines on Fig. 4.24), whose greatest intensity is in the z-direction. For the transition $^3S_1-^3P_0$, $\lambda = 404.6$ nm, only a π-component is emitted and it will not be observed in the z-direction. The emission on the third transition, $7\,^3S_1-6\,^3P_2$, $\lambda = 546$ nm, is basically of π-polarization and the fluorescence intensity in the z-direction is lower than that in the perpendicular direction. Therefore, in the z-direction the ray colour is determined primarily by the blue line with $\lambda = 435.8$ nm. The second illuminating ray had an orthogonal polarization and the pattern in it was the reverse: its colour was determined by the green line $\lambda = 546$ nm (the eye is less sensitive to the violet line $\lambda = 404.6$ nm).

The depolarization by a magnetic field of fluorescence excited in the presence of impurities and the rotation of the polarization plane were exploited by Richter in order to determine the lifetime of the $7\,^3S_1$ mercury states. However, according to Michell's analysis, the quantitative estimates were carried out incorrectly. But this work remains noteworthy as the first attempt to observe the Hanle effect upon stepwise excitation.

The experiment on mercury was repeated with great success by *Kibble* and *Pancharatnam* in the year 1965 [4.82]. The level $^3P_0$ is nondegenerate and thus the polarization of the "stepwise" emission is calculated in the same way as the polarization of the resonance fluorescence. The signal profile in these experiments is completely indistinguishable from that of the ordinary Hanle effect.

The case encountered most frequently in practice involves the transfer of the polarization moments of the excited state to the low lying states by spontaneous emission. In the work [4.83] the transfer of alignment was observed by the shift of the modulation of spontaneous radiation with respect to the modulation of the pump laser radiation. The experimental scheme is simple enough: the Q-factor of the resonator is modulated; the spontaneous radiation emerging from the side wall of the discharge tube of the laser is decomposed by a monochromator into spectral lines; light of the selected line falls on a photodetector and the alternating component of the photocurrent with the modulation frequency of the resonator's Q-factor is amplified by a phase detector. A schematic energy-level diagram of neon is sketched on Fig. 4.25. The wavy lines represent laser transitions and straight lines are spontaneous transitions. The experiment registered the radiation emitted upon the spontaneous transitions of neon atoms from the group of levels $2p\,^53p$. The detection also distinguished between light with polarization parallel and perpendicular to the discharge tube axis. In the latter case, the polarization of the observed radiation coincided with the polarization of the laser. The additional radiation on the line 614.3 nm originating from the level $2p_6$ is related to the laser transition process and it turns out to be polarized. The polarization of this line demonstrates the transfer of alignment upon spontaneous decay: the laser transition on the wavelength of 0.15 μm will deplete the population of the upper laser level; however, at the same time an alignment will be induced in it. The spontaneous transmission of excitation to the level $2p_6$ in the presence of laser generation will be passed on and the alignment will be transferred to the lower level, as a result of which the light

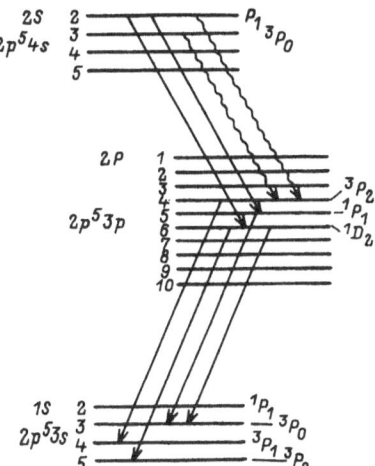

Fig. 4.25. The optical transitions in neon

intensity of one of the polarizations of the observed radiation ($\lambda = 614.3$ nm) will be decreased more than the other.

For the other lines the picture is rather complex, since their initial levels are prone to the influence of spontaneous decay, as well as directly to the laser action (generation) itself. An example of the recorded signal of the lines 626.6 and 671.7 nm is presented on Fig. 4.26. What immediately stands out is that on these two lines the intensities of the light with perpendicular polarizations are modulated in phase opposition: in one of the polarizations the light is modulated in the same way as the generation, i.e. its inensity decreases when the generation is turned off, whereas in the orthogonal polarization it increases.

The cascaded transfer of alignment from the laser levels of neon by spontaneous emission to other levels has been investigated in detail by a number of

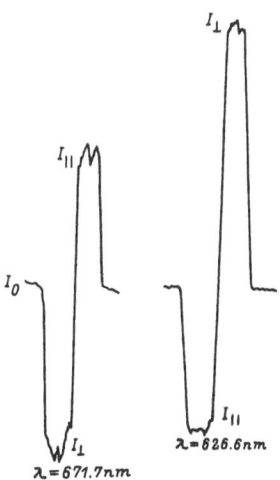

Fig. 4.26. The neon line intensity in two linear orthogonal polarizations in laser action. I_0 is the intensity of the neon line in the absence of laser generation

workers [4.84–86]. They investigated the signal profile and worked out its mathematical description.

The experiment of *Bhaskar* and *Lurio* [4.86] was designed to measure the radiation decay time of the first excited states of neon by employing the level-crossing technique. The direct optical excitation of these levels from the ground state is hampered by the fact that the resonance transitions lie in the vacuum ultraviolet region. The authors thus made use of the cascaded transfer of alignment. The experiment was designed in the following way: an atomic beam from atoms in the lower state was bombarded by an electron beam, as a result of which metastable atoms were created in the states $1s_5(^3p_2)$ and $1s_3(^3p_0)$. The beam was then transmitted through a region of uniform and controllable magnetic field that was directed along the beam or in the opposite direction. The metastable atoms were excited to one of the upper levels p by means of optical illumination from a resonance neon lamp. Upon spontaneous decay of this level, the resonance levels will be partially populated and in this process the alignment of the optically excited state p will be transferred. The intensity of the radiation from the resonance transitions $\lambda = 74.4$ and 73.6 nm was recorded as a function of the magnetic field strength.

Figure 4.27 from [4.86] shows examples of the recorded signals when illuminating the metastable atoms by the neon line 703.2 nm (a) and by the line 626.6 nm (b) allowing for the admixture of the line 621.7 nm, which could not be completely eradicated using an interference filter. The points and solid curve correspond to the experimental measurements and theoretical calculation, respectively. The lifetime of the upper state was taken from the literature and the lifetime of the lower state was varied until the best fit with the experimental curve was achieved. The lifetimes of the resonance neon levels (1.65 ns for $1s_2$ and 20.5 ns for $1s_4$), deduced from these measurements were in good agreement with the theoretical values.

The transfer of alignment by cascaded transitions accompanies almost all experiments connected with self-alignment in a plasma and the alignment

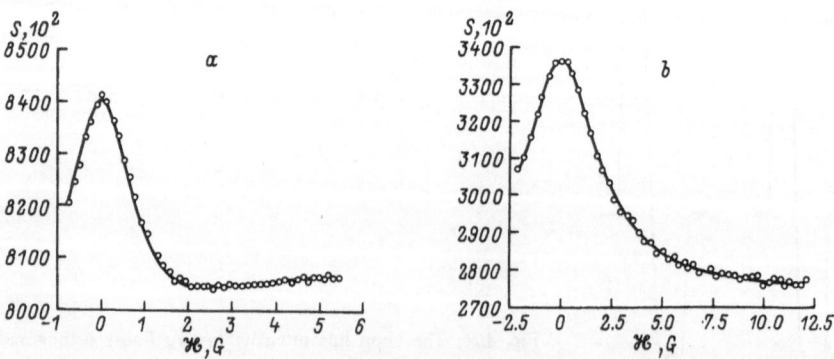

Fig. 4.27a, b. The cascade signals in the neon lines

induced by an electron beam. In both cases the excitation processes are such that the alignment is induced simultaneously on a large number of states. Upon spontaneous decay it will be transferred to a lower state, where it will add to the alignment already produced there by direct excitation.

The signal profile of cascaded transfer of alignment can be calculated on the basis of the formula (3.193)

$$\mathcal{F}_q^\kappa = (-1)^{1+\kappa} F_0 \frac{1}{\sqrt{2J_0+1}} \sum_{\kappa_0, q_0} (-1)^{q_0} (2\kappa_0 + 1)$$

$$\times \sum_{\kappa', q'} (2\kappa' + 1) \begin{pmatrix} \kappa_0 & \kappa & \kappa' \\ -q_0 & q & -q' \end{pmatrix} \begin{Bmatrix} J_1 & J_0 & 1 \\ J_1 & J_0 & 1 \\ \kappa & \kappa_0 & \kappa' \end{Bmatrix} \varrho_{q_0}^{\kappa_0} \Phi_{q'}^{\kappa'} .$$

Here $\varrho_{q_0}^{\kappa_0}$ are the moments of the initial state $|0\rangle$ and $\Phi_{q'}^{\kappa'}$ is the light tensor.

In the case of excitation transfer to another level by spontaneous emission, which is accepted to be stimulated by vacuum fluctuations, and a system possessing complete spherical symmetry, the tensor $\Phi_{q'}^{\kappa'}$ will have only one component Φ_0^0. The properties of the 3j- and 9j-symbols show that only one term will remain in the sum namely that with $\kappa = \kappa_0$ and $q = q_0$. The formula thus adopts the form

$$\mathcal{F}_q^\kappa(1) = (-1)^{J_1+J_0+\kappa} \frac{F_0}{\sqrt{3(2J_0+1)}} \begin{Bmatrix} J_1 & J_0 & 1 \\ J_0 & J_1 & \kappa \end{Bmatrix} \varrho_q^\kappa(0) \Phi_0^0 . \qquad (4.4)$$

Here $\varrho_q^\kappa(0)$ is the statistical tensor of the initial state $|0\rangle$. In a stationary case the density matrix of the final state $|1\rangle$ is given by

$$\varrho_q^\kappa(1) = \frac{\mathcal{F}_q^\kappa(1)}{\Gamma_1 - iq\Omega_1} . \qquad (4.5)$$

The last two formulae show that spontaneous emission transfers to the lower state only those components of the statistical tensor that existed in the upper level. Their relative magnitude and even the sign can be changed, but no new moments are produced. In particular, if the momentum of the final state is $J = 0$, then it is clearly impossible to transfer anything (except population) to the lower state. This is confirmed by the formula (4.4).

Let us now discuss the dependence of $\varrho_q^\kappa(1)$ on magnetic field. It is this that determines unambiguously the signal profile, observed in radiation from this level. The dependence on magnetic field is contained in the denominator of the expression for $\varrho_q^\kappa(1)$ and also in its numerator, insofar as the pump $\mathcal{F}_q^\kappa(1)$, which is related to the characteristic of the initial state $\varrho_q^\kappa(0)$, also depends on the magnetic field:

$$\varrho_q^\kappa(0) = \frac{\mathcal{F}_q^\kappa(0)}{\Gamma_0 - iq\Omega_0} . \qquad (4.6)$$

By making successive substitutions of (4.6) into (4.4) and (4.4) into (4.5), we find

$$\varrho_q^\kappa(1) = K' \frac{\mathscr{F}_q^\kappa(0)}{(\Gamma_1 - iq\Omega_1)(\Gamma_0 - iq\Omega_0)} .$$

The nature of the pump $\mathscr{F}_q^\kappa(0)$ is not important; it is significant only to know which components of the statistical tensor are present in it. Let us suppose that only $\mathscr{F}_0^0$, $\mathscr{F}_0^2$ and $\mathscr{F}_{\pm 2}^2$ are present; this describes the case of alignment reduced to the principal axes. The components $\mathscr{F}_0^0$ and $\mathscr{F}_0^2$ give rise to the components with $\kappa = 0$, $q = 0$ and $\kappa = 2$, $q = 0$ and they will be transferred to the lower state in the form of components with the same values of κ and q. Let the magnetic field be directed along the quantization axis. Then $\varrho_{q=\pm 1}^\kappa = 0$ and ϱ_0^κ do not depend on the magnetic field. Thus they do not contribute to the signal and can be neglected. The only remaining component is

$$\varrho_{q=\pm 2}^{\kappa=2} = \frac{K'}{(\Gamma_1 \mp 2i\Omega_1)(\Gamma_0 \mp 2i\Omega_0)}$$

and the part of the intensity in the corresponding polarization, which depends on the magnetic field, is given by

$$I(\mathscr{H}) = K \frac{\Gamma_0 \Gamma_1 - g_0 g_1 \mu_0^2 \mathscr{H}^2/\hbar^2}{[\Gamma_1^2 + 4(\mu_0 g_1 \mathscr{H})^2/\hbar^2][\Gamma_0^2 + 4(\mu_0 g_0 \mathscr{H})^2/\hbar^2]} .$$

Figure 4.28 shows two characteristic examples of the cascaded Hanle signal. It can be either narrower or broader than the signal produced in the absence of cascaded transition, depending on the sign of the Landé factor g. However, negative signs for this factor are encountered very rarely and the cascaded signal is almost always narrower than the direct signal.

If one considers this process in a plasma, then as well as the cascaded transfer of alignment to a certain state, there will also be an alignment produced by its direct excitation. One will observe a sum of signals due to cascaded and direct processes. These can be of different signs, depending on the sign of the alignment produced on each of the levels, as well as on the structure of the spontaneous transition in which the sign of the alignment may be changed. When the signs

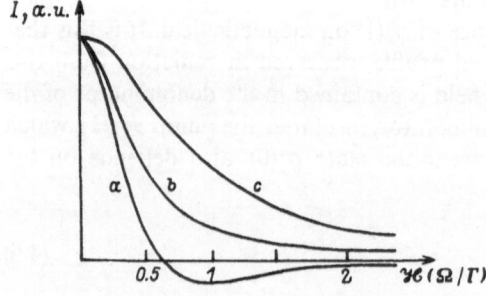

Fig. 4.28. The form of cascade signals: (a) $\Gamma_\mu = 2\Gamma_n$; $\omega_{\mu\mu'} = \omega_{nn'}$; (b) the usual Hanle signal; (c) $\Gamma_\mu = 2\Gamma_n$; $\omega_{\mu\mu'} = -\omega_{nn'}$

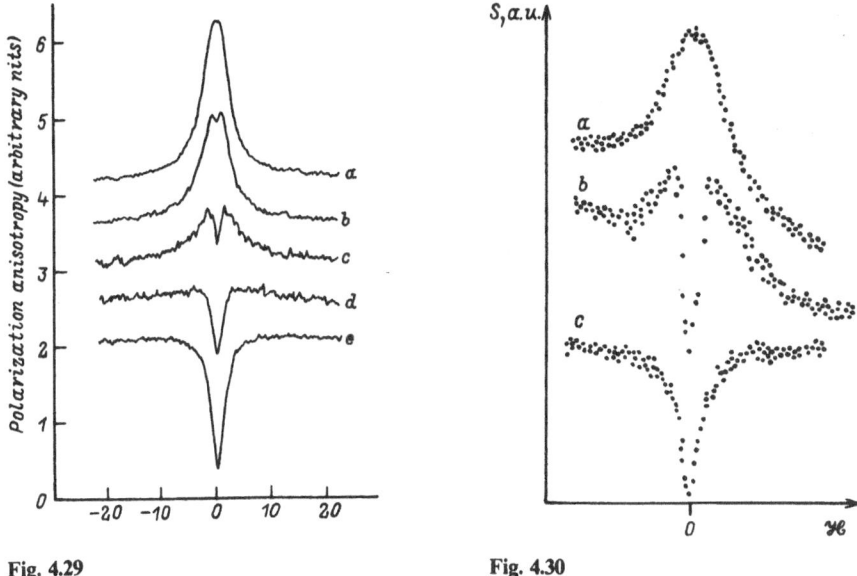

Fig. 4.29   Fig. 4.30

Fig. 4.29. Hanle effect in the neon $1s_5-2p_9$ transition (640.2 nm) showing inversion of the signal at low pressure due to the addition of argon. The partial pressures of the two gas components in micrometres of Hg are (a) Ne 125 μm, Ar 0 μm; (b) Ne 115 μm; (c) Ne 110 μm, Ar 10 μm; (d) Ne 120 μm, Ar 20 μm; (e) Ne 80 μm, Ar 25 μm. The polarization anisotropy increases when the intensity of light polarized transverse to the axis of the discharge tube increases relative to that polarized in the axial direction

Fig. 4.30. The alignment signals on the krypton line when its pressure changes (the current is 30 mA and the pressure: (a) 100 mT, (b) 30 mT, (c) 20 mT

are different and the widths of the cascaded and principal signals differ significantly, then the signals are of the type of Figs. 4.29 and 4.30 (Fig. 4.29 is adapted from [4.87] and Fig. 4.30 from [4.88]). In both cases the experiment is very simple: the spontaneous radiation in a single spectral line selected by a monochromator was measured as a function of the applied magnetic field in the observed region of the tube. In front of the monochromator was a polarizer whose axis is perpendicular to the magnetic field. The radiation was recorded in such a way that a signal is produced only when the radiation depends on the magnetic field. The signal sign change under changing discharge conditions is demonstrated on both of the figures. The signal transformation on Fig. 4.29 is related to the addition of argon to neon, on the line ($\lambda = 640.2$ nm) producing the cascaded signal. The other figure, Fig. 4.30, shows the variation of the krypton line when changing the pressure of the gas. This type of signal transformation had also been observed when varying the discharge current [4.88]. Clearly the relative contributions of the "cascaded" and "direct" Hanle signals are altered by changing the discharge conditions. This is a reason for the change in sign of the integral Hanle signal.

4.15 Diffusion of Radiation

The role of diffusion in measurements of the radiative decay time of excited states has been known for many years [4.89]. A photon, reabsorbed a number of times, survives in the volume occupied by the emitting atoms, provided that this volume is sufficiently large and indefinitely long. An observer then perceives the result of reabsorption as an increase of the excited state lifetime.

The role of radiation diffusion when observing interference phenomena is discussed in the works of *Barrat* [4.90] and *Dyakonov* and *Perel* [4.91]. The latter contains a rigorous derivation of the equation of motion of the statistical tensor of an excited state from which only one transition to the ground state is permitted. It is shown in this work, that the fluorescence decay in an infinite medium can be characterized by two factors: Γ_2, the decay of the plane polarization (or decay of the alignment) and Γ_1, the decay of the circular polarization (orientation). Of course, with these assumptions (infinite medium), the decay of population (and integral intensity) is $\Gamma_0 = 0$.

The narrowing of the level-crossing signal due to diffusion has been observed experimentally and is described in a number of works, for instance, in the fluorescence of optically excited vapour [4.92], in a discharge excited by laser radiation [4.95], and upon aligning with light from the discharge itself [4.94, 95].

We discuss the diffusion of radiation for atoms possessing only three stable states (Fig. 4.31): the ground, metastable and excited states. Let us denote the transition probability from the excited state to the ground state (resonance transition $J_1 \leftrightarrow J_0$) by d_{res}^2, and to the metastable state (observable transition $J_1 \rightarrow J_2$) by d_{obs}^2. Their sum is $d^2 = d_{\text{res}}^2 + d_{\text{obs}}^2$. It is assumed here that the metastable level is destroyed by a non-radiative transition to the lower (ground) state, thus there is no optical pumping to it, i.e. one can ignore its population. The volume occupied by the ensemble of atoms is assumed to be large, such that the absorption of resonance radiation on the resonance transition is complete. The absorption is absent on the second transition. One observes the emission exactly on this second transition. We assign to all levels certain quantum numbers J. The dependence on the magnetic field of the light intensity emitted in this transition in a given polarization is found by irradiating the ensemble of atoms with light whose frequency corresponds to the resonance transition.

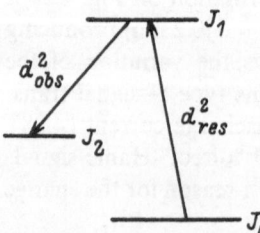

Fig. 4.31. Notation for transitions for radiation diffusion calculation

This model describes some real situations extremely well, since for complete diffusion of the radiation of the resonance transition, the signal of the second line is really observed (the whole stored energy of the excitation will be emitted exactly on this transition), whereas in the traditional problem of observing a resonance transition upon complete diffusion, the degree of polarization is equal to zero and it is impossible to observe the Hanle signal.

General Scheme for Solving the Problem. Let us examine the chain of transformations of a photon in an ensemble of atoms. The resonance radiation is incident from outside and transfers the atoms to the excited states. When this state decays there will be a certain probability of emitting a photon in the second transition (which can be registered by an observer), or light can be emitted via the resonance transition. In the latter case the light is absorbed by the ensemble of atoms and when the excited state decays, there are again two possible transitions to the two lower states. The light emitted in the former transition reaches the observer. The observer registers light from all of the excited atoms (Fig. 4.32).

The signal (i.e. the intensity as a function of magnetic field) can be calculated, provided one finds the change in the characteristics of the light field from one reabsorption event to the next. This can be treated as a two-step process. The first step is related to the transformation of light by the atom. The second step is attributed to the transversality of the light wave and it is most readily explained by means of an example: let the resonance line of the atom be subject to the normal Zeeman effect, the lower level has $J = 0$ and the upper one $J = 1$. If only a single upper sublevel $m = 0$ is excited, then such an atom will emit only on the π-transition. The radiated light propagates in various directions with different intensities (in other words the probability of photon emission is anisotropic). The light emitted perpendicular to the axis of quantization (Fig. 4.33) induces, upon its reabsorption, only π-transitions. However, light propagating in all other directions can also induce σ-transitions, since its electric field vector has a finite projection on the x–y plane. The probability of excitation of the σ-transitions can be calculated for each of the angles and then averaged over all

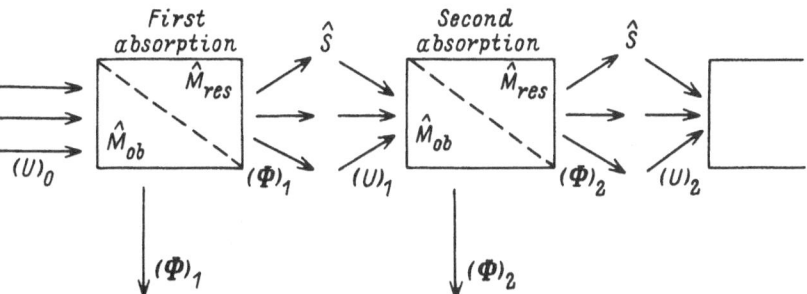

Fig. 4.32. Explanation of the calculation of alignment signals under radiation diffusion

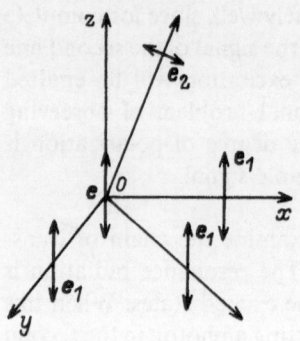

Fig. 4.33. The origin of the operator $\hat{S}$

angles. Such calculations enable one to describe the second step of transformation of the light field.

Let U be a set of quantities characterizing both the intensity and the state of polarization of the light field that illuminates the ensemble of atoms. We denote the transformation of the light by the atom by the operator $\hat{M}$. The tensors of the light emitted by the atom on the resonance transition and on the observable transition, which we will denote by Φ and Φ_{obs}, respectively, are then related to U as follows[2]

$$(\Phi)_l = (d_{res}^2/d^2)\hat{M}_{res}(U)_{l-1}, \qquad (\Phi_{obs})_l = (d_{obs}^2/d^2)\hat{M}_{obs}(U)_{l-1} \ .$$

The light emitted in the second transition emerges in its entirety from the volume, and a fraction of it reaches the observer and contributes to the signal. In contrast, the light emitted by atoms undergoing the first transition is fully reabsorbed. As discussed above, when light propagates from the radiating atom towards the absorbing one, its characteristics due to the transversality of the light wave undergo a change, and we present this transformation by way of the operator $\hat{S}$. Consequently, after the secondary absorption the light will have the characteristics $(U)_1 = \hat{S}(\Phi)_1 = (d_{res}^2/d^2)\hat{S}M_{res}(U)_0$. The transformation scheme is illustrated on Fig. 4.32. Proceeding with the same reasoning one can write the characteristics of light after any number of reabsorption events:

$$(U)_l = [(d_{res}^2/d^2)\hat{S}\hat{M}_{res}]^l(U)_0$$

$$(\Phi_{obs})_l = (d_{obs}^2/d^2)\hat{M}_{obs}[(d_{res}^2/d^2)\hat{S}\hat{M}_{res}]^{l-1}(U)_0 \ .$$

The observed signal is given by the sum of the quantities:

$$\Phi_{obs} = \sum_{l=1}^{\infty} (\Phi_{obs})_l = (d_{obs}^2/d^2)\hat{M}_{obs} \sum_{l=1}^{\infty} [(d_{res}^2/d^2)\hat{S}M_{res}]^{l-1}(U)_0 \tag{4.7}$$

---

[2] The index outside the bracket corresponds to the number of reabsorption events

from which one can easily determine the intensity in any direction and in any polarization.

Transformation Matrix M of Light Due to Atoms. The light emitted by atoms can be found through the statistical tensor ϱ_q^κ of the excited state. The latter satisfies the equation (3.173) and for the case of level crossing, where nothing depends on time: $\rho_q^\kappa = \mathscr{F}_q^\kappa/(\Gamma_\kappa - iq\Omega)$. Here $\mathscr{F}_q^\kappa$ is the exitation tensor, which is related to the characteristic of the exciting light by (3.176):

$$\mathscr{F}_q^\kappa = (-1)^{J_1 + J_0} K'' \begin{Bmatrix} 1 & 1 & \kappa \\ J_1 & J_1 & J_0 \end{Bmatrix} U_q^\kappa . \tag{4.8}$$

The intensity of spontaneous radiation can be described by the density matrix of the light, which reflects the intensity of the radiation as a function of its propagation direction and its polarization:

$$f_{q_1 q_2} = \mathscr{E}_{q_1} \mathscr{E}_{q_2}, \quad q_1, q_2 = 0, +1, -1.$$

In accordance with (3.63) the field amplitude of the spontaneous radiation in the polarization q_i is

$$\mathscr{E}_{q_i} \propto \sum_m C_m (\mathbf{d} \cdot \mathbf{e}_{q_i})_{\mu m}$$

and the density matrix is thus

$$f_{q_1 q_2} \propto \sum_{\substack{m, m' \\ \mu}} \overline{C_m C_{m'}^*} (\mathbf{d} \cdot \mathbf{e}_{q_1})_{\mu m} (\mathbf{d} \cdot \mathbf{e}_{q_2})_{\mu m'}^* = \sum_{\substack{m, m' \\ \mu}} \sigma_{mm'} (\mathbf{d} \cdot \mathbf{e}_{q_1})_{\mu m} (\mathbf{d} \cdot \mathbf{e}_{q_2})_{\mu m'}^* . \tag{4.9}$$

We transfer the density matrix of the light to the statistical tensor picture. Since we have introduced the eigenpolarizations of the dipole transitions into the density matrix, but not their projection on a chosen polarization as in (3.178), then instead of (3.179) we obtain

$$\Phi_q^\kappa = K''' \sum_{q_1, q_2} (-1)^{q_2} f_{q_1, q_2} \begin{pmatrix} 1 & 1 & \kappa \\ q_1 & -q_2 & q \end{pmatrix} . \tag{4.10}$$

Substituting $f_{q_1 q_2}$ from (4.9) and performing the transformation, similar to Sect. 3.10, we find

$$\Phi_q^\kappa = (-1)^{J_1 + J_0} K'' \begin{Bmatrix} 1 & 1 & \kappa \\ J_1 & J_1 & J_0 \end{Bmatrix} \varrho_q^\kappa . \tag{4.11}$$

Making use of the equalities (4.11), (3.173) and (4.8), we find that the l-fold reemitted light is described by

$$(\Phi_q^\kappa)_l = K \begin{Bmatrix} 1 & 1 & \kappa \\ J_1 & J_1 & J_0 \end{Bmatrix} (\Gamma - iq\Omega)^{-1} (U_q^\kappa)_{l-1} = M_q^\kappa (U_q^\kappa)_{l-1}$$

and the matrix components describing the reemission of the light by atoms, which turn out to be diagonal are given by

$$M_q^\kappa = K \begin{Bmatrix} 1 & 1 & \kappa \\ J_1 & J_1 & J_0 \end{Bmatrix} (\Gamma - iq\Omega)^{-1} .$$

The Transformation Operator $\hat{S}$. The problem of finding this operator can be divided conditionally into two parts. One can choose an arbitrary direction and finds the projections of the radiation tensor on a plane perpendicular to this direction. This set of quantities will characterize the state of polarization and intensity of the light emitted by the ensemble of atoms in the chosen direction.

One then performs averaging over angles, obtaining as a result the electromagnetic field tensor "seen" by the absorbing atom.

The first part of the problem can be solved in the following way: We transfer to a coordinate system whose z-axis makes an arbitrary angle relative to the laboratory coordinate system. Mathematically this operation is described by the rotation matrix D, which consists of the Wigner functions (Table 3.1). Then, in order to derive an explicit expression for the light component e_z, we turn to the presentation in which components of the polarization matrix are expressed as the product of the vector projections by the circular unit vectors. This transformation is described by the matrix T, whose elements are coefficients of the linear transformation which is the inverse of (4.10).

The projection operator $\hat{L}$, which exactly takes into account the transversality of the light wave, should leave unchanged all components that do not contain e_z and destroy all those that do. In the polarization matrix there are nine components of which five contain e_z. This implies that four terms of the matrix L, which is diagonal, are equal to one and five of them are equal to zero. Afterwards one needs to revert to the original presentation $(\hat{T}^{-1})$ and to the original coordinate system $(\hat{D}^{-1})$. In this way one can derive the operator, which in turn enables one to find the field tensor of the light wave radiated by a single atom and propagating to the point where another atom is located $\hat{D}^{-1}\hat{T}^{-1}\hat{L}\hat{T}\hat{D}$. The general form of this matrix is presented on Table 4.1. For the illuminating light tensor we find, upon averaging this expression over all directions,

$$S = K^S \int_0^\pi \int_0^{2\pi} \hat{D}^{-1}\hat{T}^{-1}\hat{L}\hat{T}\hat{D} \sin\beta \, d\beta \, d\gamma .$$

The transformation matrix so derived turns out to be diagonal as anticipated, since the reemission of light is a process that is physically symmetric and cannot change the symmetry of the system.

Signal Shape. Inasmuch as both the operators $\hat{S}$ and $\hat{M}$ are essentially diagonal matrices, the product of their elements in the absence of the second transition

Table 4.1. The general form of the matrix $\hat{D}^{-1}\hat{T}^{-1}\hat{L}\hat{D}$

κ	q	$\kappa'=0,\ q'=0$	$1,\ 1$	$1,\ 0$	$1,\ -1$	$2,\ 2$	$2,\ 1$	$2,\ 0$	$2,\ -1$	$2,\ -2$
0	0	$\dfrac{2}{3}$				$\dfrac{10}{\sqrt{90}}D_{02}^{2*}$	$\dfrac{10}{\sqrt{90}}D_{01}^{2*}$	$\dfrac{10}{\sqrt{90}}D_{00}^{2*}$	$\dfrac{10}{\sqrt{90}}D_{0-1}^{2*}$	$\dfrac{10}{\sqrt{90}}D_{0-2}^{2*}$
1	1		$D_{01}^{1}D_{01}^{1*}$	$D_{01}^{1}D_{00}^{1*}$	$D_{01}^{1}D_{0-1}^{1*}$					
1	0		$D_{00}^{1}D_{01}^{1}$	$D_{00}^{1}D_{00}^{1*}$	$D_{00}^{1}D_{0-1}^{1*}$					
1	-1		$D_{0-1}^{1}D_{01}^{1*}$	$D_{0-1}^{1}D_{00}^{1*}$	$D_{0-1}^{1}D_{0-1}^{1*}$					
2	2	$\dfrac{2}{\sqrt{90}}D_{02}^{2}$				S_{22}	S_{21}	S_{20}	S_{2-1}	S_{2-2}
2	1	$\dfrac{2}{\sqrt{90}}D_{01}^{2}$				S_{12}	S_{11}	S_{10}	S_{1-1}	S_{1-2}
2	0	$\dfrac{2}{\sqrt{90}}D_{00}^{2}$				S_{02}	S_{01}	S_{00}	S_{0-1}	S_{0-2}
2	-1	$\dfrac{2}{\sqrt{90}}D_{0-1}^{2}$				S_{-12}	S_{-11}	S_{-10}	S_{-1-1}	S_{-1-2}
2	-2	$\dfrac{2}{\sqrt{90}}D_{0-2}^{2}$				S_{-22}	S_{-21}	S_{-20}	S_{-2-1}	S_{-2-2}

$S_{qq'} = D_{2q}^{2}D_{2q'}^{2*} + D_{-2q}^{2}D_{-2q'}^{2*} + \tfrac{1}{3}D_{0q}^{2}D_{0q'}^{2*}.$

$(d_0 = 0)$ can be given by

$$(SM)^{\kappa = 0} = 1$$

$$(SM)^{\kappa = 1} = 0.5(\Gamma_1 - iq\Omega)^{-1} \begin{Bmatrix} 1 & 1 & 1 \\ J_1 & J_1 & J_0 \end{Bmatrix}^2 \begin{Bmatrix} 1 & 1 & 0 \\ J_1 & J_1 & J_0 \end{Bmatrix}^{-2} \Gamma_0$$

$$(SM)^{\kappa = 2} = 0.7(\Gamma_2 - iq\Omega)^{-1} \begin{Bmatrix} 1 & 1 & 2 \\ J_1 & J_1 & J_0 \end{Bmatrix}^2 \begin{Bmatrix} 1 & 1 & 0 \\ J_1 & J_1 & J_0 \end{Bmatrix}^{-2} \Gamma_0 .$$

Having an explicit form of the operator $\hat{S}\hat{M}$, it is not difficult to find the components of the observed light tensor: One makes use of the formula (4.7), in which the operator $\hat{M}_{obs}$ is also an operator $\hat{M}_{res}$, but in which J_0 is replaced by J_2.

Let us explore the signal shape for the most standard conditions of the experiment with illumination by linearly polarized light whose polarization vector is perpendicular to the magnetic field. The non-zero components $(u)_0$ will be $(u_0^0)_0$ and $(u_{\pm 2}^2)_0$, and the observation tensor will contain only these components. The signal that depends on the magnetic field will have the form

$$I_{obs} = K_2 \frac{(1 - \alpha_2)\Gamma_2}{[\Gamma_2(1 - \alpha_2)]^2 + 4\Omega^2} ,$$

where

$$\alpha_2 = 0.7 \frac{d_{res}^2}{d^2} \begin{Bmatrix} 1 & 1 & 2 \\ J_1 & J_1 & J_0 \end{Bmatrix}^2 \begin{Bmatrix} 1 & 1 & 0 \\ J_1 & J_1 & J_0 \end{Bmatrix}^{-2} .$$

We can define $\Gamma_2' = \Gamma_2(1 - \alpha_2)$.

As is seen from the formula, the Hanle signal again has a Lorentzian form; however, its width is determined not by the decay time $1/\Gamma_2$ but by the coherence time, which differs by the factor $(1 - \alpha_2)^{-1}$. If there is diffusion of the light on a transition with a normal Zeeman effect ($J_1 = 1$, $J_0 = 0$) and in the absence of the second decay channel ($d_{obs} = 0$) one obtains the well known increase of life "time" $\Gamma_2'/\Gamma_2 = 0.3$ [4.90, 91]. In the work [4.96] the signal has been used to determine the ratio of probabilities of the two transitions.

The signal form of the crossing with $\Delta m = 1$, which is observed in a light of circular polarization when the excitation is also due to light of circular polarization (both of the beams are perpendicular to the magnetic field) is determined by the components

$$I_{obs} = K_1 \frac{(1 - \alpha_1)\Gamma_1}{[\Gamma_1(1 - \alpha_1)]^2 + \Omega^2} ,$$

$$\alpha_1 = 0.5 \frac{d_{res}^2}{d^2} \begin{Bmatrix} 1 & 1 & 1 \\ J_1 & J_1 & J_0 \end{Bmatrix}^2 \begin{Bmatrix} 1 & 1 & 0 \\ J_1 & J_1 & J_0 \end{Bmatrix}^{-2} ,$$

$$\Gamma_1' = \Gamma(1 - \alpha_1) .$$

The signal width decreases by a factor of $(1 - \alpha_1)$. The mean intensity

$$\bar{I} = (\Phi_0^0)_{obs} = \frac{d_{obs}^2}{d^2} (M_0^0)_{obs} \sum_l \frac{d_{res}^2}{d^2} (U_0^0)_0 = (M_0^0)_{obs} (U_0^0)_0$$

behaves as if, in the absence of diffusion, the whole decay of the excited state had taken place via the second transition.

The above method of deriving the expression for the level-crossing signal upon diffusion of radiation allows one to make some (approximate) conclusions about the influence of the radiation diffusion on the signal in a finite chamber.

Let us return to the transformation matrix $\hat{S}$ of the light. It is this that contains the difference from the previous case. It is worked out in the same way as in the previous case, but before averaging over angles, one has to multiply every element of the matrix (Table 4.1) by a coefficient that takes into account the fact that the radiation emerges from the volume. This coefficient is smaller than unity and equal to $(1 - k)$, where k is the absorption and is a function of angle. In the general case this leads to the fact that off-diagonal elements of $\hat{S}$ do not integrate to zero. This in turn means that, upon radiation diffusion, one component gives rise to the others, and can, moreover, induce components of a tensor of different rank. Hence, for example, the intensity (U_0^0) gives rise to an alignment – this is a self-alignment (Sect. 4.10).

It is quite interesting to emphasize that an orientation ($\kappa = 1$), if it is absent in the illuminating light, cannot arise under any conditions and thus cannot transform to an alignment and influence the population: the matrix elements corresponding to this transformation are always equal to zero.

In a volume with axial symmetry, the coefficient depends only on the angle β. All the off-diagonal elements, when integrated with respect to another angle, become zero in this case and the matrix $\hat{S}$ remains diagonal. However, its elements are no longer equal to one another for every q. This implies that the longitudinal and transverse alignments each decays with its own time constant.

The whole situation will be very complex if one takes into account that the properties of the light field and the concentration of the excited atoms vary within the volume occupied by the vapour [4.97].

4.16 Influence of the Laboratory Magnetic Field on the Hanle Signal Shape. False Hanle Signals

The laboratory magnetic field (due to the Earth's field and other sources) is one of the obvious factors which distorts the anticipated form of the recorded interference signals. By superimposing on the controlled magnetic field, it changes the magnitude and direction of the latter, and thus also influences the registered light intensity. It is quite obvious that it will influence all types of interference signals, including those whose controlling parameter is an electric field. The laboratory magnetic field induces the smallest distortion in the case where its direction coincides with that of the external variable magnetic field. Then it leads merely to a variation of the field magnitude without changing its direction. Therefore, it results only in a shift of the registered curve on the magnetic field scale.

We examine a similar influence of the laboratory magnetic field on the example of the Hanle effect [4.98]. Upon investigating this influence it has been found that the intensity depends on the applied external controlled magnetic field even when the Hanle signal (in the conventional understanding of this word) is absent (for instance, when observing light of linear polarization along the controlled magnetic field). The profiles of these dependences are highly complex and their width in terms of the magnetic field scale does not necessarily depend on the lifetime of the emitting state. The law which governs the intensity changes is determined mainly by the "rate" of variation of the direction of the total magnetic field, but not by the coherence time of the radiating state. Therefore, signals of this origin were called false alignment signals (or false Hanle signals) by analogy to the "ghosts" of diffraction spectra.

The signal profiles calculated with inclusion of the influence of the laboratory fields are depicted in Figs. 4.34 and 4.35. In these, the unit of magnetic field is taken to be the field in which the Zeeman splitting is equal to the width of the level Γ. All results of the calculations on Figs. 4.34 and 4.35 are given on this scale. The reference point on the ordinate (relative intensities) corresponds to the signal on the transition $J = 1 \rightarrow J = 0$ at the maximum value of the alignment.

Figure 4.34 includes all possible combinations of orthogonal and coinciding directions of linear polarization of the exciting light, the external magnetic field and polarization of the observed light (also linear). Altogether there are nine. However, since for $\theta = 0$ there is no dependence on the angle ϕ, one can confine oneself to the single value $\phi = 0$. Figure 4.35 shows the signals for some other directions of the laboratory field relative to the tensor axes of the excitation and observation.

The table of curves, given in Fig. 4.34 is symmetric with respect to the diagonal, since the curves do not change upon interchanging the angles $\theta \rightleftarrows \theta_{obs}$ and $\phi \rightleftarrows \phi_{obs}$, i.e. upon interchanging the values Φ and Φ_{obs}. For this reason they are presented only for one set of angles. The curves in the second and third columns of Figs. 4.35 and 4.36 illustrate the Hanle signal distortion by the laboratory field. They differ strongly. If the polarizations of the exciting and observed light are perpendicular to one another and the laboratory magnetic field coincides with one of these directions, then it will be manifested only in broadening of the signal; nevertheless, its form remains Lorentzian. If however, the polarization of the illuminating and observed light coincide, and the laboratory field is perpendicular to them, then the latter simply decreases the signal amplitude and does not change its width; however, if the laboratory field, polarization of the exciting and observation light are all parallel to one another, then the signal shape will be distorted; its amplitude will grow and the width may decrease.

The curves in the first columns of Figs. 4.35 and 4.36 illustrate the false Hanle signals: The polarization of excitation is parallel to the controlled magnetic field and in the absence of the laboratory magnetic field, the intensity does not depend on the strength $\mathcal{H}_z$. The profile of the false Hanle signals is quite varied and depends, for a weak laboratory magnetic field, on $\mathcal{H}_y/\Gamma$.

On Fig. 4.36 is depicted the experimentally measured intensity dependence of a radiation with a polarization along the z-axis on the field strength $\mathcal{H}_z$ [4.98]. The observation has been carried for the line $\lambda = 441.6$ nm of the cadmium ion, emitted by a direct current discharge in cadmium vapour. In the discharge tube, in which the temperature is maintained at a value of 290°C, the cadmium vapour pressure was 26.5×10^{-3} torr; the internal tube diameter was 3.5 mm and the discharge current 30 mA. An alignment was induced on the state that emitted this line in the discharge tube itself, with the axis of alignment parallel to the discharge tube axis. The theoretical dependence of $I(\mathcal{H}_z)$ is given by the solid line. Furthermore, the parameter $\mathcal{H}_y/\Gamma$ that enters this dependence, as well as the scale along the abscissa and ordinate, are fitted to yield the best coincidence with the experimental points. The deviation of the curves at large $\mathcal{H}_z$ is ascribed by the authors to the influence of the magnetic field on the characteristics of the discharge. The best agreement of the curves occurred when $\mathcal{H}_y/\Gamma = 1.5$. The perpendicular component of the magnetic field is known to be $\mathcal{H}_y = 0.45$ Oe, then $\Gamma = \mathcal{H}_y/1.5 = 0.3$ Oe, which upon the Landé factor $g = 1.2$ gives $\tau = 160$ ns. The coherence time of this level, measured from the "normal" Hanle signal is equal to $\tau = 140$ ns ($\Delta\mathcal{H} = 0.35$ Oe). The agreement between these numbers confirms the informativeness of the false Hanle signals. The curve depicted on Fig. 4.36 for $\mathcal{H}_y/\Gamma = 1.0$ enables one to judge the precision of such an estimation. In contrast, when the decay constant is known, the false Hanle signals can be used in order to assess the laboratory magnetic field (its transverse component).

From the above discussion it follows that precautionary measures (the degree of compensation of the transverse component of the magnetic field) which are sufficient for observing uniaxial alignment signals may turn out to be insufficient when the alignment is biaxial. Actually, for a uniaxial alignment, the observed signal is slightly distorted even by weak (up to $\mathcal{H}_y \approx \Gamma$) laboratory magnetic fields, as is seen from the two curves on the right-hand side of Figs. 4.34 and 4.35. Upon biaxial alignment, the observed signal consists of an admixture of the signal due to the laboratory field (or, more precisely, its component perpendicular to $\mathcal{H}_z$) and the false Hanle signal from the alignment with perpendicular axis. This is particularly important if the latter is greater in magnitude than the observed signal. An example is the observation of the radial alignment in a cylindrical discharge tube, to which the transverse laboratory field adds a false signal from the axial alignment. The latter signal is usually stronger than the former.

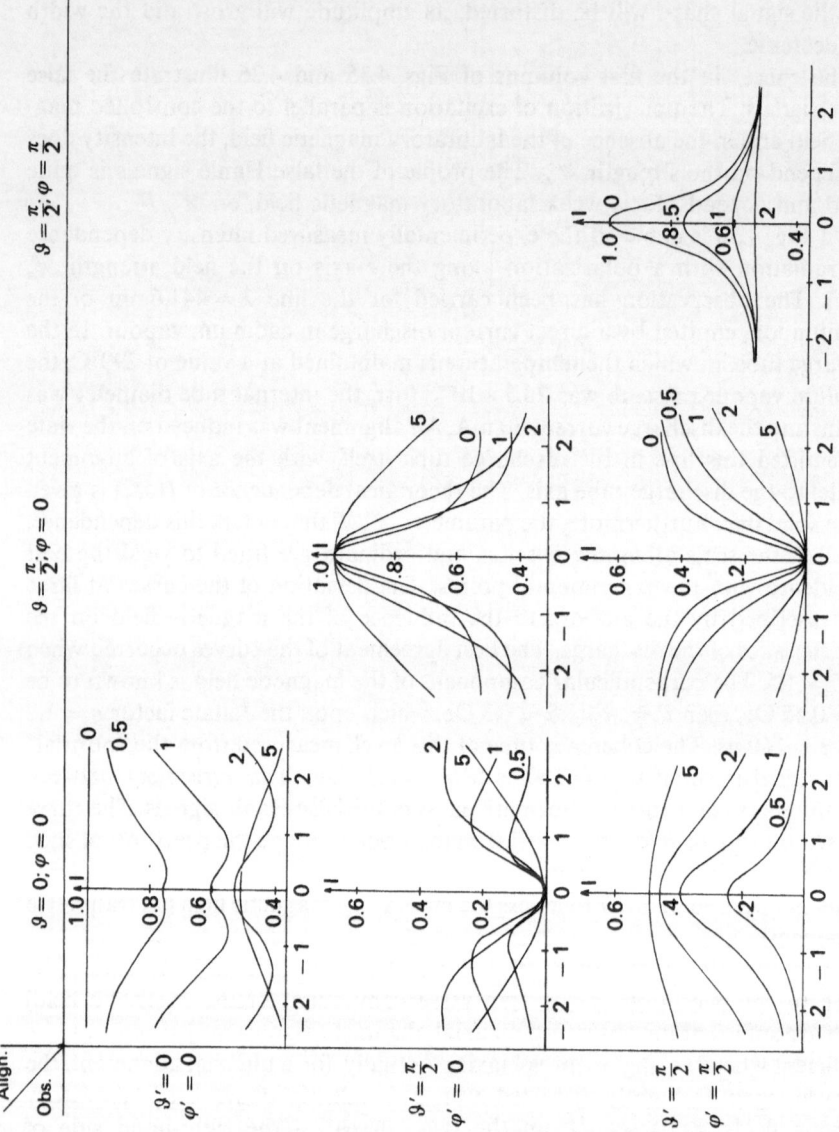

Fig. 4.34. The magnetic dependences of the light intensity for different angles between the direction on magnetic field (↑) alignment axis (‖) and observation polarization (⫴). The ordinate is the light intensity and the abscissa is H_z/Γ. The numbers on the curves are the parameters H_y/Γ

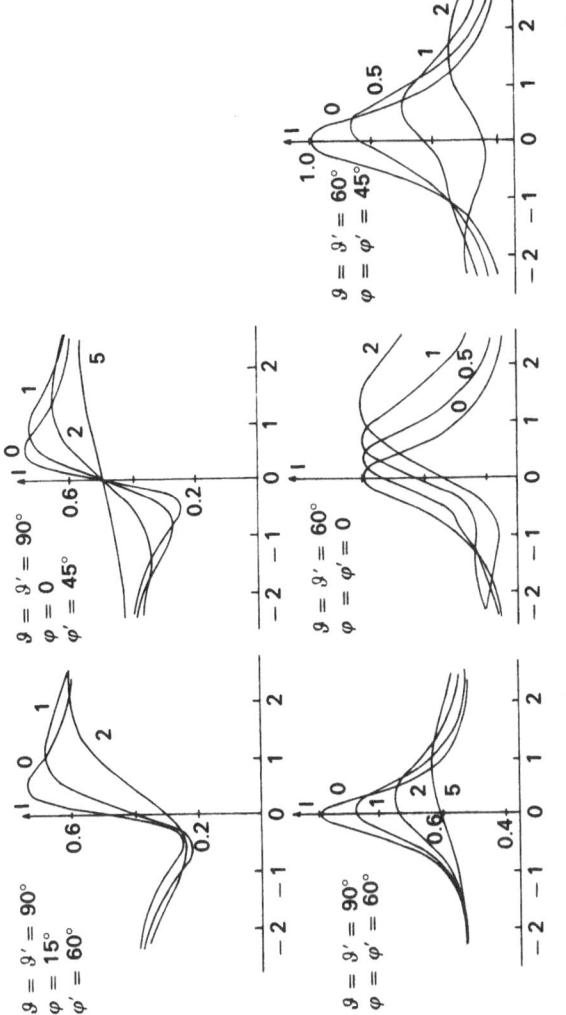

Fig. 4.35. As Fig. 4.34 but for other angles. The superscript on H_y is the value of angle $(-\phi)$. The z-axis is perpendicular to the other vectors

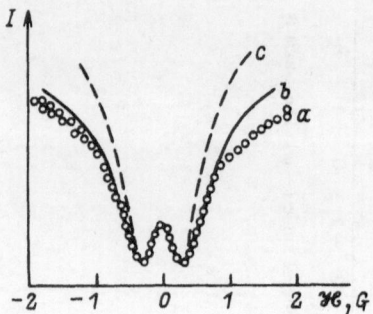

Fig. 4.36. The false Hanle signal observed on the cadmium line $\lambda = 441.5$ nm (a). (b) Calculation for $H_y/\Gamma = 1.5$; (c) calculation at $H_y/\Gamma = 1$

4.17 Spectral Content of the Exciting Light and Absorption Line Profile

In all measurement techniques that exploit level-crossing, one is interested in the position of the crossing signals in terms of the magnetic field scale and in their width. If all the crossing signals are well resolved from one another, as is the case in caesium [4.99, 100], then their relative magnitudes do not play any role. It is quite different when the signal is a complex curve due to the overlap of signals of each individual pair of the crossing levels. In this case the most reliable technique to extract the necessary information is to compare the experimental curve with the calculated one. The latter depends significantly on the assumed relative excitation rate of each of the hyperfine sublevels in the theoretical calculation. A consideration of the relative intensities of the hyperfine components was the decisive factor in achieving a high accuracy of determination of the hyperfine interaction constant of the level $^2P_{3/2}$ of sodium [4.101] and rubidium [4.102].

The other difficulty is to distinguish the spector of the exciting light from the white light that is related to the variation of the absorption lineshape of vapour in a magnetic field. Under the action of a magnetic field the absorption line splits and its components shift in opposite directions from the line centre. Provided the Zeeman splitting of a line is much smaller than the Doppler absorption linewidth or the width of the illuminating line, the absorption will remain constant. However, when the broader of these turns out to be comparable with the natural width in the same magnetic fields (which do not yet fully depolarize the fluorescence) the integral intensity of the fluorescence will change noticeably. This is because both the absorption lineshape and the absorption strength will change. As an example one can consider the Hanle effect on the level $^3p_4$, excited by laser radiation of wavelength 3.39 µm [4.103]. The Doppler linewidth of the absorption on this transition is close to the natural linewidth. The dependence of the observed fluorescence intensity on magnetic field (Fig. 4.37) is determined by two effects: by the Hanle signal, which dominates in the region $\mathcal{H} = 0$ and by

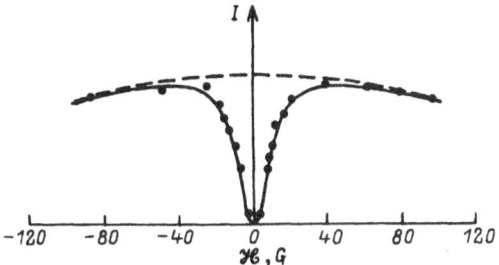

Fig. 4.37. The Hanle effect on the neon level $3p_4$ under exitation by neon line $\lambda = 3.39\ \mu m$

the decrease of the integral intensity, which is significant at the edges of the Hanle effect.

Flichtner et al. [4.104] observed anomalies in the Hanle signal on the level $^2P_{3/2}$ of rubidium-87 upon increasing the temperature of the cell and they ascribed them to the variation of absorption in the magnetic field. In order to take into account the error related to this, the authors described the variation of the excitation in a magnetic field by an empirical formula, whose parameters were fitted from the experiment.

The illumination of an atomic beam by monochromatic light can serve as a pictorial demonstration of the influence of the spectral content of a light on the Hanle signal profile. Let us assume that the frequency of the light coincides with the resonance frequency of the atom. Let us direct the incident light perpendicular to the atomic beam and we will change the external magnetic field, which is parallel to the atomic beam motion, in the vicinity of zero. In a magnetic field, the Zeeman sublevels change their energy and if the initial frequency of the illuminating light coincides with the frequency of the atomic transition, then upon applying a magnetic field this coincidence will be lifted. The frequency of the light wave will no longer be in resonance with the atomic transition and the absorption of the light will drop. If the excited line is broad, the number of exited atoms does not depend on magnetic field, and the fluorescence intensity variation is related to the interference effects only. When the exiting line is narrow, the integral intensity depends on the external field and, in this case, the observed signal is attributed to both effects.

In Chap. 3 we derived the expression for the density matrix of an ensemble of atoms excited by a monochromatic light (3.18)

$$C_n C_k^* = \frac{\mathscr{E}^2 (d \cdot e_\lambda)_{n0} (d \cdot e_\lambda)_{k0}^*}{\hbar^2 (\omega - \omega_n + i\Gamma/2)(\omega - \omega_k - i\Gamma/2)} \ .$$

We confine ourselves to the case of the presence of only two sublevels in the excited state and the simplest organization of the observation: illumination by a linearly polarized light and observation of light of the same polarization. We define

$$\omega_n = \omega_0 + \Omega, \qquad \omega_k = \omega_0 - \Omega \ ,$$

where Ω is the Larmour frequency. The intensity of the emitted light is

$$I = K \frac{\mathscr{E}^2}{\hbar^2} \left\{ \frac{|(\boldsymbol{d}\cdot\boldsymbol{e})_{no}|^4}{(\omega - \omega_0 - \Omega)^2 + \Gamma^2/4} + \frac{|(\boldsymbol{d}\cdot\boldsymbol{e})_{ko}|^4}{(\omega - \omega_0 + \Omega)^2 + \Gamma^2/4} \right.$$

$$\left. + \frac{2|(\boldsymbol{d}\cdot\boldsymbol{e})_{no}|^2|(\boldsymbol{d}\cdot\boldsymbol{e})_{ko}|^2[(\omega-\omega_0)^2 - \Omega^2 + \Gamma^2/4]}{[(\omega-\omega_0)^2 - \Omega^2 + \Gamma^2/4]^2 + \Gamma^2\Omega^2} \right\}.$$

The form of the signal will be highly simplified when, in the absence of a magnetic field, the frequency of the monochromatic light coincides with the frequency of the atomic transition, i.e. $\omega = \omega_0$. We then have:

$$I = K \frac{4\mathscr{E}^2}{\hbar^2\Gamma^2} \left\{ \frac{|(\boldsymbol{d}\cdot\boldsymbol{e})_{no}|^4 + |(\boldsymbol{d}\cdot\boldsymbol{e})_{ko}|^4}{1 + 4\Omega^2/\Gamma^2} \right.$$

$$\left. + \frac{2|(\boldsymbol{d}\cdot\boldsymbol{e})_{no}|^2|(\boldsymbol{d}\cdot\boldsymbol{e})_{ko}|^2(1 - 4\Omega^2/\Gamma^2)}{(1 + 4\Omega^2/\Gamma^2)^2} \right\}.$$

In this expression the first term is diagonal and, unlike the excitation by broad lines, the diagonal terms depend on the magnetic field. The form of the interference part of the signal will have the same shape as the signal profile for the cascaded resonance (Sect. 4.14). The signal shape will be very complex when the frequency of the illuminating light is detuned from the frequency of the atomic transition.

In Chap. 3 it has already been said that, if the absorbing atoms are in random motion and are distributed over wide velocity ranges such that the Doppler linewidth is far greater than the natural width, then the form of the signal will be the same as in the case of excitation by a broad line. This applies not only to the Hanle signal, but also to the other interference phenomena, such as the beat resonance and the parametric resonance.

4.18   Faraday Rotation

When light is transmitted through a vapour its properties will change not only as a result of the absorption process, but also due to the dispersion behaviour of the vapour. The dispersion is also a function of magnetic field and can therefore influence the signal profile in question.

When light is propagating through a vapour in a direction *perpendicular* to the magnetic field, the two linear polarizations parallel and perpendicular to the magnetic field will be the characteristic polarizations. The difference in the refractive indices is relatively small. However, it can be significant when observing along the magnetic field. The characteristic polarizations in a vapour for light propagating *parallel* to the magnetic field are the right- and left-hand circular polarizations. The difference in the refractive indices for these leads to

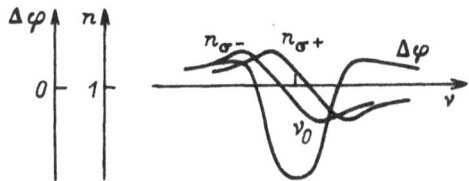

Fig. 4.38. Rotation of the plane of light
polarization as a function frequency separation from the spectral line centre

the rotation of the plane of polarization of linearly polarized light – the well-known Faraday effect. The rotation of the polarization plane is a function of the wavelength of the light. An example of this dependence in the vicinity of the spectral transition is given in Fig. 4.38. Conversion of the sign of rotation within the Doppler absorption line is known as the Macaluzo–Corbino effect.

For a low vapour density and thin vapour layer through which the observed fluorescence passes from the place of excitation to the output of the cell, the rotation of the polarization plane is small and its influence is not prominent. It is manifest when the absorption is appreciable. The radiation emerging from the cell contains all frequencies, but each of them, upon passing through the vapour layer, will experience its own additional rotation of the polarization plane, such that light from the central part of the line rotates in one direction and that from the edge of the line in the other. When observing the whole fluorescence light without spectral decomposition, the plane of its polarization may turn out to be rotated by some angle and this angle of rotation increases with the magnetic field. We note that different angles and signs of rotation for different spectral harmonics, besides rotating the polarization plane, lead to the depolarization of light. The polarization plane of fluorescence under the influence of Faraday rotation turns out to be rotated by an angle other than that of the true Hanle effect. This produces in the signal the same qualitative change as changing the duration of coherency, i.e. the measured lifetime will turn out to be different from the radiation lifetime.

Such is the qualitative picture of the influence of the Faraday rotation on the Hanle effect. The phenomenon is treated in more detail in the work of *Lecler* [4.105]. This gives the fundamental theory of the phenomenon; numerical computations are performed for the simplest special case, and the influence of Faraday rotation on the resonance fluorescence of natural mercury on the line 185 nm $(6\,^1S_0-6\,^1P_1)$ is investigated experimentally. The resonance cell contained three input windows for illumination by light and three output windows for observing the fluorescence (Fig. 4.39). This experiment showed that the signal form does not depend on the path length that the light has traversed in the cell from the entrance window to the place of observation; the signal becomes distorted when the path length from the location of excitation to the exit window is increased. The intensity difference for light with linear polarizations inclined at angles of 45° to the polarization plane of the exciting light (Fig. 4.39) was taken as the signal. Such a signal has an antisymmetric form.

The theoretical and experimental curves are in good agreement with one another and they plainly show that the signal drops in magnitude and decreases

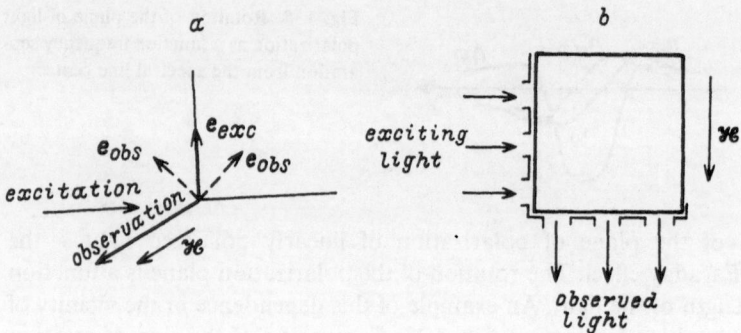

Fig. 4.39a, b. The Faraday rotation experiment. From [4.105]

in width upon increasing the thickness of the vapour layer traversed by the fluorescence light. For a qualitative assessment we present here data obtained in [4.78]: If one takes the signal width as the initial signal width at a mercury vapour concentration of $2 \cdot 10^{11}$ cm^{-3} and with a distance between the entrance window and exit window of 0.5 cm (τ of the excited state ~ 1 ns), then, when increasing this distance to 1.5 cm, the signal width will decrease by 10–20%. Faraday rotation will distort not only the Hanle signals, but also the level-crossing signals. Then, in the neighbourhood of the level-crossings the fluorescence is once more polarized. As passing through the vapour layer, both the degree of polarization as well as its direction will change. Consequently, the signal registered will also change. As far as the signal variation in one or the other direction from its maximum are different due to the different values of the magnetic field, one can anticipate the signal shift and the existence of its asymmetry.

4.19 Hanle Effect Due to Excitation That Is Random with Time

In the previous section we discussed the Hanle effect for time-independent and pulsed excitations with a specified choice of the observation time. One can also consider [4.106] a somewhat different experimental design: We assume now that the vapour is excited by some random sequence of light pulses with an interval between them that is significantly longer than the lifetime of the atom. We will register not the intensity at each given time instant, but will store the fluorescence energy until it decays out. Such measurements are practically quite feasible (in some respects they are even simpler, since in reality one cannot measure the "instantaneous" intensity because of the finite response time of any recording device). The calculation of the signal due to each of the pulses is not a difficult task and one obtains

$$\Delta I = K' \int_{t_1}^{t_2} \sum_{m,m'} G_{m'm}\sigma_{mm'}(t)\,dt \ ,$$

where $\sigma_{mm'}$ is given by the expression (3.95). Assuming the pulse to be single, one can change the limit of integration to infinity; after performing the integration we then find

$$\Delta I = K \sum_{m,m'} \frac{G_{m'm}F_{mm'}}{\Gamma + i\omega_{mm'}} \ .$$

If the pulses are so weak that they do not perturb the ground state and do not induce transitions from the excited states to the lower, i.e. if they remain within the limit of the weak field approximation, then one can assume the signals from different pulses of excitation to be independent. Even when the pulses succeed one another sufficiently rapidly that the next pulse excites atoms at an instant when the fluorescence excited by the previous pulse has not yet decayed, the observed signal will be a simple sum of signals from the individual pulses. For this to hold there is no restriction on the time interval between the pulses. One thing alone is important – the fluorescence energy excited by all pulses must have been received in full by the photodetector. Consequently, if, within some time interval, the system of atoms is illuminated by light of broader spectral content, arbitrarily changing with time, and if the signal is stored from the start of the illumination over a time greater than the time of illumination at least by a few τ, then the signal dependence on the magnetic field will have the customary profile for the level-crossing phenomenon.

4.20   Polarization of Atomic Fluorescence in a Flame

In the fluorescence technique of spectral analysis the solution to be analysed is injected into the flame of a burner. The flame is illuminated by light from a spectral lamp and the fluorescence intensity is measured. It has been observed [4.6] that the fluorescence resonance is polarized. The analysis was conducted in the flame of a hydrogen-air burner at atmospheric pressure. The cadmium solution was injected into the flame and the flame was illuminated by light from a resonance cadmium lamp through a polarizer, the radiation of the flame was also directed through a polarizer into a monochromator. The latter selected the cadmium line $\lambda = 228.8$ nm. The authors reported that they observed 100% polarization of this cadmium line. *Lombardi* and *Chenevier* [4.107] repeated the experiment [4.6]. In their experiment the polarization of this line did not exceed 35%. They showed that the polarization of the resonance fluorescence is produced by the alignment, but not by the instrumental error. They applied a magnetic field to the flame and recorded the dependence of the fluorescence

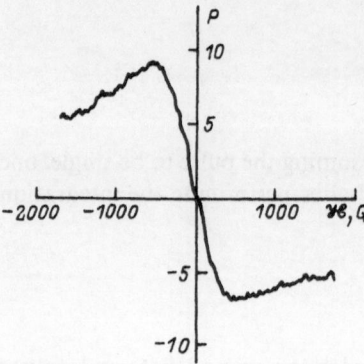

Fig. 4.40. The Hanle effect in a flame. From [4.107]

intensity on the magnetic field, as illustrated in Fig. 4.40. As is seen from the figure, strong fields of the order of kiloGauss destroy the polarization. The presence of an alignment at atmospheric pressure and the broad width of the Hanle signal are attributed to the very short lifetime of the emitting state formed under the influence of quenching collisions.

The experimental results, i.e. the width of the Hanle signal and the degree of polarization, were used by the authors to evaluate the cross-section of the quenching and depolarizing collisions ($\sigma^2_{depol} + \sigma^2_{qu} = 50\,\text{Å}^2$ and $\sigma^2_{qu} \leq 7\,\text{Å}^2$ for the gas, which the authors assumed to be a mixture of 30% water vapour and 70% molecular nitrogen).

4.21 Detection of the Polarization Moments by Radioactivity

The investigation of radioactive isotopes is complicated, above all due to the small available sizes of the samples. The technique of recording the orientation and alignment induced by optical radiation using the anisotropy of the radioactive decay was developed in approximately 1975 [4.108–112]. From the present-day point of view, the registration of radioactivity has no advantages with respect to sensitivity over the registration of atoms by means of optical emission: On the one hand, the radioactive emission is far more energetic than optical emission and the quantum efficiency of the detectors for the former is higher than for the latter. On the other hand, however, a nucleus can emit only a single radioactive particle, after which it transforms to another nucleus and will make no further contribution. When using intense illumination, an atom can emit several photons corresponding to one and the same atomic transition, such that the sensitivity of observation is much enhanced. However, a detailed comparison of these techniques would divert us from the problems we wish to treat. Therefore, we discuss this method in general terms as it is published in the literature. We will describe the method based on the example of its application to measure the isotope shift on spectral lines of radioactive isotopes of mercury ^{199M}Hg [4.108]. The hyperfine structure of the isotope of mercury is significantly broader than the Doppler width of the line $\lambda = 253.7$ nm. The authors of this work illuminated the sample of mercury vapour using a lamp containing the mercury isotope ^{204}Hg. A controlled magnetic field was applied to the lamp. The illuminating lines were split into three components, the frequencies of the 6-components depend on the magnetic field strength. Certain magnetic fields gave rise to coincidence between the frequencies of one of the Zeeman components of the illumination line and one of the hyperfine components of the mercury isotope under investigation. At the point of coincidence of these frequencies, an optical pumping mechanism is switched on and, under its influence, the mercury groud state is oriented or aligned. The presence of polarization moments indicates a deviation of the angular momentum distribution from spherical symmetry. The angular momentum of an atom with a hyperfine structure is determined by the electron orbit and by the nucleus of the atom. Since the electron orbit of an atom of mercury possesses no angular momentum in the ground state, the angular momentum of the atomic state is fully given by the nuclear moment. This way it is possible to align or orient the nucleus of atoms by optical means of excitation, thanks to the interaction of the electron shell with the nucleus. In the case of radioactive nuclei, their alignment is accompanied by an anisotropy in the radioactive emission. In such a way, the measured intensity difference in the radioactive emission in the directions parallel and perpendicular to the polarization of the illuminating light (or direction of the illumination) enables one to observe the coincidence of the resonance frequencies of the ^{204}Hg atom in a magnetic field and the hyperfine transition in the examined sample. On the left-hand side of Fig. 4.41 are sketched the levels of ^{204}Hg the hfs centre of gravity ^{199M}Hg and at the top right are the Zeeman sublevels of ^{204}Hg in a magnetic field and the hyperfine components ^{199M}Hg. The lower part shows the experimental anisotropy of the radioactive radiation as a function of the magnetic field. The signal peak corresponds to the coincidence of the frequency components ^{204}Hg and ^{199M}Hg. These measurements, excluding the isotropic shifts, were used to determine the coefficient of quadrupole interaction and from this the nuclear quadrupole moment.

The nuclear properties of the radioactive isotopes of the alkali metals have been determined by a similar technique [4.109, 110]. Via optical pumping, circularly polarized illumination causes, as a rule, the orientation of the ground state. This orientation affects the distribution of the nuclear angular momenta. The latter is determined from the anisotropy of β radiation. When a rf field of the corresponding frequency is applied, transitions will take place between the hyperfine levels of the Zeeman components and the orientation decreases. The point of coincidence of the frequency of the rf field with the frequency of the interatomic transitions is detected by the decrease of the anisotropy of the β decay. The resonance transition frequencies are the main information required to determine the nuclear constants.

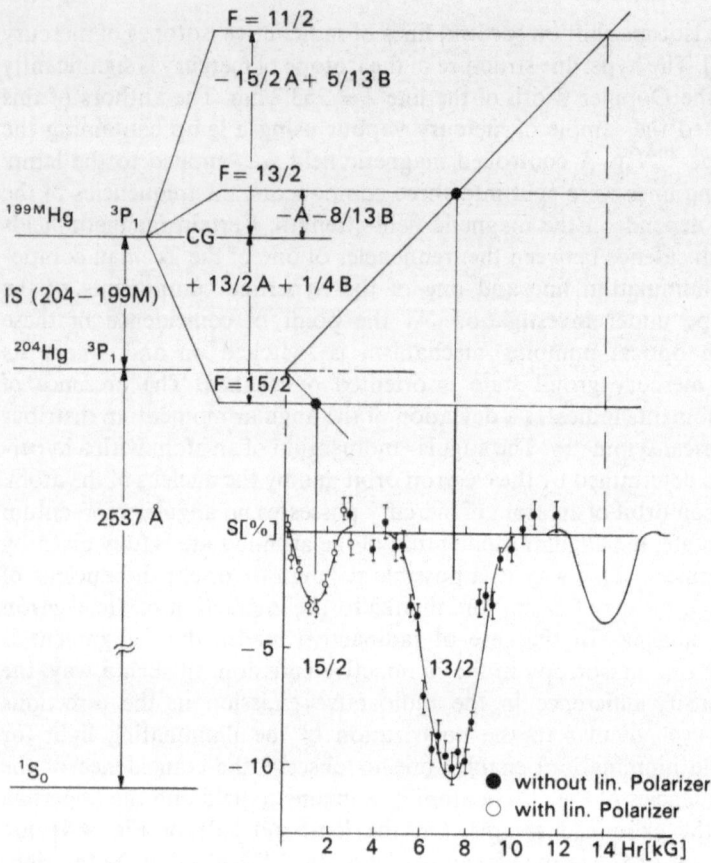

Fig. 4.41. Zeeman scanning of ^{199M}Hg. The top of the figure shows the $^3P_1$ state of ^{199M}Hg and ^{204}Hg. The latter is split into its Zeeman sublevels. The anisotropy obtained in the M4 transition is shown at the bottom of the figure as a function of the magnetic field over the ^{204}Hg spectral lamp

4.22 Use of the Polarization Moments for Improving the Accuracy of Nonlinear Spectroscopic Techniques

The production of polarization moments is used in nonlinear laser spectroscopy in order to increase sensitivity. We recap here on the essence of saturation absorption techniques. An intense monochromatic laser radiation interacts with atoms or molecules possessing "resonance" with the frequency of the laser radiation. The resonance will occur only for the portion of atoms with a specified ray velocity (i.e. with a given velocity projection on the ray direction). Under the influence of the radiation, atoms leave the lower level and make a transition to the excited state. From here they reach, via various processes, states which do not interact with the laser radiation. A "dip" will thus arise in the velocity distribution of atoms. Its width is determined by the natural linewidth. Frequency tunable probe radiation can detect this distribution. If, as probe radiation, one uses radiation from the very same laser, but propagating in the opposite direction to the saturating beam, then the absorption decrease will arise at the laser frequency, coinciding with the resonance frequency of the stationary atoms. This permits one to identify the resonance frequency of atoms with high precision, certainly higher than is achieved by measuring the Doppler line centre.

The accuracy of the measurement is ultimately limited by noise, including shot noise. The latter is proportional to the square root of the light intensity. The induction of polarization moments allows one to reduce shot noise. We will demonstrate the organization of the saturation absorption signal by means of the work reported in [4.113], in which the wavelength $H_\alpha D_1$, of the components $(2S_{1/2}-3P_{1/2})$ was measured with high precision. A single frequency dye laser (Coherent Inc. Model 599, Rhodamin 101 dye), pumped by the line $\lambda = 568.2$ nm of the krypton ion laser, generated a power of up to 50 mW near 656 nm with an approximate bandwidth of 1 MHz. A fraction of the laser radiation was divided into the probe beam (0.5 mW) and saturating beam (15 mW), which were directed in almost opposite directions in the positive column of a discharge of 55 cm length. The beams intersect at an angle of 1–5 millirad, and the absorption near the resonance was approximately 40%. The telescopic system expanded the beam to 7 mm in order to avoid the power shift. The probe beam was passed through linear polarizers, placed at the input and output of the cell, which were misaligned by 2° from the exact crossing position. In such a design, less than 0.05% of the probe beam intensity will reach the photodetector. A circularly polarized saturating beam will induce slight gyroscopic birefringence in the sample gas, which could rotate the probe beam polarization. The light flux reaching the detector is a very sensitive indicator of any variation in the polarization. A resonance signal of the dispersion type was observed at the centre of the Doppler-broadened line, where both beams interacted with the same atoms. (A further subject of this investigation was a measurement of the frequency difference of the laser and the reference radiation).

Figure 4.42 assists one in understanding the reason behind the rotation of the polarization plane and its dependence on frequency. The upper part of the figure shows the spectral line of absorption with a dip at the frequency of the saturating laser beam. However, within this region the laser beam has removed only those atoms that are capable of absorbing radiation not only of the laser frequency, but also the laser polarization and, as a consequence, the refractive indices of the right and left circularly polarized waves is different. This difference is depicted in the lower part of Fig. 4.42 by the solid line. The broken line shows the refractive index, which is for both circular polarizations of the unperturbed Doppler line. The authors recorded the derivative of this difference, by modulating the frequency of the laser radiation with a modulation depth of 8 MHz and

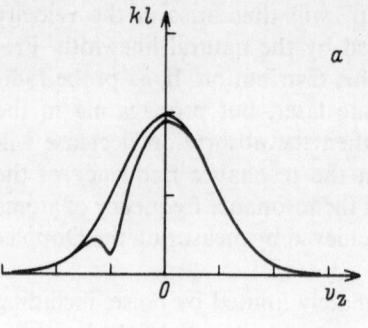

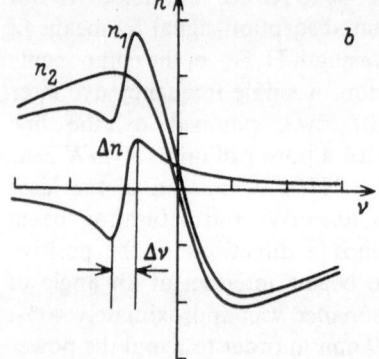

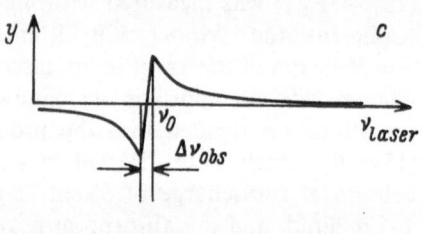

Fig. 4.42. Explanation of the origins of the signals reported in [4.113]. (a) The form of the absorption line in an intense light field; (b) the refraction coefficient: n_1 for one of the circular polarizations and Δn the difference between the refraction coefficients of the two circular polarizations. (c) The angle of rotation of the polarization for the wave propagating in the direction opposite to the saturating beam. Δv is the frequency width of the anomaly and Δv_{obs} the width of observed effect: $\Delta v_{obs} = \frac{1}{2} \Delta v$

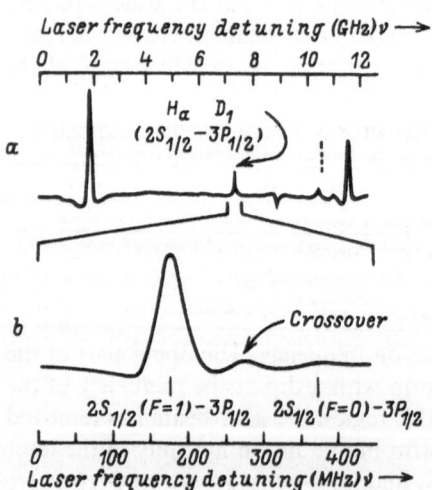

Fig. 4.43. (a) Polarization spectrum of H_α. The dashed line indicates the position of the iodine reference line. (b) Polarization spectrum of the $H_\alpha D_1$ ($2S_{1/2}-3P_{1/2}$) fine-structure component

modulation frequency of 2.3 kHz, using a phase detector. The registered signal is illustrated on Fig. 4.43. Its width is half that of the function $\Delta n_\pm$ on Fig. 4.42, since the frequency of the probe beam and the anomaly in the refractive index are moving towards one another along the line contour.

4.23 Conclusion

In this chapter, we have attempted to characterize experimental work on interference phenomena, concentrating on effects that either hinder information extraction or open new possibilities. Therefore, the content of this chapter is necessarily somewhat arbitrary and the picture is far from complete.

The shortcomings of our presentation can be compensated to some extent by reference to [4.11, 114–120], as well as the following (far from complete) bibliographical works of recent years: The works [4.121–131] are devoted to the Hanle effect; of these, [4.130, 131] are considered the Hanle effect in the ground state; level anti-crossing is discussed in [4.132, 133]; the quantum beat is treated in the works [4.134–138]; the influence of cascaded transitions is discussed in [4.139–142]. In Ref. [4.143] level-crossing in forward scattered light is investigated. Alignment detection by observation of the optogalvanic effect is reviewed in [4.144]. A few other studies connected with the topic of this chapter are reported in [4.145–158].

5. Calculation of Interference Signals

In Chap. 3 the description of interference phenomena was presented in quantum mechanical language. An interference signal is recorded as a change of parameters of the observed light as a function of the external field strength. The form of the signal depends on the atomic characteristics and polarization of the radiation. The atomic characteristics are represented by the quantities d – the matrix elements – and Ω – the energy gap between the interfering levels. Both these groups of quantities depend on the external fields. The explicit dependences will be discussed at the end of this chapter. This knowledge permits one to relate the observed interference signals to the fundamental atomic characteristics, such as the constant of magnetic dipole interaction between an electron orbital and a nucleus, from which the nuclear magnetic moment and the constant of quadrupole interaction are calculated, and thence the nuclear quadrupole moment. However, if the relationship between the signal profile and the fundamental quantity is known, then the latter can be derived from the interference signals.

5.1 An Atom in a Magnetic Field

An atom will interact with a magnetic field provide it itself possesses a magnetic moment. The atomic magnetic moment is formed from the orbital and spin momenta of the electrons. (The nuclear magnetic moment is quite small and we will not take into consideration for the time being.)

The magnetic moment is directly linked to the mechanical momentum and, due to the fact that the electron charge is negative, the directions of the momentum vector and the magnetic moment are antiparallel

$$\mu_l = -\mu_0 L, \qquad \mu_s = -2\mu_0 S . \tag{5.1}$$

μ_0 is the well known Bohr magneton:

$$\mu_0 = \frac{e\hbar}{2mc}$$

(in some books the minus sign is included in the quantity μ_0, i.e., the value μ_0 is a negative quantity and in the formula (5.1) the minus sign does not appear).

Defining the atomic magnetic moment

$$\mu_{at} = \mu_{el} = \mu_l + \mu_s \tag{5.2}$$

through its total mechanical momentum[1]

$$J = L + S$$

we get

$$\mu_{at} = -\mu_0 g_J J \ . \tag{5.3}$$

As is well known g_J is called the Landé factor. Transferring from vectors to operators, we can find the numerical value of g_J: Considering (5.1) and (5.2), we have

$$\mu_0 \hat{L} + 2\mu_0 \hat{S} = \mu_0 g_J \hat{J} \ ,$$

and multiplying both sides of this equation by $\hat{J}$, we get

$$\hat{L} \cdot \hat{J} + 2\hat{S} \cdot \hat{J} = g_J \hat{J}^2 \ . \tag{5.4}$$

Making use of the obvious equalities

$$\hat{S}^2 = (\hat{J} - \hat{L})^2 = \hat{L}^2 + \hat{J}^2 - 2\hat{J} \cdot \hat{L} \ ,$$
$$\hat{L}^2 = (\hat{J} - \hat{S})^2 = \hat{J}^2 + \hat{S}^2 - 2\hat{J} \cdot \hat{S} \ , \tag{5.5}$$

and replacing them by their eigenvalues:

$$\hat{J}^2 \to J(J+1), \qquad \hat{L}^2 \to l(l+1), \qquad \hat{S}^2 \to S(S+1)$$

from (5.4) we obtain

$$g_J = \frac{3J(J+1) - l(l+1) + S(S+1)}{2J(J+1)} \ . \tag{5.6}$$

The interaction energy of the magnetic moment with a magnetic field of strength $\mathscr{H}$ is

$$\Delta W = -\mu \cdot \mathscr{H} \tag{5.7}$$

and the corresponding energy operator is

$$\hat{H}_M = -\hat{\mu} \cdot \hat{\mathscr{H}} = \mu_0 g_J \hat{J} \cdot \hat{\mathscr{H}} \ . \tag{5.8}$$

The exact form of the operator $\hat{H}_M$ depends on the choice of coordinate axes. Let

---

[1] Strictly speaking, the quantities J, l and S (as well as I) are not mechanical momenta, but the quantum numbers of these momenta. However, because of the proportionality between these quantities, it is accepted practice to refer to them as momenta.

us choose, as is usually done for convenience, the z-axis, called the axis of quantization, along the magnetic field $\mathscr{H}$. Hence, for weak fields $\mathscr{H}$, in which one can assume J to be a good quantum number,

$$\hat{H}_{M} = \mu_{0} g_{J} \mathscr{H} \hat{J}_{z} , \tag{5.9}$$

where $\hat{J}_{z}$ is the projection of $\hat{J}$ on the z-axis. The total Hamiltonian of an atom in a magnetic field contains the Hamiltonian of a free atom $\hat{H}_{0}$:

$$\hat{H} = \hat{H}_{0} + \hat{H}_{M} . \tag{5.10}$$

In the absence of a magnetic field, there exist $2J + 1$ different wavefunctions corresponding to the operator $\hat{H} = \hat{H}_{0}$. These describe the degenerate states with a given value of J and corresponding to different orientations of the momentum J in space, or, in other words, to different projections of the vector J on the z-axis.

The wavefunction of an atom can be represented as the product of individual probability amplitudes that depend on the distance from the nucleus r and on the angles θ and ϕ:

$$\Psi = \Psi(r) \Psi(\theta, \phi) .$$

The operator $\hat{H}_{0}$ acts only on the radial part of the function. Indeed, the atom's energy cannot depend on its orientation in space, provided it is isotropic, i.e. does not contain any field. The operator $\hat{H}_{M}$, in contrast, acts only on the angular component of the wavefunction $\Psi(\theta, \phi)$. In reality, it does not change the distance between the electron and the nucleus (from the classical point of view), but the interaction with the magnetic field depends on the relative direction of $\mathscr{H}$ and the magnetic moment μ_{at}, i.e. exactly on the angles θ and ϕ. Therefore, in the problem under consideration, one can neglect the radial part of the function and in determining the interaction of an atom with an external magnetic field restrict oneself to the angular components, which, as is known, are generalized spherical functions $\Psi_{JM}(\theta, \phi)$ [5.1].

Inasmuch as these functions are completely determined by the quantum numbers J and M, we will use the expression Ψ_{JM}. The quantum number M is the eigenvalue of the operator $\hat{J}_{z}$:

$$\hat{J}_{z} \Psi_{JM} = M \Psi_{JM} . \tag{5.11}$$

The operator of the magnetic interaction has been defined through the total atomic moment (5.2). Taking into account the equation (5.3), we easily find the eigenvalues of the magnetic interaction operator

$$\hat{H}_{M} \Psi_{JM} = \mu_{0} g_{J} \mathscr{H} M \Psi_{JM} , \tag{5.12}$$

that is,

$$\Delta W = \mu_{0} g_{J} \mathscr{H} M . \tag{5.13}$$

The last expression is the well-known formula for the Zeeman splitting: the degeneracy with respect to the quantum numbers is lifted, the splitting is linear with respect to the magnetic field strength $\mathscr{H}$ and the levels are equally spaced.

5.2 The Hyperfine Structure

The hyperfine structure of atomic levels results from the interaction between the magnetic moment of the electron orbital and the magnetic moment of the nucleus. The electric quadrupole interaction of the orbitals with the nucleus also contributes to the hyperfine structure. This interaction is somewhat weaker than the magnetic dipole interaction and it is manifest in atomic states with moments greater than 1/2.

The nuclear magnetic moment can be written as:

$$\mu_I = \mu_0 g_I I \ . \tag{5.14}$$

Here I is the nuclear spin and g_I is a constant. As a rule, this quantity is positive; however, there are some atoms for which g_I is negative (for example O^{17}, K^{40} and He^3). For such atoms the magnetic moment is in the opposite direction to the nuclear spin I.

The nuclear magnetic moment is three orders of magnitude smaller than the electronic moments. For measuring the nuclear magnetic moment one can use either the Bohr magneton or the nuclear magneton

$$\mu_{nuc} = e\hbar/2Mc \ , \tag{5.15}$$

where M is the proton mass. The relationship between these units is obvious $\mu_{nuc} = (m/M)\mu_0$.

If the magnetic moment is calculated in terms of the Bohr magneton, then the g factor will be denoted by g_I:

$$\mu_I = \mu_0 g_I I \ , \tag{5.16}$$

and if in terms of the nuclear magneton, then by

$$\mu_I = \mu_{nuc} g_I' I \ ,$$

where of course

$$\frac{g_I'}{g_I} = \frac{M}{m} \ . \tag{5.17}$$

The additional energy resulting from the interaction of the two magnetic moments μ_{el} and μ_I can be written as

$$\Delta W = -K(\mu_{el} \cdot \mu_I) \ ,$$

and considering the expressions (5.3) and (5.14) we find

$$\Delta W = A \mathbf{I} \cdot \mathbf{J} .\tag{5.18}$$

The hyperfine interaction operator is found, in accordance with the rules of quantum mechanics, by replacing the vectors with the corresponding operators

$$\hat{H}_{\text{hfs}} = A \hat{\mathbf{I}} \cdot \hat{\mathbf{J}},\tag{5.19}$$

where A is the same constant as in (5.18). It is called the magnetic dipole interaction constant. The total atomic Hamiltonian can be written

$$\hat{H} = \hat{H}_0 + A \hat{\mathbf{I}} \cdot \hat{\mathbf{J}}.\tag{5.20}$$

Let us denote the eigenfunction of the total Hamiltonian $\hat{H}$ by Ψ. The sum of the moments $\mathbf{I}$ and $\mathbf{J}$ form the total mechanical momentum of an atom $\mathbf{F}$:

$$\mathbf{J} + \mathbf{I} = \mathbf{F}\tag{5.21}$$

$$\hat{\mathbf{F}} = \hat{\mathbf{J}} + \hat{\mathbf{I}} .\tag{5.22}$$

In order to find the operator $\hat{\mathbf{I}} \cdot \hat{\mathbf{J}}$ we will utilize the following relation (as in Sect. 5.1):

$$\hat{\mathbf{F}}^2 = \hat{\mathbf{J}}^2 + \hat{\mathbf{I}}^2 + 2\hat{\mathbf{I}} \cdot \hat{\mathbf{J}},\tag{5.23}$$

from which

$$A \hat{\mathbf{I}} \cdot \hat{\mathbf{J}} = A \frac{\hat{\mathbf{F}}^2 - \hat{\mathbf{J}}^2 - \hat{\mathbf{I}}^2}{2} .\tag{5.24}$$

Consequently, the eigenfunctions of the operator $A \hat{\mathbf{I}} \cdot \hat{\mathbf{J}}$ must simultaneously be eigenfunctions of the three operators $\hat{\mathbf{F}}^2, \hat{\mathbf{J}}^2$ and $\hat{\mathbf{I}}^2$. As is known, a state with a given angular momentum F describes the eigenfunctions of an operator of the square of the momentum $\hat{\mathbf{F}}^2$ with the eigenvalues $F(F + 1)$:

$$\hat{\mathbf{F}}^2 \Psi = F(F + 1) \Psi ,$$

$$\hat{\mathbf{J}}^2 \Psi = J(J + 1) \Psi ,\tag{5.25}$$

$$\hat{\mathbf{I}}^2 \Psi = I(I + 1) \Psi .$$

In view of this it is quite simple to find the eigenvalue of the total Hamiltonian

$$\hat{H}\Psi = \left(E_0 + A \frac{F(F + 1) - J(J + 1) - I(I + 1)}{2} \right) \Psi\tag{5.26}$$

or, in the notation frequently adopted,

$$\hat{H}\Psi = \left(E_0 + \frac{AC}{2} \right) \Psi .$$

From (5.26) one sees that a state with given values of J and I splits into a series of levels, called hyperfine levels, each of which is characterized by its own value of F. The number of sublevels is equal to $2I + 1$ when $J > I$ and $2J + 1$ when $J < I$.

Equation (5.26) also yields the rule of intervals; the distances $\Delta W_{F_k - F_{k-1}} : \Delta W_{F_{k-1} - F_{k-2}} : \ldots$ between the sublevels are related to one another by

$$F_{\max} : (F_{\max} - 1) : \cdots : (F_{\min} + 1) \ .$$

Thus the energies of the states are determined. Let us now find the magnetic moment of an atom whose nucleus possesses a spin I. It is given by the sum of moments of the nucleus μ_I and electron μ_e

$$\mu_F = \mu_l + \mu_s + \mu_I = \mu_e + \mu_I \ .$$

Taking into account the relationship between the magnetic moments and the mechanical momenta, we obtain

$$\mu_F = -\mu_0 g_J J + \mu_0 g_I I \ .$$

On the other hand, the magnetic moment μ_F is related to the mechanical momentum of the atom F by

$$\mu_F = -\mu_0 g_F F \ ,$$

where g_F is simply a coefficient. We find it from the equation

$$-\mu_0 g_F F = -\mu_0 g_J J + \mu_0 g_I I \ ,$$

both sides of which are multiplied in a scalar form by F

$$g_F F^2 = g_J J \cdot F - g_I I \cdot F \ .$$

We then proceed to the operator form of the expression

$$g_F \hat{F}^2 = g_J \hat{J} \cdot \hat{F} - g_I \hat{I} \cdot \hat{F} \ .$$

Using the equation (5.22) and performing the transformation as in (5.23, 24) for determining the product of operators, we get

$$g_F = \frac{F(F + 1) + J(J + 1) - I(I + 1)}{2F(F + 1)} g_J$$

$$- \frac{F(F + 1) - J(J + 1) + I(I + 1)}{2F(F + 1)} g_I \ . \tag{5.27}$$

It is known from the theory of moments [5.1, 2] that the eigenfunctions of the operators of the squared momenta are simultaneously the eigenfunctions of the operator of the momentum projection onto the axis of quantization

$$\hat{F}_z \Psi = m_F \Psi \ . \tag{5.28}$$

The eigenvalues m_F of the operator $\hat{F}_z$ take values from F to $-F$ in unit intervals and are called the magnetic quantum numbers of the hyperfine sublevels.

The hyperfine splitting varies from some ten thousand MHz down to very small values. The classic technique for investigating hyperfine structures involves spectral decomposition by devices of high resolving power, in particular by the Fabry–Perot interferometer. Experimental studies have shown that the interval rule is not always satisfied. The deviation is explained by yet another interaction of the valence electron with the nucleus, related in this case to the charge distribution within the nuclear volume. If this distribution deviates from spherical symmetry, then the nucleus will have an electric quadrupole moment. But before treating the quadrupole interaction, we will discuss the relation between the nuclear properties and the magnetic dipole interaction constant A. This relationship enables one to find the nuclear magnetic moment from the experimentally determined quantity A.

5.3   The Magnetic Dipole Interaction Constant

In order to determine the relationship between the nuclear magnetic moment μ_I and the constant A let us examine the motion of an electron in a nuclear field following the scheme used by *Landau* and *Lifshitz* [5.3].

The electron motion is given by the equation

$$\hat{H}\Psi = \left[\frac{1}{2m} \left(\hat{p} + \frac{e}{c} \mathscr{A} \right)^2 + \frac{e\hbar}{mc} \hat{S} \cdot \mathscr{H} + U(r) \right] \Psi$$

$$= \left[\frac{1}{2m} \hat{p}^2 + \frac{e}{mc} \hat{p} \cdot \mathscr{A} + \frac{e^2}{2mc^2} \mathscr{A}^2 + \frac{e\hbar}{mc} \hat{S} \cdot \mathscr{H} + U(r) \right] \Psi \ , \tag{5.29}$$

where $\mathscr{A}$ is the magnetic vector potential, $\mathscr{H}$ the magnetic field strength at the position of the electron due to the nucleus, $\hat{S}$ the spin operator and $\hat{p}$ the momentum operator.

The term $\frac{1}{2}\hat{p}^2/m$ in the operator describes the free motion of an electron (in classical mechanics, this corresponds to an inertial motion). The term $U(r)$ is the electron potential energy due to fields of non-magnetic origin. The remaining terms constitute the operator of the interaction between nuclear and electronic magnetic moments. Moreover, the term

$$\hat{H}_1 = \frac{e}{mc} \hat{p} \cdot \hat{\mathscr{A}} + \frac{e^2}{2mc^2} \hat{\mathscr{A}}^2 \tag{5.30}$$

is responsible for the interaction of the nucleus with the orbital magnetic moment μ_l, and

$$\hat{H}_2 = \frac{eh}{mc} \hat{S} \cdot \mathcal{H} \tag{5.31}$$

describes the interaction with the spin magnetic moment of the electron.

The expression (5.30), in turn, contains two terms, of which the second is the diamagnetic response of the electron orbital to the magnetic field of the nucleus which is quadratic in $\mathcal{A}$. Because this interaction is weak, the second term is small compared to the first and one can neglect it. In this way, the Hamiltonian of the hyperfine magnetic dipole interaction contains altogether two terms

$$\hat{H} = \frac{e}{mc} [\hat{p} \cdot \mathcal{A} + h(\hat{S} \cdot \mathcal{H})] \; . \tag{5.32}$$

As can be inferred from electrodynamics [5.4],

$$\mathcal{A} = \frac{\mu_I \times r}{r^3} \; ; \qquad \mathcal{H} = \frac{3r(\mu_I \cdot r) - \mu_I r^2}{r^5} \; , \tag{5.33}$$

where r is the distance between the nucleus and the electron. Substituting (5.33) into (5.32), we find

$$\hat{H} = \frac{eh}{mc} \frac{\hat{p} \cdot \hat{\mu}_I \times \hat{r}}{r^3} + \frac{eh}{mc} \left(\frac{3\hat{r}(\hat{\mu}_I \cdot \hat{r}) - \hat{\mu}_I r^2}{r^5} \right) \cdot \hat{S} \; . \tag{5.34}$$

Let us use the known relationship $A \cdot B \times C = A \times B \cdot C$, together with the definition

$$\hat{L} = \frac{1}{h} \hat{r} \times \hat{p}$$

to rewrite the expression (5.34) in the form

$$\hat{H} = \frac{2\mu_0^2 g_I \hat{I}}{r^3} \cdot [\hat{L} + 3(\hat{S} \cdot n')n' - \hat{S}] \; , \tag{5.35}$$

where

$$n' = \frac{\hat{r}}{|r|} \; ; \qquad \mu_0 = \frac{eh}{2mc} \; ; \qquad \hat{\mu}_I = \mu_0 g_I \hat{I} \; .$$

The quantity $1/r^3$ depends on the state in question, i.e. on all of the quantum numbers, including the angular momentum J. If we determine all the quantum numbers J, l, I and S, then we can average over r^{-3}, and in the formula (5.35) replace r^{-3} by the factor $\overline{r^{-3}}$. Besides averaging over r^{-3} we also average over the direction of the vector r, i.e. with respect to n'. After averaging, the operator

(5.35) acquires the following form:

$$\hat{H} = 2\mu_0^2 g_I \left(\frac{1}{r^3}\right) \hat{I} \cdot \left[\hat{L} - \frac{6(\hat{S} \cdot \hat{L})\hat{L} - 2\hat{L}^2\hat{S}}{(2l-1)(2l+3)}\right] . \qquad (5.36)$$

Comparing the formulae (5.36) and (5.19) we see that

$$\hat{H} = 2\mu_0^2 g_I \overline{r^{-3}} \hat{I} \cdot g\hat{J} .$$

We find the quantity g in the following way: we multiply by the operator $\hat{J}$ both sides of the equation

$$\left[L - \frac{6(\hat{S} \cdot \hat{L})\hat{L} - 2\hat{L}^2\hat{S}}{(2l-1)(2l+3)}\right] = g\hat{J} .$$

We then replace the operators $(\hat{J} \cdot \hat{S})$, $(\hat{J} \cdot \hat{L})$, $\hat{L}^2$ and $\hat{J}^2$ by their eigenvalues

$$(\hat{S} \cdot \hat{J}) \rightarrow \frac{J(J+1) + S(S+1) - l(l+1)}{2}$$

$$(\hat{L} \cdot \hat{J}) \rightarrow \frac{J(J+1) + l(l+1) - S(S+1)}{2}$$

$$\hat{L}^2 \rightarrow l(l+1)$$

$$\hat{J}^2 \rightarrow J(J+1)$$

$$\hat{S}^2 \rightarrow S(S+1)$$

and

$$\hat{L} \cdot \hat{S} = \hat{J} \cdot \hat{L} - \hat{L}^2 ,$$

taking into account, that $S = 1/2$. This yields

$$g = \frac{l(l+1)}{J(J+1)} .$$

It is now clear that the magnetic dipole interaction constant A is equal to

$$A = \frac{2\mu_0^2 g_I \overline{r^{-3}} l(l+1)}{J(J+1)} . \qquad (5.37)$$

A somewhat stricter treatment that takes relativistic corrections into account leads to the appearance of an additional factor f, whose value is quite close to unity:

$$A = \frac{2\mu_0^2 g_I \overline{r^{-3}} l(l+1)}{J(J+1)} f . \qquad (5.38)$$

This form of the magnetic dipole interaction constant of the nuclear with the electron orbital moment is also derived by *Kopferman* [5.5].

5.4   Quadrupole Interaction Between a Nucleus and an Electron Shell

The quadrupole interaction caused by the non-spherical charge distribution in the volume of a nucleus has already been mentioned above. In the presence of a magnetically split hyperfine structure, the quadrupole moment merely shifts the energy of the sublevels of given moments F and does not result in additional splitting.

It is clear that the nuclear quadrupole moment can interact with an electron state that does not possess spherical symmetry, i.e. an electron shell that itself possesses a quadrupole moment. For such a state the total moment J has to be greater than 1/2.

Let us turn to the determination of the electronic energy level shift in an atom under the influence of a nuclear quadrupole moment. For this purpose it is necessary to find the additional interaction energy of the charges of the nucleus with the electron shell in terms of the deviation from spherical symmetry. The Coulomb interaction between point charges q_i and q_k has the well-known form

$$W_{ik} = \frac{q_i q_k}{r_{ik}}$$

or

$$W_{ik} = \phi(r_{ik}, q_i)q_k = \phi(r_{ik}, q_k)q_i \; , \tag{5.39}$$

where $\phi(r_{ik}, q_i)$ is the potential produced by the charge q_i at the location of the charge q_k. If the charges are distributed somehow in space, then the interaction energy between the charge distribution q_i and the charge distribution q_k is determined by integrating over W. We introduce the notation with help of Fig. 5.1. We place the origin of the coordinate system at the charge centre of the nucleus, given by analogy with the definition of centre of mass. The vector I determines the nuclear moment. The charge density within the nucleus is denoted by ϱ and its distribution is not so far specified. r_i and r_k are points in space and r_{ik} is the distance between the points i and k. The potential $d\phi$ at point k produced by an elemental charge ϱdV located at the point i is equal to

$$d\phi_k = \frac{\varrho(r_i)\,dV}{r_{ik}} \; . \tag{5.40}$$

The point k can equally well be inside the nucleus. The point i is not restricted to the nuclear volume, but when it is outside one has $\varrho = 0$. The total potential ϕ at

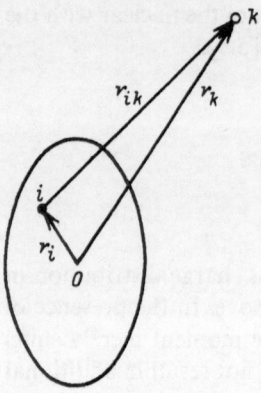

Fig. 5.1. Explanation of the definition of potential at a point outside a non-spherical nucleus

the point k is thus

$$\phi_k = \int \frac{\varrho(r_i)\,dV}{r_{ik}} . \tag{5.41}$$

The integration extends over all space, but since $\varrho(r) = 0$ for r_i outside the nucleus, the integral can be taken over the nuclear volume. If the entire charge of the electron shell were concentrated at the point k, then the nuclear–electron shell interaction would be given by

$$W = -e\phi_k .$$

However, the electron charge is spread out in space. The electron charge density in our notation is $\varrho(r_k)$. We note that the integrals over all space are obviously

$$\int \varrho(r_k)\,dV = -e \tag{5.42}$$

and

$$\int \varrho(r_i)\,dV = Z , \tag{5.43}$$

where Z is the nuclear charge.

Taking into account the distribution of charges the interaction energy becomes

$$W = \int \phi_k \varrho(r_k)\,dV = \iint \frac{\varrho(r_i)\,\varrho(r_k)\,dV_i\,dV_k}{r_{ik}} . \tag{5.44}$$

In the above expression, the volume integral is performed over the part of space that is occupied by the nucleus $(\varrho(r_i))$ and by the charge of the electron shell $(\varrho(r_k))$. Such notation is instructive; nonetheless both of the integrations can equally well span an infinite space with

$$dV_i = r_i^2 \sin\theta \, d\theta \, d\phi \, dr_i = dx \, dy \, dz \equiv dx_1 \, dx_2 \, dx_3 \ ,$$

$$dV_k = r_k^2 \sin\theta \, d\theta \, d\phi \, dr_k = dX \, dY \, dZ \equiv dX_1 \, dX_2 \, dX_3 \ . \tag{5.45}$$

We expand the factor $1/r_{ik}$ into a series with respect to r_i. This can be done provided $r_i \ll r_k$. Such an inequality applies for all electron orbitals, except the spherically symmetric orbitals S and $P_{1/2}$ (penetrating orbitals). However it is exactly for such spherical symmetry that the quadrupole interaction is absent. For the remaining cases r_i is of the order of 10^{-4} nm, and r_k is of the order of a few angstrom and the inequality is valid. The expansion has the form

$$\frac{1}{r_{ik}} = \frac{1}{r_k} - r_i \nabla \frac{1}{r_k} + \frac{1}{2} \sum_{\alpha,\beta} x_\alpha x_\beta \frac{\partial^2}{\partial X_\alpha \partial X_\beta} \frac{1}{r_k} + \cdots \ ,$$

where x_α and x_β are the projections of r_i on the cartesian coordinate axes, oriented in space arbitrarily. We neglect terms whose order in r_i is higher than two (r_i^3 etc.) due to their smallness. After substituting into (5.44) we get

$$W = \int\limits_{V_i V_k} dV_i dV_k \varrho(r_i) \varrho(r_k)$$

$$\times \left[\frac{1}{r_k} - r_i \nabla \frac{1}{r_k} + \frac{1}{2} \sum_{\alpha,\beta} x_\alpha x_\beta \frac{\partial^2}{\partial X_\alpha \partial X_\beta} \frac{1}{r_k} + \cdots \right]. \tag{5.46}$$

The first term is the usual nucleus–electron interaction and in the present case this is of no interest. The second term is identically equal to zero from symmetry considerations. Actually, it is proportional to the nuclear dipole moment, but in the chosen coordinate system (origin at the charge centre), this is by definition equal to zero.

The quadrupole interaction in which we are interested is given by the last term. We write it as a product of two integrals

$$\Delta W = \frac{1}{2} \sum_{\alpha,\beta} \int dV_i \varrho(r_i) x_\alpha x_\beta \int dV_k \varrho(r_k) \frac{\partial^2}{\partial X_\alpha \partial X_\beta} \frac{1}{r_k} \ , \tag{5.47}$$

each of which is a second rank tensor

$$\Delta W = \sum_{\alpha,\beta} \tfrac{1}{2} Q'_{\alpha\beta} D_{\alpha\beta} \ . \tag{5.48}$$

In this expression all the nuclear characteristics are contained in the symbol $Q'_{\alpha\beta}$, and those of the atomic shell in $D_{\alpha\beta}$. The problem consists in finding the nuclear quadrupole moment Q_0 from experimental data, i.e. based on the measured values of ΔW. For this reason one has to isolate the quantity Q_0 from the right-hand side of (5.47). Let us rewrite the expression (5.47) in a more conveni-ent form for further transformation. In order to do this we add in the integral the term $r^2\delta_{\alpha\beta}$, which is permissible since

$$\mathrm{Tr}\{D_{\alpha\beta}\} = K\left(\frac{\partial^2}{\partial X^2} + \frac{\partial^2}{\partial Y^2} + \frac{\partial^2}{\partial Z^2}\right)\frac{1}{r_k} = 0 \ .$$

Then

$$\Delta W = \frac{1}{6}\sum_{\alpha,\beta}\int\mathrm{d}V_i\,\varrho(r_i)(3x_\alpha x_\beta - r^2\delta_{\alpha\beta})\int\mathrm{d}V_k\,\varrho(r_k)\frac{\partial^2}{\partial X_\alpha\partial X_\beta}\frac{1}{r_k} \ . \qquad (5.49)$$

Denoting the first integral by $Q_{\alpha\beta}$ (without prime), we have

$$\Delta W = \tfrac{1}{6}\sum_{\alpha,\beta}Q_{\alpha\beta}D_{\alpha\beta} \ . \qquad (5.50)$$

The quantities $Q_{\alpha\beta}$ constitute the so-called quadrupole moment tensor.

The naturally selected direction in the nucleus is the direction of its spin. One can attach the coordinate system to this direction. The expression (5.50) is valid in any cartesian coordinate system whose origin is at the centre of the charge, including the one in which $z\|I$. Clearly the direction of I is also the symmetry axis of the charge distribution, since all deviations are averaged as the nucleus "spins" about this axis. This means, in the new coordinate system, that the tensor $Q_{\alpha\beta}$ is reduced to the principal axes such that terms with $\alpha \neq \beta$ become zero, and, moreover, $Q_{xx} = Q_{yy}$. Thus the tensor $Q_{\alpha\beta}$ is now characterized by only two numbers Q_{xx} and Q_{zz}. The projection $Q_{zz} \equiv Q_0$ is called the observable quadrupole or simply the nuclear quadrupole moment. In accordance with (5.49)

$$Q_0 = \int\varrho\,\mathrm{d}V(3z^2 - r^2) \ . \qquad (5.51)$$

We denote the integrals by

$$\int\mathrm{d}V_i\,\varrho(r_i)x_\alpha^2 = \overline{x_\alpha^2} \ .$$

It is obvious that for spherical symmetry

$$\overline{x^2} = \overline{y^2} = \overline{z^2} = \tfrac{1}{3}\overline{r^2} \ .$$

Then $Q_0 = \int(3z^2 - r^2)\varrho\,\mathrm{d}V = 0$ and there is no shift of the hyperfine sublevels, i.e. $\Delta W = 0$. If the nucleus is elongated along the spin direction I, then $\overline{z^2} > \tfrac{1}{3}\overline{r^2}$. The expression (5.51) will then be greater than zero, and the quadrupole moment will be positive; if the nucleus is flattened along the direction I, then the quadrupole moment Q_0 is negative.

In accordance with (5.50) the interaction energy is determined by the product of the tensors $Q_{\alpha\beta}$ and $D_{\alpha\beta}$. The principal axes of these tensors do not coincide. Their angular distribution in space obeys the rules of quantum mechanics. Therefore one must refer to the operators.

Let us turn to the definition of the tensor $D_{\alpha\beta}$. We rewrite it as

$$D_{\alpha\beta} = \int (-e)\,\varrho'(r_k)\,\frac{3X_\alpha X_\beta - r_k^2\delta_{\alpha\beta}}{r_k^5}\,dV_k$$

$$= \int (-e)\,\varrho'(r_k)\,\frac{3n_\alpha n_\beta - \delta_{\alpha\beta}}{r_k^3}\,dV_k \ . \tag{5.52}$$

Here $(-e)\,\varrho'(r_k) = \varrho(r_k)$, and $n_{\alpha,\beta}$ are components of a unit vector, directed along r_k. The density $\varrho'(r_k)$ is simply the modulus of the wavefunction, which we denoted by $\Psi(r)$ (Sect. 5.1)

$$\varrho'(r_k) = |\Psi(r)|^2 \ . \tag{5.53}$$

We have seen that one can represent $\Psi(r)$ as a product of radial and angular parts of the wavefunction

$$\Psi(r) = \Psi(r)\,\Psi(\theta,\phi) \ . \tag{5.54}$$

In view of this, the integral (5.52) can be given as a product of integrals

$$D_{\alpha\beta} = (-e)\int \frac{|\Psi(r_k)|^2}{r_k^3}\,r_k^2\,dr_k$$

$$\times \int |\Psi(\theta,\phi)|^2\,(3n_\alpha n_\beta - \delta_{\alpha\beta})\sin\theta\,d\theta\,d\phi \ . \tag{5.55}$$

According to the definition of mean values

$$\int \frac{|\Psi(r_k)|^2 r_k^2\,dr_k}{r_k^3} = \overline{r_k^{-3}}$$

and

$$\int |\Psi(\theta,\phi)|^2\,(3n_\alpha n_\beta - \delta_{\alpha\beta})\sin\theta\,d\theta\,d\phi = \overline{3n_\alpha n_\beta - \delta_{\alpha\beta}} \ . \tag{5.56}$$

According to *Landau* [5.3], for a state with a given quantum number J, the latter expression corresponds to the operator

$$- 3a_{Jl}\,\frac{\hat{J}_\alpha\hat{J}_\beta + \hat{J}_\beta\hat{J}_\alpha - \frac{2}{3}J(J+1)\delta_{\alpha\beta}}{(2l+3)(2l-1)} \ . \tag{5.57}$$

Here a_{Jl} is the eigenvalue of the operator

$$\hat{a}_{Jl} = \frac{3\hat{\boldsymbol{L}}\cdot\hat{\boldsymbol{J}}(2\hat{\boldsymbol{L}}\cdot\hat{\boldsymbol{J}} - 1) - 2\hat{\boldsymbol{L}}^2\cdot\hat{\boldsymbol{J}}^2}{l(2l-1)(J+1)(2J+3)} \ . \tag{5.58}$$

It is completely determined by the quantum numbers. We note that for states with $J = 1/2$ (for these states $l = 0$ or $l = 1$) $a_{Jl} = 0$. This indicates that $D_{\alpha\beta}$ for these states is equal to zero and hence the quadrupole interaction is absent. On the other hand, the states with $J = 1/2$ are spherically symmetric and for symmetry reasons the quadrupole interaction must be absent (this has already been mentioned above).

In the same way one can deduce the known [5.3] expression for the nuclear quadrupole moment operator through the nuclear spin projection $\hat{I}_{\alpha,\beta}$ on the axes x, y and z:

$$Q_{\alpha\beta} = \frac{3}{2} Q_0 \frac{\hat{I}_\alpha \hat{I}_\beta + \hat{I}_\beta \hat{I}_\alpha - \frac{2}{3}\hat{I}^2 \delta_{\alpha\beta}}{I(2I-1)} . \tag{5.59}$$

Let us return to the definition of the interaction energy (5.50). This corresponds to an operator which we denote by $\hat{B}$:

$$\hat{B} = \tfrac{1}{6} Q_{\alpha\beta} D_{\alpha\beta} ; \tag{5.60}$$

$$\hat{B} = \frac{3}{4} eQ_0 \overline{r^{-3}} \frac{a_{JI}I}{(2I+3)J(2J-1)I(2I-1)}$$

$$\times (\hat{I}_\alpha \hat{I}_\beta + \hat{I}_\beta \hat{I}_\alpha - \tfrac{2}{3}\delta_{\alpha\beta}\hat{I}^2)(\hat{J}_\alpha \hat{J}_\beta + \hat{J}_\beta \hat{J}_\alpha - \tfrac{2}{3}\delta_{\alpha\beta}\hat{J}^2) . \tag{5.61}$$

The operators $\hat{I}_{\alpha,\beta}$ and $\hat{J}_{\alpha,\beta}$ commute, and because of this we obtain, after multiplying out the expressions in the brackets,

$$\hat{B} = eQ_0 \overline{r^{-3}} \left(a_{JI} \frac{2I}{2I+3} \right) \frac{3(\hat{\boldsymbol{I}} \cdot \hat{\boldsymbol{J}})^2 + \frac{3}{2}(\hat{\boldsymbol{I}} \cdot \hat{\boldsymbol{J}}) - I(I+1)J(J+1)}{2J(2J-1)I(2I-1)} . \tag{5.62}$$

The coefficient

$$eQ_0 \overline{r^{-3}} a_{JI} \frac{2I}{2I+3} \equiv B \tag{5.63}$$

is known as the quadrupole interaction constant. For a state with a given value of the quantum number F, the operator $\hat{B}$ is a characteristic operator (as in the case of determination of the magnetic dipole interaction) with eigenvalues given by

$$\Delta W = B \frac{\frac{3}{4}C(C+1) - I(I+1)J(J+1)}{2J(2J-1)I(2I-1)} , \tag{5.64}$$

where C is given, as in (5.26), by

$$C = F(F+1) - I(I+1) - J(J+1) . \tag{5.64a}$$

One can determine the quadrupole interaction constant B from the experimentally measured hyperfine splitting. From this constant one can also determine the quadrupole moment. The most difficult task in this calculation is the determination of the quantities $\overline{r^{-3}}$. However, if the magnetic dipole moment and the magnetic dipole interaction constant A are known, then it is convenient to utilize the relationship

$$\frac{B}{A} = \frac{J(J+1)}{(I+1)(2I+3)} a_{JI} \frac{eQ_0}{\mu_0^2 g_I} , \qquad (5.65)$$

which does not contain $\overline{r^{-3}}$ but nonetheless enables one to find the quadrupole moment Q_0. This "observable" quadrupole moment Q_0 is the time averaged one. The nucleus rotates about an axis that does not coincide with the symmetry axis of the charge distribution, thereby creating the "true" quadrupole moment, known as the quadrupole moment of a stationary nucleus Q_s.

The relationship between Q_0 and Q_s is given by the following expression

$$Q_0 = Q_s \frac{(2I-1)I}{(2I+3)(I+1)} . \qquad (5.66)$$

From here, one sees in particular that when $I = 1/2$ there is no observable quadrupole moment.

5.5 Transition Matrix Elements of the Electric Dipole Moment

The probability amplitude of the dipole transitions between the states Ψ_2 and Ψ_1 is proportional to the matrix elements of the dipole moment $d = -er$.

The pump matrix F and observation matrix G contain the matrix elements of the dipole moment projections on the direction of the electric vector of the illuminating electromagnetic field e_λ

$$\langle 2|d \cdot e_\lambda|1 \rangle = (d \cdot e_\lambda)_{21}$$

and on the direction of the electric vector of the observed light

$$\langle 1|d \cdot e_r|2 \rangle = (d \cdot e_r)_{12} .$$

Since the vectors e_λ and e_r have no influence on the wavefunctions, one can "transfer" them through $|1\rangle$ or $|2\rangle$ and consequently the following equalities hold

$$\langle 1|d \cdot e_r|2 \rangle = \langle 1|d|2 \rangle \cdot e_r \equiv d_{12} \cdot e_r ,$$

$$\langle 2|d \cdot e_\lambda|1 \rangle = \langle 2|d|1 \rangle \cdot e_\lambda \equiv d_{21} \cdot e_\lambda .$$

For definiteness it is assumed that the transition from the state $|1\rangle$ to the state $|2\rangle$ takes place with the absorption of energy and the transition from $|2\rangle$ to $|1\rangle$ corresponds to spontaneous emission.

Let us turn to the calculation of the matrix element d_{12}. The matrix elements d_{12} [5.6] are called the matrix elements of the dipole moment of the transition $|2\rangle \rightarrow |1\rangle$

$$d_{12} = \langle 1|-er|2\rangle = -\int \Psi_1^* er\Psi_2 \, dV \ . \tag{5.67}$$

Because $\Psi(r, \theta, \phi) = \Psi(r)\Psi(\theta, \phi)$ equation (5.67) can be rewritten as

$$d_{12} = \int_0^\infty \Psi_1^*(r)(-er)\Psi_2(r)r^2 \, dr$$

$$\times \int_0^{2\pi} d\phi \int_0^\pi d\theta \sin\theta \, \Psi_1^*(\theta, \phi)n'\Psi_2(\theta, \phi) \ , \tag{5.68}$$

where $n' = r/|r| = r/r$. The first cofactor in (5.68) is the radial part of the matrix element

$$-e \int_0^\infty \Psi_1^*(r)r\Psi_2(r)r^2 \, dr = R_{12}$$

and does not depend on the magnetic quantum numbers. It is therefore identical for all Zeeman components. It is determined by the quantum numbers l, S, and J. In problems in which these numbers are the same for all components of the structure, the quantity R_{12} is a kind of constant. We will not be concerned with the radial part of the matrix element, but pay attention instead to the calculation of the angular part of the matrix element of the dipole moment

$$d_{12} = R_{12} \int_0^{2\pi} d\phi \int_0^\pi d\theta \sin\theta \Psi_1^*(\theta, \phi)n'\Psi_2(\theta, \phi) \ . \tag{5.69}$$

The angular parts of the wavefunctions are described by spherical tensors [5.1], and they can be represented as a superposition of products of the spherical functions $Y_{lm}(\theta, \phi)$ responsible for the orbital properties of the electron shell, with the spin function Y_{Sm_S}, which depends on the spin S and on its projection on the z-axis (m_S):

$$\Psi(l, S, J, M) = \sum_{m, m_s} C_{lmSm_s}^{JM} Y_{lm}(\theta, \phi) Y_{Sm_s} \ . \tag{5.70}$$

Here $C_{lmSm_s}^{JM}$ are coefficients of the vector summation (Klebsch–Gordon coefficients).

The wavefunction Ψ can be expanded in terms of the products $Y_{lm} Y_{Sm_s}$ with the coefficients $C_{Sm_slm}^{JM}$. This merely corresponds to a different choice of phase.

From the properties of the Klebsch–Gordon coefficients, it follows that $m + m_s = M$, as a consequence of which the double sum in (5.70) becomes in practice a sum over one index. Since the spherical functions are variously defined in the literature (they differ in phase), we introduce here a definition which will be used throughout this work and which corresponds to the definition given in [5.2]:

$$Y_{lm}(\theta, \phi) = (-1)^m \sqrt{\frac{(2l + 1)(l - m)!}{4\pi(l + m)!}} \sin^m \theta \frac{d}{d \cos \theta} P_l(\cos \theta) e^{im\phi} , \qquad (5.71)$$

where $P_l(\cos \theta)$ is the Legendre polynomial.

As is well known, this system of spherical functions is orthonormal, i.e.

$$\int_0^{2\pi} d\phi \int_0^{\pi} d\theta \sin \theta Y^*_{l_i m_i}(\theta, \phi) Y_{l_k m_k}(\theta, \phi) = \delta_{l_i, l_k} \delta_{m_i, m_k} . \qquad (5.72)$$

The spin functions have the properties of unit vectors:

$$Y^*_{S m_s} Y_{S' m'_s} = \delta_{S, S'} \delta_{m_s, m'_s} . \qquad (5.73)$$

For single-electron systems $S = 1/2$ and $m_s = \pm 1/2$.

Expressing the wavefunction through the expansion (5.70), from (5.69) we find for the matrix element

$$d_{12} = R_{12} \sum_{\substack{m_1, m_2, \\ m_{s_1}, m_{s_2}}} \int_0^{2\pi} d\phi \int_0^{\pi} d\theta \sin \theta Y^*_{l_1 m_1}(\theta, \phi) Y^*_{S_1 m_{s_1}}$$

$$\times n' Y_{l_2 m_2}(\theta, \phi) Y_{S_2 m_{s_2}} C^{J_1 M_1}_{l_1 m_1 S_1 m_{s_1}} C^{J_2 M_2}_{l_2 m_2 S_2 m_{s_2}} . \qquad (5.74)$$

Here, J, M, S, m_S, l, m are the angular momentum quantum numbers of the states with the corresponding indices.

The operator n' is a unit vector, and therefore it does not act on the spin functions. As a result of (5.73), the expression (5.74) will have the following form

$$d_{12} = R_{12} \sum_{m_1, m_2} C^{J_1 M_1}_{l_1 m_1 S m_s} C^{J_2 M_2}_{l_2 m_2 S m_s}$$

$$\times \int_0^{2\pi} d\phi \int_0^{\pi} d\theta \sin \theta Y^*_{l_1 m_1}(\theta, \phi) n' Y_{l_2 m_2}(\theta, \phi) . \qquad (5.75)$$

In order to calculate d_{12} it is convenient to replace n' in (5.75) by its expansion over circular unit vectors e_0, e_+, e_- [Chap. 2; (2.6–9)]

$$n' = \cos \theta \, e_0 - \frac{\sin \theta e^{-i\phi}}{\sqrt{2}} e_+ + \frac{\sin \theta e^{i\phi}}{\sqrt{2}} e_- \qquad (5.76)$$

because the circular projections $\cos \theta$, $\mp \sin e^{\mp i\phi}$, multiplied by some kind of spherical function, transform this to other spherical functions with known coefficients. Given below are these recurrence relations [5.2, 3]

$$\cos\theta\, Y_{lm} = Y_{l+1\,m}\sqrt{\frac{(l-m+1)(l+m+1)}{(2l+1)(2l+3)}} + Y_{l-1\,m}\sqrt{\frac{(l-m)(l+m)}{(2l-1)(2l+1)}}$$

$$e^{i\phi}\sin\theta\, Y_{lm} = -Y_{l+1\,m+1}\sqrt{\frac{(l+m+1)(l+m+2)}{(2l+1)(2l+3)}}$$

$$+ Y_{l-1\,m+1}\sqrt{\frac{(l-m)(l-m-1)}{(2l-1)(2l+1)}} \tag{5.77}$$

$$e^{-i\phi}\sin\theta\, Y_{lm} = Y_{l+1\,m-1}\sqrt{\frac{(l-m+1)(l-m+2)}{(2l+1)(2l+3)}}$$

$$- Y_{l-1\,m-1}\sqrt{\frac{(l+m)(l+m-1)}{(2l-1)(2l+1)}} \ .$$

Now, due to the orthonormality of the functions Y_{lm} the quantity d_{12} is obtained immediately (for instance, by allowing the operator n' to act on $Y_{l_2 m_2}$),

$$d_{12} = R_{12} \sum_{m_1,\,m_2} C_{l_1 m_1 S m_s}^{J_1 M_1} C_{l_2 m_2 S m_s}^{J_2 M_2}$$

$$\times \left\{ \left[\sqrt{\frac{(l_2-m_2+1)(l_2+m_2+1)}{(2l_2+1)(2l_2+3)}}\, \delta_{l_1,\,l_2+1} \right. \right.$$

$$+ \sqrt{\frac{l_2^2-m_2^2}{(2l_2-1)(2l_2+1)}}\, \delta_{l_1,\,l_2-1} \Bigg]\delta_{m_1,\,m_2}\boldsymbol{e}_0$$

$$+ \left[-\sqrt{\frac{(l_2-m_2+1)(l_2-m_2+2)}{(2l_2+1)(2l_2+3)}}\, \delta_{l_1,\,l_2+1} \right.$$

$$+ \sqrt{\frac{(l_2+m_2)(l_2+m_2-1)}{(2l_2-1)(2l_2+1)}}\, \delta_{l_1,\,l_2-1} \Bigg]\delta_{m_1,\,m_2-1}\frac{\boldsymbol{e}_+}{\sqrt{2}}$$

$$+ \left[-\sqrt{\frac{(l_2+m_2+1)(l_2+m_2+2)}{(2l_2+1)(2l_2+3)}}\, \delta_{l_1,\,l_2+1} \right.$$

$$+ \sqrt{\frac{(l_2-m_2)(l_2-m_2-1)}{(2l_2-1)(2l_2+1)}}\, \delta_{l_1,\,l_2-1} \Bigg]\delta_{m_1,\,m_2+1}\frac{\boldsymbol{e}_-}{\sqrt{2}} \Bigg\} \ . \tag{5.78}$$

For convenience of presentation, and primarily for facilitating analysis, we transform the expression to another form. We introduce the notation

$$\boldsymbol{e}_0 = \boldsymbol{e}_{\nu=0}; \qquad \boldsymbol{e}_+ = \boldsymbol{e}_{\nu=1}; \qquad \boldsymbol{e}_- = \boldsymbol{e}_{\nu=-1} \ ,$$

such that the values of the roots in the square brackets coincide with the coefficients of the vector addition [5.2]. Then

$$d_{12} = R_{12} \left[\sqrt{\frac{l_2}{2l_2+1}} \delta_{l_2,\, l_2-1} - \sqrt{\frac{l_2+1}{2l_2+1}} \delta_{l_1,\, l_2+1} \right]$$

$$\times \sum_{m_1,\, m_2,\, \nu} C^{J_1M_1}_{l_1m_1Sm_s} C^{J_2M_2}_{l_2m_2Sm_s} C^{l_2m_2}_{l_1m_1\nu} e_\nu \ . \tag{5.79}$$

From the theory of angular momenta, we know the relation coupling the product of the 3j-symbols with the 6j-symbols [5.1]:

$$\sum_{\alpha,\, \beta,\, \delta} C^{c\gamma}_{a\alpha b\beta} C^{e\varepsilon}_{d\delta b\beta} C^{d\delta}_{a\alpha f\phi} = (-1)^{b+c+d+f} \sqrt{(2c+1)(2d+1)} C^{e\varepsilon}_{c\gamma f\phi} \begin{Bmatrix} a & b & c \\ e & f & d \end{Bmatrix} \ . \tag{5.80}$$

Making use of these relationships, as well as the symmetry properties of the 6j-symbols, we transform the formula (5.79) to the following form

$$d_{12} = (-1)^{S+J_1} \Pi' \begin{Bmatrix} S & l_1 & J_1 \\ 1 & J_2 & l_2 \end{Bmatrix} \sqrt{2J_1+1} C^{J_2M_2}_{J_1M_1\nu} e_\nu \ . \tag{5.81}$$

Here, for brevity, we introduce the coefficient Π', equal to

$$\Pi' = R_{12}(-1)^{l_2+1}(\sqrt{l_2}\delta_{l_1,\, l_2-1} - \sqrt{l_2+1}\delta_{l_1,\, l_2+1}) \ .$$

The expression for d_{21} can be derived from (5.81) by exchanging the indices $1\leftrightarrows2$. Upon this $\nu \to \nu' = -\nu$, since $\nu = M_2 - M_1$ and $\nu' = M_1 - M_2$. As a result we obtain (because $R_{12} = R_{21}$):

$$d_{21} = (-1)^{S+J_2} \Pi' \begin{Bmatrix} S & l_2 & J_2 \\ 1 & J_1 & l_1 \end{Bmatrix} \sqrt{2J_2+1} C^{J_1M_1}_{J_2M_21\nu'} e_{\nu'} \ . \tag{5.82}$$

Using the following property of the Klebsch–Gordon coefficients:

$$C^{J_1M_1}_{J_2M_21\nu'} = (-1)^{J_1-J_2+\nu'} \sqrt{\frac{2J_1+1}{2J_2+1}} C^{J_2M_2}_{J_1M_11-\nu'}$$

together with the equality

$$e^*_\nu = (-1)^\nu e_{-\nu}; \qquad \begin{Bmatrix} S & l_1 & J_1 \\ 1 & J_2 & l_2 \end{Bmatrix} = \begin{Bmatrix} S & l_2 & J_2 \\ 1 & J_1 & l_1 \end{Bmatrix} ,$$

we can rewrite (5.82) as

$$d_{21} = \left[(-1)^{S+J_1} \Pi' \begin{Bmatrix} S & l_2 & J_2 \\ 1 & J_1 & l_1 \end{Bmatrix} \sqrt{(2J_2+1)(2J_1+1)} \right] \frac{C^{J_2M_2}_{J_1M_11\nu}}{\sqrt{2J_2+1}} e^*_\nu \ . \tag{5.83}$$

Comparing (5.83) and (5.81) it is seen that $d_{21} = d_{12}^*$. This reflects the fact that the operator d is Hermitian.

The quantities within the square brackets in (5.83) are called the reduced matrix elements of the transition $|1\rangle \to |2\rangle$. The reduced matrix element does not depend on M_1 and M_2 (the whole dependence on M_1 and M_2 is determined by Klebsch–Gordon coefficient).

Denoting the reduced matrix element by $\| d_{21} \|$, we rewrite (5.83) as

$$d_{21} = \| d_{21} \| \frac{C_{J_1 M_1 1 v}^{J_2 M_2}}{\sqrt{2J_2 + 1}} e_v^*$$

$$= \| d_{21} \| (-1)^{J_2 - M_2} \begin{pmatrix} J_2 & 1 & J_1 \\ -M_2 & v & M_1 \end{pmatrix} e_v^* = d_{21} e_v^* , \qquad (5.83a)$$

where

$$\| d_{21} \| = \Pi'(-1)^{S + J_1} \begin{Bmatrix} S & l_2 & J_2 \\ 1 & J_1 & l_1 \end{Bmatrix} \sqrt{(2J_2 + 1)(2J_1 + 1)} .$$

By exchanging $1 \leftrightarrows 2$ we obtain a similar expression for d_{12}:

$$d_{12} = \| d_{12} \| (-1)^{J_1 - M_1} \begin{pmatrix} J_1 & 1 & J_2 \\ -M_1 & v' & M_2 \end{pmatrix} e_{v'}^* = d_{12} e_{v'}^* , \qquad (5.83b)$$

where

$$\| d_{12} \| = \Pi'(-1)^{S + J_2} \begin{Bmatrix} S & l_2 & J_2 \\ 1 & J_1 & l_1 \end{Bmatrix} \sqrt{(2J_2 + 1)(2J_1 + 1)} .$$

From (5.83a) and (5.83b) it follows that

$$\| d_{21} \| = \| d_{12} \| (-1)^{J_2 - J_1}; \qquad d_{21} = d_{12}(-1)^v .$$

It is also obvious that

$$d_{12} = d_{21}^* = \| d_{21} \| (-1)^{J_2 - M_2} \begin{pmatrix} J_2 & 1 & J_1 \\ -M_2 & v & M_1 \end{pmatrix} e_v . \qquad (5.81a)$$

The formulae (4.83a, b) are convenient for calculating the matrix elements d required in order to calculate the pump and observation matrices. If the relative values of the probabilities of the Zeeman components are required, then one can drop the reduced matrix element, since it has the same value for all magnetic components of a particular spectral line. The values of the Klebsch–Gordon coefficients in the symbols are presented in Tables 5.1 and 5.2.

Some of the relationships which we now introduce are useful for monitoring and carrying out the calculations. Since

$$\sum_{M_1} (C_{J_1 M_1 1 v}^{J_2 M_2})^2 = 1$$

Table 5.1. Klebsch–Gordon coefficients $C_{b\beta 1\mu}^{a\alpha}$

$\quad\mu$ a	$+1$	0	-1
$b+1$	$\sqrt{\dfrac{(b+\alpha)(b+\alpha+1)}{2(2b+1)(b+1)}}$	$\sqrt{\dfrac{(b-\alpha+1)(b+\alpha+1)}{(2b+1)(b+1)}}$	$\sqrt{\dfrac{(b-\alpha)(b-\alpha+1)}{2(2b+1)(b+1)}}$
b	$-\sqrt{\dfrac{(b+\alpha)(b-\alpha+1)}{2b(b+1)}}$	$\dfrac{\alpha}{\sqrt{b(b+1)}}$	$\sqrt{\dfrac{(b-\alpha)(b+\alpha+1)}{2b(b+1)}}$
$b-1$	$\sqrt{\dfrac{(b-\alpha)(b-\alpha+1)}{2b(2b+1)}}$	$-\sqrt{\dfrac{(b-\alpha)(b+\alpha)}{b(2b+1)}}$	$\sqrt{\dfrac{(b+\alpha+1)(b+\alpha)}{2b(2b+1)}}$

Table 5.2

$$\begin{Bmatrix} a & b & c \\ 1 & c-1 & b-1 \end{Bmatrix} = (-1)^s \sqrt{\frac{s(s+1)(s-2a-1)(s-2a)}{(2b-1)2b(2b+1)(2c-1)2c(2c+1)}}$$

$$\begin{Bmatrix} a & b & c \\ 1 & c-1 & b \end{Bmatrix} = (-1)^s \sqrt{\frac{2(s+1)(s-2a)(s-2b)(s-2c+1)}{2b(2b+1)(2b+2)(2c-1)2c(2c+1)}}$$

$$\begin{Bmatrix} a & b & c \\ 1 & c-1 & b+1 \end{Bmatrix} = (-1)^s \sqrt{\frac{(s-2b-1)(s-2b)(s-2c+1)(s-2c+2)}{(2b+1)(2b+2)(2b+3)(2c-1)2c(2c+1)}}$$

$$\begin{Bmatrix} a & b & c \\ 1 & c & b \end{Bmatrix} = (-1)^{s+1} \frac{2[b(b+1)+c(c+1)-a(a+1)]}{\sqrt{2b(2b+1)(2b+2)2c(2c+1)(2c+2)}}$$

$$s = a+b+c$$

it is quite simple to write the sum of transition probabilities from the initial state $|2\rangle$ to all possible final states $|1\rangle$:

$$\sum_{M_1} |d_{12}|^2 = \sum_{M_1} |d_{21}|^2 = \frac{\|d_{21}\|^2}{2J_2+1} \; .$$

The absence of the quantum number M_2 on the right-hand side is noteworthy and implies that the sum of probabilities of all Zeeman transitions from any Zeeman sublevel will not depend on its magnetic quantum number M_2. In particular, this indicates that the radiative decay probabilities of all of the sublevels are identical, and similarly that their lifetimes are identical.

In the same manner, one can find the sum of probabilities over the initial states: Using

$$\sum_{M_2} (C_{J_1 M_1 1\nu}^{J_2 M_2})^2 = \frac{2J_2+1}{2J_1+1}$$

we find

$$\sum_{M_2} |d_{12}|^2 = \frac{\|d_{21}\|^2}{2J_1 + 1} .$$

From this it follows that the sum of probabilities does not depend on the magnetic quantum number of the final state M_1. In particular this means that the probability for the atom to be in any of the final magnetic sublevels is identical, provided, of course, all the initial Zeeman states are identically populated. Moreover, the summation of probabilities of all transitions for any initial state leads to one and the same value and the summation over all final states also results in a single value for all final sublevels. These sums themselves are related to one another by

$$\frac{\sum_{M_1} |d_{12}|^2}{\sum_{M_2} |d_{12}|^2} = \frac{2J_1 + 1}{2J_2 + 1} .$$

5.6   Eigenpolarizations of Transitions

In accordance with (5.83a) and (5.81a),

$$d_{12} = \|d_{21}\| (-1)^{J_2 - M_2} \begin{pmatrix} J_2 & 1 & J_1 \\ -M_2 & v & M_1 \end{pmatrix} e_v = d_{21} e_v ,$$

$$d_{21} = \|d_{21}\| (-1)^{J_2 - M_2} \begin{pmatrix} J_2 & 1 & J_1 \\ -M_2 & v & M_1 \end{pmatrix} e_v^* = d_{21} e_v^* .$$

The vector e_v is known as the eigenpolarization of a transition from the state $|2\rangle$ to the state $|1\rangle$. Correspondingly, e_v^* is the eigenpolarization of a transition from the state $|1\rangle$ to the state $|2\rangle$. (We recall that $e_v^* = (-1)^v e_{-v}$). The eigenpolarizations of the transitions allow one to find the polarization of the emitted and absorbed light. The latter is defined as the polarization of such light, whose absorption is maximal. We write the electric field strength of the incident monochromatic wave in the following way:

$$E = \mathscr{E}_0 (e_\lambda \exp(-i\omega t + i\phi) + \text{c.c.}) .$$

The absorption amplitude of such light by an atomic system is proportional to the matrix element:

$$\langle 2 | d \cdot e_\lambda | 1 \rangle = d_{21} \cdot e_\lambda$$

or, isolating from d_{21} the characteristic polarization vector

$$\langle 2 | d \cdot e_\lambda | 1 \rangle = d_{21} e_v^* \cdot e_\lambda .$$

It is seen that the wave with the polarization $e_\lambda = e_v$ is absorbed most strongly. If, for example, $v = +1$, then the field strength of the light wave will have the form[2]

$$E = \mathcal{E}_0(e_+ \exp(-i\omega t + i\phi)) + e_+^* \exp(i\omega t - i\phi)$$

$$= -\mathcal{E}_0[e_x \cos(\omega t - \phi) + e_y \sin(\omega t - \phi)] \ . \tag{5.84}$$

According to the correspondence principle [5.2] the polarization is given by the matrix element d:

$$E \propto \mathrm{Tr}\{\hat{\sigma} d\} = \sigma_{12} d_{21} + \sigma_{21} d_{12} \ .$$

Substituting in this expression σ_{12} and σ_{21} from Sect. 3.3 we have

$$E = K_1(d_{21} \exp(i\omega t - i\phi) + d_{12} \exp(-i\omega t + i\phi)) \ ,$$

where ω is the frequency radiated in the transition from $|2\rangle$ to $|1\rangle$. This formula can be rewritten as

$$E = K(e_v^* \exp(i\omega t - i\phi) + e_v \exp(-i\omega t + i\phi)) \ .$$

Let $M_2 = M_1 + 1$, such that $v = 1$. Whereas the polarization of the emitted light propagating along the z-axis is a right-handed screw, i.e. if one observes along the direction of propagation of the light, then the electric vector will rotate in the clockwise direction (from the x-axis to the y-axis):

$$E = \mathcal{E}_0(e_+ \exp(-i\omega t + i\phi) + e_+^* \exp(+i\omega t - i\phi))$$

$$= -\mathcal{E}_0[e_x \cos(\omega t - \phi) + e_y \sin(\omega t - \phi)] \ . \tag{5.85}$$

The light propagating in the opposite direction will be a left-handed screw and its electric vector for an observer looking along the propagation direction of the light will rotate anticlockwise. (For completeness we note that for an observer looking as before along the z-axis, i.e. counter to the light, the electric vector will rotate in a clockwise direction.)

One can also solve another problem: namely, given the propagation direction of the light n, one can find its polarization. It is quite clear that when the vector n deviates from the z-axis, the circular polarization becomes an elliptic one, turning into a linear one when $n \perp z$. Thereafter it becomes elliptic again with different sign and then acquires circular polarization of opposite direction when the light propagates along the negative z-axis.

Let us excite an atomic system by a light wave of right-handed polarization, propagating along the positive z-axis. Such a polarization, according to (5.84),

---

[2] This light is directed along the z-axis or counter to it. In the first case, it is polarized in the right circular sense when observed in the direction of propagation (right-handed screw), and in the second case in the left circular sense (left-handed screw).

excites the transition from $|1\rangle$ to $|2\rangle$ with $M_2 = M_1 + 1$. The spontaneous emission upon the transition from $|2\rangle$ to the initial state $|1\rangle$ is also directed along the z-axis and, in accordance with (5.85), also possesses right-hand polarization. Such a coincidence always holds true, and the polarization of the absorbed light always coincides with that of the spontaneous radiation between the same states (provided their direction of propagation is the same).

We note that the radiation amplitude in any arbitrary polarization is found as the projection of the vector E, given in (5.85), on $e_{r'}$

$$E = \mathscr{E}_0[(e_+ \cdot e_{r'}^*)e_{r'}e^{-i\omega t} + (e_+^* \cdot e_{r'})e_{r'}^* e^{i\omega t}] \; . \tag{5.85a}$$

From a comparison of (5.85a) with the expression for the operator of spontaneous radiation (3.72), one can see that $e_{r'} = e_r^*$.

5.7 Matrix Elements of the Dipole Transition Between States with Hyperfine Structure

The state wavefunctions of atoms possessing nuclear spin – the so-called hyperfine wavefunctions – can be defined in terms of an expansion similar to (5.70)

$$\Psi_{Fm_F} = \sum_{m_I} C_{JMIm_I}^{Fm_F} \Psi_{JM} Y_{Im_I} \; . \tag{5.86}$$

The function Ψ_{Fm_F} is an eigenfunction of the operators $\hat{J}^2$, $\hat{I}^2$ and $\hat{F}^2$, as well as of the operator $\hat{H}_{\text{hfs}} = A\hat{I}\cdot\hat{J}$ with the eigenvalues $AC/2$ (Sect. 5.2). The functions Ψ_{JM} were examined immediately above (5.70). They are eigenfunctions of the operator $\hat{H}_0$ and of $\hat{J}^2$ and $\hat{J}_z$. The function Y_{Im_I} is an eigenfunction of the operators $\hat{I}^2$ and $\hat{I}_z$. The product of the last two functions characterizes the state of systems whose atomic orbitals (Ψ_{JM}) do not interact with the state of the atomic nucleus (Y_{Im_I}); the latter gives the magnitude and orientation of the nuclear spin in space. Because of this, the functions

$$\Phi_{JMIm_I} = \Psi_{JM} Y_{Im_I} \tag{5.87}$$

are called the decoupled functions or strong field functions. Referring to the equation (5.86), one can obtain the expansion of these functions in terms of the hyperfine wavefunctions Ψ_{Fm_F}:

$$\Phi_{JMIm_I} = \sum_F C_{JMIm_I}^{Fm_F} \Psi_{Fm_F} \; . \tag{5.88}$$

It should be emphasized that the wavefunction can be defined both in terms of these functions Φ, or using another method slightly different to (5.86):

$$\Psi_{Fm_F} = \sum_{m_I} C_{Im_I JM}^{Fm_F} Y_{Im_I} \Psi_{JM} \; . \tag{5.89}$$

The terms in these two expansions (5.86, 89) differ by the factor $(-1)^{I+J-F}$. However, since for the calculation of the wavefunctions in a magnetic field we adopted a given order of summation of moments, here we are obliged to take the expansion (5.86). Using a different one may lead to serious errors when calculating the interference signals from atomic systems in external fields.

Let us return to the determination of matrix elements of the dipole transitions. The dipole transition operator contains the vector $n' = r/|r|$ (5.68). This operator does not act on the spin function. The remaining terms in the operator are only coefficients and all these quantities can be taken out of the integral, as was done above in deriving the formula (5.75). As in the previous case, we denote the magnetic sublevels as states $|1\rangle$ and $|2\rangle$ (note that these are now magnetic sublevels of the hyperfine structure between which the dipole moment is determined). Moreover, the state $|2\rangle$ can be any sublevel with the quantum number J_2 (the initial or primary group of states), and $|1\rangle$ is any of the sublevels with the quantum number J_1 (final state). The dipole matrix element between the states $|2\rangle$ and $|1\rangle$ can be written as

$$
d_{12} = \langle l_1 J_1 F_1 m_{F_1} | n' | l_2 J_2 F_2 m_{F_2} \rangle R_{12}
$$

$$
= R_{12} \sum_{m_l, M_1, M_2} C^{F_1 m_{F_1}}_{J_1 M_1 I m_l} C^{F_2 m_{F_2}}_{J_2 M_2 I m_l}
$$

$$
\times \int_0^{2\pi} d\phi \int_0^{\pi} d\theta \sin\theta\, \Psi^*_{J_1 M_1}\, n'\, \Psi_{J_2 M_2} \, . \tag{5.90}
$$

In this equation the spin functions are absent, since, due to their orthonormality, their product is equal to either one or zero. The latter case also leads to the absence of a summation over the second set m'_l. The integral in (5.90) is the quantity that we derived in the previous section, formula (5.81). Substituting the expression for this integral into (5.90), we find

$$
d_{12} = (-1)^{S+J_1} \Pi' \sqrt{2J_1+1} \sum_{M_1, M_2, m_l} \begin{Bmatrix} S & l_1 & J_1 \\ 1 & J_2 & l_2 \end{Bmatrix} C^{J_2 M_2}_{J_1 M_1 1\nu}
$$

$$
\times C^{F_1 m_{F_1}}_{J_1 M_1 I m_l} C^{F_2 m_{F_2}}_{J_2 M_2 I m_l} e_\nu \, . \tag{5.91}
$$

The three factors can, in accordance with (5.80), be replaced by two factors

$$
d_{12} = (-1)^{I+S+J_1+J_2+F_1+1} \sqrt{(2J_1+1)(2J_2+1)}\, \Pi' \begin{Bmatrix} S & l_1 & J_1 \\ 1 & J_2 & l_2 \end{Bmatrix}
$$

$$
\times \sqrt{2F_1+1}\, C^{F_2 m_{F_2}}_{F_1 m_{F_1} 1\nu} \begin{Bmatrix} I & J_1 & F_1 \\ 1 & F_2 & J_2 \end{Bmatrix} e_\nu \, . \tag{5.92}
$$

As in the previous case, for convenience we denote by Π the series of factors common to all components of the hyperfine structure of an isolated spectral line:

$$
\Pi = (-1)^{I+S+J_1+J_2+1} \sqrt{(2J_1+1)(2J_2+1)}\, \Pi' \begin{Bmatrix} S & l_1 & J_1 \\ 1 & J_2 & l_2 \end{Bmatrix} . \tag{5.93}
$$

Then

$$d_{12} = (-1)^{F_1} \Pi \sqrt{2F_1 + 1} \, C^{F_2 m_{F_2}}_{F_1 m_{F_1} 1 \nu} \begin{Bmatrix} I & J_1 & F_1 \\ 1 & F_2 & J_2 \end{Bmatrix} e_\nu \ . \tag{5.94}$$

In another, symmetrical representation

$$d_{12} = (-1)^{F_2 + \nu} \Pi \sqrt{2F_2 + 1} \, C^{F_1 m_{F_1}}_{F_2 m_{F_2} 1 - \nu} \begin{Bmatrix} I & J_1 & F_1 \\ 1 & F_2 & J_2 \end{Bmatrix} e_\nu \ . \tag{5.95}$$

In interference phenomena involving levels with hyperfine structure, all the hf components of the spectral line usually play a role. Therefore here we will not distinguish the matrix elements. They will be different for every pair of F_1 and F_2. The general formula (5.94) or (5.95) correctly determines not only the amplitudes of transitions, but also their relative sign. The latter, as is clear from previous chapters, is of great importance when calculating interference signals.

The quantities entering into these formulae can be found from Tables 5.1 and 5.2. The 6j-symbols entering the expressions (5.91–94) can be interpreted as quantities proportional to the probabilities of the hyperfine transitions. This turns out to be obvious when one finds the sum of transition probabilities over all polarizations, i.e. over all possible finite magnetic sublevels:

$$\sum_{m_{F_1}} |d_{12}|^2 = \Pi^2 (2F_1 + 1) \begin{Bmatrix} I & J_1 & F_1 \\ 1 & F_2 & J_2 \end{Bmatrix}^2 \sum_{m_{F_1}} (C^{F_2 m_{F_2}}_{F_1 m_{F_1} 1 \nu})^2 \ . \tag{5.96}$$

We recall that

$$\sum_{m_{F_1}} (C^{F_2 m_{F_2}}_{F_1 m_{F_1} 1 \nu})^2 = 1 \ .$$

Thus

$$\sum_{m_{F_1}} |d_{12}|^2 = \Pi^2 (2F_1 + 1) \begin{Bmatrix} I & J_1 & F_1 \\ 1 & F_2 & J_2 \end{Bmatrix}^2 \ . \tag{5.97}$$

We thus realize that the sum of transition probabilities from every magnetic sublevel within the range of one hyperfine component is one and the same. This result is similar to that obtained in the previous section in relation to the dipole transitions in atoms without nuclear spin.

The expression (5.95) enables one to readily carry out the summation over m_{F_2}:

$$\sum_{m_{F_2}} |d_{12}|^2 = \Pi^2 (2F_2 + 1) \begin{Bmatrix} I & J_1 & F_1 \\ 1 & F_2 & J_2 \end{Bmatrix}^2 \ . \tag{5.98}$$

By analogy to the relationship (5.85), we find

$$\frac{\sum\limits_{m_{F_1}} |d_{12}|^2}{\sum\limits_{m_{F_2}} |d_{12}|^2} = \frac{2F_1 + 1}{2F_2 + 1} \ . \tag{5.99}$$

This relationship is valid if one takes into consideration only a single hyperfine transition. Let us look at the sum of probabilities for transitions from one initial state to other final states within one spectral line possessing a hyperfine structure

$$\sum_{m_{F_1}, F_1} |d_{12}|^2 = \Pi^2 \sum_{F_1} (2F_1 + 1) \begin{Bmatrix} I & J_1 & F_1 \\ 1 & F_2 & J_2 \end{Bmatrix} = \frac{\Pi^2}{2J_2 + 1} .$$ (5.100)

On the other hand,

$$\sum_{m_{F_2}, F_2} |d_{12}|^2 = \frac{\Pi^2}{2J_1 + 1} .$$ (5.101)

Let us examine the relationship of transition probabilities for two different hyperfine final states with the quantum numbers F_1 and F'_1:

$$\frac{\sum_{m_{F_1}} |d_{12}|^2}{\sum_{m_{F'_1}} |d_{12}|^2} = \frac{2F_1 + 1}{2F'_1 + 1} \frac{\begin{Bmatrix} I & J_1 & F_1 \\ 1 & F_2 & J_2 \end{Bmatrix}^2}{\begin{Bmatrix} I & J_1 & F'_1 \\ 1 & F_2 & J_2 \end{Bmatrix}^2} = \frac{W^2(IJ_1 1F_2; F_1 J_2)}{W^2(IJ_1 1F_2; F'_1 J_2)} .$$ (5.102)

Here $W(IJ_1 1F_2; F_1 J_2)$ are the so called *Racah* coefficients [5.1]:

$$W(IJ_1 1F_2; F_1 J_2) = (-1)^{1+F_2+I+J_1} \sqrt{(2F_1 + 1)(2J_2 + 1)} \begin{Bmatrix} I & J_1 & F_1 \\ 1 & F_2 & J_2 \end{Bmatrix} .$$ (5.103)

The Racah coefficients determine the relative probabilities of the hyperfine transitions: the relationship of transition probabilities from one hyperfine sub-level to all other possible final states. We note here the well-known relationship

$$\sum_{F_1} W^2(IJ_1 1F_2; F_1 J_2) = 1 .$$ (5.104)

In addition, it is not difficult to find the relative probabilities of the hyperfine transitions within the whole range of the hyperfine structure of a line. Let us denote these by the symbol $W^2(F_2 F_1)$. The probability amplitudes given in such a way will be equal to

$$W^2(F_2 F_1) = (2F_1 + 1)(2F_2 + 1) \begin{Bmatrix} I & J_1 & F_1 \\ 1 & F_2 & J_2 \end{Bmatrix}^2$$

$$= \frac{2F_2 + 1}{2J_2 + 1} W^2(IJ_1 1F_2; F_1 J_2) = \frac{2F_1 + 1}{2J_1 + 1} W^2(IJ_2 1F_1; F_2 J_1) . ,$$ (5.105)

These transition probabilities determine the relative intensity of the components of the hyperfine structures.

From (5.104, 105) we find that

$$\frac{\sum\limits_{F_1} W^2(F_2 F_1)}{\sum\limits_{F_1} W^2(F_2' F_1)} = \frac{2F_2 + 1}{2F_2' + 1} . \tag{5.106}$$

This formula reflects the well-known rule of intensities.

One can set up one more useful relationship:

$$\frac{\sum\limits_{F_1} W^2(F_2 F_1)}{\sum\limits_{F_2} W^2(F_2 F_1)} = \frac{2F_2 + 1}{2F_1 + 1} \frac{2J_1 + 1}{2J_2 + 1} . \tag{5.107}$$

The relationships (5.106, 107) are very useful for numerical calculations of the transition probabilities, since they serve as a good test. However, errors of sign in d_{12} will not be manifested. To check that the sign is correct, one may look at the symmetry of the dipole transition matrix, which is quite easy to deduce when the number of components is large.

5.8 The Stark Effect

An external electric field acting on the electron orbital of an atom has the effect of mixing the energy sublevels and partially lifting the degeneracy. In the literature it is usual to distinguish the linear and quadratic Stark effects. The linear Stark effect is the interaction of the electric dipole moment of an electron orbital of a quantum system with the electric field. However, atoms do not possess characteristic electric dipole moments except in rare cases. Therefore, the linear Stark effect occurs predominantly in molecules and in atoms only for hydrogen. For these systems the additional energy is

$$\Delta W = - \boldsymbol{d} \cdot \boldsymbol{\mathscr{E}} , \tag{5.108}$$

where $\boldsymbol{d}$ is the dipole moment of the electron orbital (as in Chap. 2, $\boldsymbol{d} = -e\boldsymbol{r}$, where the radius vector originates from the atomic nucleus) and $\boldsymbol{\mathscr{E}}$ is the electric field vector.

However, an electric field can also induce a dipole moment in an electrically neutral system. In a first-order approximation this is proportional to $\boldsymbol{\mathscr{E}}$: i.e. $\boldsymbol{d} = K\boldsymbol{\mathscr{E}}$. Then

$$\Delta W = - \boldsymbol{d} \cdot \boldsymbol{\mathscr{E}} \propto \kappa \mathscr{E}^2 . \tag{5.109}$$

In quantum mechanics the interaction of an atom with the electric field is given by an operator

$$\hat{V} = \hat{H}_e = - \hat{\boldsymbol{d}} \cdot \boldsymbol{\mathscr{E}} . \tag{5.110}$$

Unlike the magnetic field operator, which is diagonal in the eigenfunctions of the Hamiltonian H_0 (not including the interaction with the nucleus), this operator mixes atomic states with different l and the diagonal matrix elements of $\hat{V}$ are equal to zero:

$$\langle \Psi_l|\hat{H}_0 + \hat{V}|\Psi_l\rangle = \langle \Psi_l|\hat{H}_0|\Psi_l\rangle + \langle \Psi_l|\hat{V}|\Psi_l\rangle$$
$$= \langle \Psi_l|\hat{H}_0|\Psi_l\rangle = E \ . \tag{5.111}$$

We expand the eigenfunctions Ψ'' of the operator $\hat{H}_0 + \hat{V}$ in terms of the eigenfunctions of the operator $\hat{H}_0$

$$\Psi'' = \sum_{k''} C_{k''}\Psi_{k''} \ .$$

Among the functions $\Psi_{k''}$ there may be functions that are degenerate in energy, but which differ in magnetic quantum number. The functions Ψ'' satisfy the equation

$$(\hat{H}_0 + \hat{V})\Psi'' = E''\Psi'' \ ;$$

this means

$$(\hat{H}_0 + \hat{V})\sum_{k''} C_{k''}\Psi_{k''} = E'' \sum_{k''} C_{k''}\Psi_{k''} \ .$$

We will be interested in some kind of degenerate states, whose wavefunctions are denoted by the indices p. We divide the sum over k'' into two sums over k' and over p. Then

$$(\hat{H}_0 + \hat{V})\left(\sum_{k'} C_{k'}\Psi_{k'} + \sum_p C_p\Psi_p\right) = E''\left(\sum_{k'} C_{k'}\Psi_{k'} + \sum_p C_p\Psi_p\right) . \tag{5.112}$$

We multiply from the left-hand side by the function Ψ_k^* belonging to the set $\Psi_{k'}^*$, and we integrate over the whole volume. In a first-order approximation ($V_{kk'} = 0$) we obtain a system of equations

$$C_k E_k + \sum_p C_p V_{kp} = E''C_k \ . \tag{5.113}$$

Multiplying (5.112) by $\Psi_{p'}^*$ in the same way, we find the system

$$C_{p'} E_{p'} + \sum_{k'} C_{k'} V_{p'k'} = E''C_{p'} \ ,$$

which can be rewritten as

$$C_{p'}(E'' - E_{p'}) = \sum_{k'} C_{k'} V_{p'k'} \ . \tag{5.114}$$

We recall that here E'' is the energy due to the perturbation and $E_{p'}$ is the energy of the unperturbed atomic state p'. All E_p are identical and therefore one can assume the quantity $\varepsilon = E'' - E_{p'}$ to be the energy shift of the state p under the action of the perturbation $\hat{V}$.

From (5.113) we find

$$C_k = \frac{\sum\limits_p C_p V_{kp}}{E'' - E_k} .$$

We substitute the expression derived into (5.114):

$$C_{p'}\varepsilon = \sum_k \sum_p C_p \frac{V_{kp} V_{p'k}}{E'' - E_k} .$$

From the sum over p we then isolate the term with $p = p'$

$$C_{p'}\varepsilon = \sum_k \left(\sum_{p \neq p'} C_p \frac{V_{kp} V_{p'k}}{E'' - E_k} + C_{p'} \frac{V_{kp'} V_{p'k}}{E'' - E_k} \right) .$$

Let us rewrite this same equation in another form:

$$C_{p'} \left(\varepsilon - \sum_k \frac{V_{kp'} V_{p'k}}{E'' - E_k} \right) - \sum_{k, p \neq p'} C_p \frac{V_{kp} V_{p'k}}{E'' - E_k} = 0 . \tag{5.115}$$

In this way we have obtained a system of linear equations, whose number is equal to the number of different indices p, for the coefficients C. One can solve the system provided its determinant is equal to zero. From this latter condition for solving the secular equation one finds all possible values ε. Knowing these, one can find the expansion coefficients C.

The solution sought will be simplified if one takes the z-axis along the electric field $\mathscr{E}$. In this case $e_v = e_0$ and the matrix elements of the dipole transition (5.81) will be different from zero only if the magnetic quantum numbers of the coupled states are equal. When they are not equal, the coefficient C in (5.81) will be equal to zero. This indicates, that, if the index k (which also contains the value of the magnetic quantum number) is given, then the index p can take only one value; all the remaining matrix elements become zero and, as a result of this, the second sum in (5.115) becomes zero:

$$C_p \left(\varepsilon - \sum_k \frac{V_{kp} V_{pk}}{E'' - E_k} \right) = 0 . \tag{5.116}$$

The system of equations

$$\varepsilon = \sum_k \frac{V_{kp} V_{pk}}{E'' - E_k} , \tag{5.117}$$

which is the secular equation of the system (5.116), allows one to find the shift in all of the energies, for the states p as well as k. For every energy value there exists a corresponding wavefunction, given by a set of coefficients C_p and C_k:

$$\Psi'' = \sum_p C_p \Psi_p + \sum_{k,p} C_p \frac{V_{kp}}{E'' - E_k} \Psi_k \ . \tag{5.118}$$

The coefficients C can be found from the system of equations (5.114) by substituting the values ε_p of the roots of the equation (5.117). If the root ε_p is nondegenerate, then the sums over p will contain only by a single term each. Then, with the normalization condition

$$\sum_k C_k^2 + C_p^2 = 1 \ ,$$

we have

$$C_p = \left(1 + \sum_k \frac{|V_{kp}|^2}{(E'' - E_k)^2} \right)^{-1/2} \tag{5.119}$$

and

$$\Psi'' = \left(1 + \sum_k \frac{|V_{kp}|^2}{(E'' - E_k)^2} \right)^{-1/2} \Psi_p$$
$$+ \sum_k \left(1 + \sum_{k'} \frac{|V_{k'p}|^2}{(E'' - E_{k'})^2} \right)^{-1/2} \frac{V_{kp}}{E'' - E_k} \Psi_k \ . \tag{5.118a}$$

These expressions are exact in a first approximation; however, each of (5.117) is an equation of higher degree[3] dependent on the number of terms in the sum over k and, as a rule, cannot be solved analytically. Nevertheless, if ε is small in comparison to $E'' - E_k$, then its presence in the denominator can be neglected and it is simple to get an approximate value of ε:

$$\varepsilon_p = \sum_k \frac{V_{kp} V_{pk}}{E_p - E_k} \ . \tag{5.119a}$$

The degree of approximation can be quite easily evaluated from the same equation (5.117).

With the assumption $|\varepsilon| \ll |E_p - E_k|$, where k is arbitrary, one has

$$\sum_k \frac{|V_{kp}|^2}{(E_p - E_k)^2} < \frac{\varepsilon}{(E_p - E_k)_{min}} \ll 1 \tag{5.120}$$

---

[3] This is evident from expression (5.113).

as a consequence of which the sum under the square root in (5.119) can be neglected, and

$$\Psi_p'' = \Psi_p + \sum_k \frac{V_{kp}}{E_p - E_k} \Psi_k \ . \tag{5.121}$$

In most problems involving the interference of states, it turns out that one can also neglect the factors $V_{kp}/(E_p - E_k)$ and use the unperturbed functions.

The elements V_{pk} include the matrix elements of the electric dipole transitions. Their values are determined with the help of the formula (5.82). We substitute the expression for V_{pk}, assuming $v = 0$ ($\mathscr{E} \parallel z$),

$$V_{pk} = - d_{pk}\mathscr{E} = (-1)^{S+J_p+l_p} R_{pk}$$

$$\times (\sqrt{l_p}\, \delta_{l_k, l_p-1} - \sqrt{l_p+1}\, \delta_{l_k, l_p+1})\sqrt{2J_p + 1}\, C^{J_k M_p}_{J_p M_p 10}$$

$$\times \begin{Bmatrix} S & l_p & J_p \\ 1 & J_k & l_k \end{Bmatrix} \mathscr{E} \tag{5.122}$$

into the formula for the energy shift ε (5.117). We express the coefficient C^2 in the explicit form

$$(C^{J_k M_p}_{J_p M_p 10})^2 = \frac{(J_p + 1)^2 - M_p^2}{(2J_p + 1)(J_p + 1)} \delta_{J_p, J_k-1}$$

$$+ \frac{M_p^2}{J_p(J_p + 1)} \delta_{J_p J_k} + \frac{J_p^2 - M_p^2}{J_p(2J_p + 1)} \delta_{J_p, J_k+1} \ , \tag{5.123}$$

and multiplying by

$$\frac{2J_k + 1}{2J_k + 1} = (2J_k + 1)\left(\frac{\delta_{J_p, J_k-1}}{2J_p + 3} + \frac{\delta_{J_p, J_k}}{2J_p + 1} + \frac{\delta_{J_p, J_k+1}}{2J_p - 1} \right)$$

we isolate the terms containing the quantum numbers M. From the remainder we isolate the sum, which can be treated as the shift of the centre of gravity of a term of quantum number $J = J_p$. The resulting expression can be written as

$$\varepsilon_p = \alpha_p \mathscr{E}^2 + \beta_p \left(M_p^2 - \frac{J_p(J_p + 1)}{3} \right) \mathscr{E}^2 \ , \tag{5.124}$$

where

$$\alpha_p = \sum_k \frac{|R_{pk}|^2}{3(E_p - E_k)} (l_p \delta_{l_k, l_p-1} + (l_p + 1)\delta_{l_k, l_p+1})$$

$$\times (2J_k + 1) \begin{Bmatrix} S & l_p & J_p \\ 1 & J_k & l_k \end{Bmatrix}^2 = \sum_k \frac{\Pi^2_{J_p J_k}}{3(E_p - E_k)} \tag{5.125}$$

and

$$\beta_p = \sum_k \frac{|R_{pk}|^2}{E_p - E_k} (l_p \delta_{l_k, l_p - 1} + (l_p + 1)\delta_{l_k, l_p + 1})(2J_k + 1)$$

$$\times \left\{ \begin{matrix} S & l_p & J_p \\ 1 & J_k & l_k \end{matrix} \right\}^2 \left[-\frac{\delta_{J_p, J_k + 1}}{J_p(2J_p - 1)} + \frac{\delta_{J_p, J_k}}{J_p(J_p + 1)} - \frac{\delta_{J_p, J_k - 1}}{(J_p + 1)(2J_p + 3)} \right].$$

$$(5.126)$$

We note that the sum over M of the quantities in brackets in (5.124) is equal to zero, which also indicates that $\alpha_p \mathscr{E}^2$ is the displacement of the centre of gravity of the term p.

The expression (5.124) shows that the energy level as a whole shifts in proportion to the square of the electric field strength; moreover, it splits into $J + 1$ sublevels for integer J and into $J + 1/2$ sublevels for half-integral J. The square of the magnetic quantum number will thus take this many different values. From here one can deduce that there is no splitting in an electric field for the levels with $J = 1/2$ and $J = 0$. For any J the electric field partially lifts the degeneracy, but levels with $+ M$ and $- M$ will remain degenerate. The levels will not be equally spaced and the distance between them increases in proportion to $\mathscr{E}^2$.

Let us introduce an operator whose eigenfunctions coincide with the eigenfunctions of the operator $\hat{H}_0$ and whose eigenvalues describe the energy variation of the sublevels with the quantum numbers J and M

$$\hat{V} = \alpha \mathscr{E}^2 + \beta \left(\hat{J}_z^2 - \frac{J(J + 1)}{3} \right) \mathscr{E}^2 .$$

$$(5.127)$$

If the operators are replaced by their eigenvalues, one obtains the formula (5.124).

The operator $\hat{V}$ has the form (5.127) in the coordinate system with $z \parallel \mathscr{E}$. If one reverses the coordinate axis, then the expression derived in the reversed coordinate will have the most general form. It is obvious that the first term in (5.127) – the shift of the centre of gravity of a term, and the term $\beta \mathscr{E}^2 J(J + 1)/3$ do not depend on the coordinate system and that their forms are always the same. The operator $\hat{J}_z$ in the reversed coordinate system possesses components $\hat{J}_x$, $\hat{J}_y$ and $\hat{J}_z$, and the operator $\hat{J}_z^2$ is a component of the tensor $\hat{J}_i \hat{J}_j$. The tensor $\mathscr{E}_i \mathscr{E}_j$ possesses same the type of components. Taking into consideration all of the above, for an arbitrary choice of the coordinate system relative to the direction of the field $\mathscr{E}$, the operator for the interaction with the electric field has the form

$$\hat{V} = \alpha \mathscr{E}^2 + \beta \hat{J}_i \hat{J}_j \mathscr{E}_i \mathscr{E}_j - \beta \frac{J(J + 1)}{3} \mathscr{E}^2$$

$$= \alpha \mathscr{E}^2 + \beta \left(\hat{J}_i \hat{J}_j - \frac{J(J + 1)}{3} \delta_{i,j} \right) \mathscr{E}_i \mathscr{E}_j .$$

$$(5.128)$$

As usual, the summation convention applies to repeated indices.

5.9 Atoms with Nonzero Nuclear Spin in External Fields

In Sect. 5.1 we implicitly assumed that the interaction between the angular momenta L and S leading to the fine splitting is far greater than the magnetic interaction. Therefore the L–S interaction was included in the fundamental Hamiltonian.

In the presence of a nuclear spin yet another interaction will arise – the hyperfine interaction. This can also be an interaction with external fields. Therefore, as in many other problems of quantum mechanics, the description of an atom in external fields will be based on the methods of perturbation theory. One can assume that the problem is solved when one has found the wavefunctions Ψ' that are eigenfunctions of the operator $\hat{H}$ composed of the basic operator H_0 and the perturbation operator $\hat{V}$. The operator H_0 has the eigenfunctions Ψ and eigenenergies E

$$H_0 \Psi_n = E_n \Psi_n \ .$$

We are interested in a single atomic state with given values of the quantum numbers l and J. Thus, we confine ourselves to a set of Ψ_n for which all $E_n = E$. We represent the perturbed wavefunction in terms of an expansion over the eigenfunctions Ψ:

$$\Psi' = \sum_n C_n \Psi_n \ . \tag{5.129}$$

The Schrödinger equation for the function Ψ' can be written

$$\hat{H}\Psi' = (E + \varepsilon)\Psi' = E'\Psi' \ . \tag{5.130}$$

Substituting (5.129) into (5.130), multiplying both sides of the equation from the left subsequently by Ψ_n^* and integrating, we find, with the notation $V_{ik} = \int \Psi_i^* \hat{V} \Psi_k dv$, the matrix element of the perturbation operator:

$$C_1(V_{11} - \varepsilon) + C_2 V_{12} + \cdots = 0$$

$$C_1 V_{21} + C_2(V_{22} - \varepsilon) + \cdots = 0$$

$$\vdots \tag{5.131}$$

$$C_1 V_{n1} + \cdots + C_n(V_{nn} - \varepsilon) = 0 \ .$$

In order to determine the energy of an atom it is sufficient to define the conditions for solving the system (5.131) with respect to the coefficients C, i.e. to find the roots ε of the secular equation:

$$\begin{vmatrix} V_{11} - \varepsilon & V_{12} & V_{13} & \cdots & V_{1n} \\ V_{21} & V_{22} - \varepsilon & V_{23} & \cdots & V_{2n} \\ \vdots & \vdots & \vdots & & \vdots \\ V_{n1} & V_{n2} & V_{n3} & \cdots & V_{nn} - \varepsilon \end{vmatrix} = 0 \ . \tag{5.132}$$

The known values ε enable one to solve the system (5.131) with respect to the coefficients C and in this way to solve the problem of determining of the unknown functions Ψ'. The ease or complexity of this procedure is determined first of all by the choice of the operators $\hat{H}_0$ and $\hat{V}$ and the basis functions Ψ, which must be known explicitly.

In formulating the problem (all $E_n = E$), the operator has in fact already been defined. The additional terms $\hat{H}_{hfs}$ and $\hat{H}_{out}$ together constitute the perturbation operator

$$\hat{V} = \hat{H}_{hfs} + \hat{H}_{ext} \; . \tag{5.133}$$

The perturbations $\hat{H}_{hfs}$ and $\hat{H}_{ext}$ are both weak in comparison with $\hat{H}_0$; however, their relation to one another is to some extent arbitrary. The relationship between them is determined by the most rational method of solution. The possibilities will be explored in the following sections.

5.10   Perturbation Operators and Their Matrix Elements

Let us introduce here the perturbation operators, which have already been discussed above.

The hyperfine structure operator, i.e. the operator of the magnetic dipole and electric quadrupole between the electron and nucleus is

$$\hat{H}_{hfs} = \hat{A} + \hat{B} \; . \tag{5.134}$$

The two terms are the magnetic dipole interaction operator (Sect. 5.2, eq. (5.19)):

$$\hat{A} = A\hat{\boldsymbol{I}} \cdot \hat{\boldsymbol{J}} \; ,$$

and the operator of the electric quadrupole – nucleus interaction (5.62, 63):

$$\hat{B} = B \frac{3(\hat{\boldsymbol{I}} \cdot \hat{\boldsymbol{J}})^2 + \frac{3}{2}(\hat{\boldsymbol{I}} \cdot \hat{\boldsymbol{J}}) - \hat{I}^2 \hat{J}^2}{2J(2J-1)I(2I-1)} \; .$$

Both of these operators are diagonal in the hyperfine wavefunctions, which turn out to be their eigenfunctions. The quadrupole interaction operator does not produce an additional splitting of the hyperfine sublevels, but merely shifts their energies slightly.

The operator for the interaction with the magnetic field can be found in the same way as was done in (5.8, 9). The magnetic moment of an atom is a vector sum $\mu_e + \mu_I$ (Sect. 5.2), and therefore the operator contains two summands reflecting the interaction with the electron orbital and with the nucleus of the atom:

$$\begin{aligned}
\hat{H}_M &= \mu_0 g_J \hat{\boldsymbol{J}} \cdot \mathcal{H} - \mu_0 g_I \hat{\boldsymbol{I}} \cdot \mathcal{H} \\
&= \mu_0 g_J (\hat{J}_x \mathcal{H}_x + \hat{J}_y \mathcal{H}_y + \hat{J}_z \mathcal{H}_z) - \mu_0 g_I (\hat{I}_x \mathcal{H}_x + \hat{I}_y \mathcal{H}_y + \hat{I}_z \mathcal{H}_z) \; .
\end{aligned} \tag{5.135}$$

Here μ_0 is the Bohr magneton, g_J and g_I are the electron and nuclear Landé factors, respectively, and $\mathscr{H}$ is the magnetic field strength. In the general case, it is not possible to represent the operator $\hat{H}_M$ in terms of g_F. This can only be done when F is a "good" quantum number, i.e. when the system of functions with particular values of F is a complete set for the given problem. In a coordinate system with the quantization axis directed along the magnetic field, one has

$$\hat{H}_M = \mu_0 g_J \hat{J}_z \mathscr{H} - \mu_0 g_I \hat{I}_z \mathscr{H} . \tag{5.136}$$

The next perturbation is the perturbation by an electric field. The magnetic dipole and electric quadrupole moments of the nucleus do not interact with the electric field, and therefore their presence will have no influence on the interaction operator with the electric field, which is determined by the previously given formulae (5.110) and (5.128):

$$\hat{H}_e = \alpha \mathscr{E}^2 + \beta \left(\hat{J}_i \hat{J}_j - \frac{J(J+1)}{3} \delta_{ij} \right) \mathscr{E}_i \mathscr{E}_j .$$

Here $\mathscr{E}_{ij}$ are components of the electric field vector and $\hat{J}_{i,j}$ are operators of the projection of the angular momentum $(i, j = x, y, z)$. In the coordinate system in which the quantization axis is in the direction of the electric field (5.127):

$$\hat{H}_e = \alpha \mathscr{E}^2 + \beta \left(\hat{J}_z^2 - \frac{J(J+1)}{3} \right) \mathscr{E}^2 .$$

One can choose as basis functions either the hyperfine functions Ψ_{Fm_F} or the broken link functions Φ_{JMIm_I}. In both cases, the off-diagonal matrix elements of the operator $\hat{V}$ are nonzero. However, in the first case they are attributed only to the operator $\hat{H}_{out}$, whereas in the second case they result from the operator $\hat{H}_{hfs}$ and the operator $\hat{H}_{out}$, depending on the characteristics of the perturbation and on the choice of coordinate system, may or may not contribute to the off-diagonal matrix elements of $\hat{V}$.

As can be seen, all the perturbation operators are defined through the angular momentum operator. Therefore, by determining their matrix elements, one can also find the matrix elements of any operator V, within the scope of the above-discussed interactions. The angular momentum operators satisfy the following equations [5.7]:

$$\hat{J}_z \Psi_{JM} = M \Psi_{JM} ,$$

$$\hat{J}_+ \Psi_{JM} = (\hat{J}_x + i\hat{J}_y) \Psi_{JM} = \sqrt{(J-M)(J+M+1)} \Psi_{J,M+1} ,$$

$$\hat{J}_- \Psi_{JM} = (\hat{J}_x - i\hat{J}_y) \Psi_{JM} = \sqrt{(J+M)(J-M+1)} \Psi_{J,M-1} , \tag{5.137}$$

$$\hat{J}^2 \Psi_{JM} = J(J+1) \Psi_{JM} ,$$

from which one can obtain matrix elements in the decoupled functions

$$\langle J, M+1|\hat{J}_x|J, M\rangle = \frac{1}{2}\sqrt{(J-M)(J+M+1)}$$

$$\langle J, M-1|\hat{J}_x|J, M\rangle = \frac{1}{2}\sqrt{(J+M)(J-M+1)}$$

$$\langle J, M+1|\hat{J}_y|J, M\rangle = -\frac{i}{2}\sqrt{(J-M)(J+M+1)}$$ (5.138)

$$\langle J, M-1|\hat{J}_y|J, M\rangle = \frac{i}{2}\sqrt{(J+M)(J-M+1)}$$

(by making the replacements $\hat{J}\to\hat{I}$, $J\to I$ and $M\to m_I$, one can similarly derive the equations for the nuclear moments).

The matrix elements of the products of the operators $V_{nn'}$ are

$$V_{nn'} = \langle n|\hat{V}^{(1)}\hat{V}^{(2)}|n'\rangle = \sum_k \langle n|\hat{V}^{(1)}|k\rangle\langle k|V^{(2)}|n'\rangle \ . \tag{5.139}$$

By using this rule it is not difficult to work out the matrix elements of the operators $\hat{J}_x^2$, $\hat{J}_x\hat{J}_y$, $\hat{I}\hat{J}$, etc. For example,

$$\langle J, M+1|\hat{J}_x^2|J, M\rangle = \sum_k \langle J, M+1|\hat{J}_x|J, k\rangle\langle J, k|\hat{J}_x|J, M\rangle \ .$$

Because either the first or the second factor must be zero, we have

$$\langle J, M+1|\hat{J}_x^2|J, M\rangle = 0 \ .$$

5.11   The Zeeman Effect in Atoms with Hyperfine Structure

When treating atoms in only a magnetic field, it is quite natural to choose the quantization axis along the direction of the magnetic field. The Hamiltonian of the atom can then be written

$$\hat{H} = \hat{H}_0 + A\,\hat{I}\cdot\hat{J} + \hat{B} + g_J\mu_0\mathcal{H}\hat{J}_z - g_I\mu_0\mathcal{H}\hat{I}_z \ . \tag{5.140}$$

The most interesting case is when $\hat{H}_{\text{hfs}}$ and $\hat{H}_{\text{out}}$ are of the same order of magnitude (intermediate fields). It is precisely in this perturbation range that one has the level degeneracy that is responsible for the interference phenomena. When solving such a problem, one includes in the perturbation operator both the operators $\hat{H}_{\text{out}}$ and $\hat{H}_{\text{hfs}}$:

$$\hat{H} = \hat{H}_0 + \hat{V} \ ,$$

$$\hat{V} = A\,\hat{\boldsymbol{I}}\cdot\hat{\boldsymbol{J}} + \hat{B} + g_J\mu_0\mathcal{H}\hat{J}_z - g_I\mu_0\mathcal{H}\hat{I}_z \ . \tag{5.141}$$

Equation (5.131) contains the matrix elements V_{ik}, which are conveniently rewritten

$$V_{ik} = \langle i|A\,\hat{\boldsymbol{I}}\cdot\hat{\boldsymbol{J}} + \hat{B}|k\rangle + \langle i|g_J\mu_0\mathcal{H}\hat{J}_z - g_I\mu_0\mathcal{H}\hat{I}_z|k\rangle \ . \tag{5.142}$$

The operator $\hat{H}_{\text{ext}}$ is diagonal in terms of the decoupled function, and the hyperfine operator $\hat{H}_{\text{hfs}}$ is expressed in terms of the hyperfine functions. If the hyperfine functions are chosen as basis functions, then one can solve the problem by replacing the hyperfine functions in the second term by their expansions in terms of the functions Φ_{JMIm_I}. If, however, the basis is taken as the decoupled functions, then it is natural to replace Φ_{JMIm_I} in the first term by its expansion (5.88):

$$V_{ik} = \sum_{F,F'} C^{Fm_F}_{JM^iIm_I^i} C^{F'm_{F'}}_{JM^kIm_I^k}\langle Fm_F|A\,\hat{\boldsymbol{I}}\cdot\hat{\boldsymbol{J}} + \hat{B}|F'm_{F'}\rangle$$

$$+ \langle JIM^im_I^i|g_J\mu_0\mathcal{H}\hat{J}_z - g_I\mu_0\mathcal{H}\hat{I}_z|JIM^km_I^k\rangle \ . \tag{5.143}$$

Now, since in the expression for V_{ik} there remain only diagonal matrix elements, these can be replaced by the eigenvalues to yield

$$V_{ik} = \sum_F C^{Fm_F}_{JM^iIm_I^i} C^{Fm_F}_{JM^kIm_I^k}\left(\frac{AC}{2} + B\frac{\frac{3}{4}C(C+1) - I(I+1)J(J+1)}{2I(2I-1)J(2J-1)}\right)$$

$$+ \mu_0\mathcal{H}(g_JM^i - g_Im_I^i)\delta_{ik} \ ; \tag{5.144}$$

$$C = F(F+1) - I(I+1) - J(J+1) \ .$$

The other method of tackling the problem involves directly solving the matrix elements of the operators $\hat{A}$ and $\hat{B}$ in the functions Φ_{JMIm_I}; this can be done using the formulae given in the previous section.

Thanks to the properties of the matrix elements, the system of equations (5.131) decomposes into a series of independent subsystems, each of which links states with the same quantum number $m = M + m_I$. The derived systems of equations can rarely be solved completely in analytical form. In most cases the problem is solved with the help of a computer.

An analytical solution can be found for an atom with arbitrary nuclear spin, but only for the level with $J = 1/2$ (or vice versa). We will discuss this case in detail, and, as will be seen here below, we will obtain the well-known Breit–Rabi formula. Let us take the functions Φ_{JMIm_I} as basis functions. There will be $(2J+1)(2I+1)$, i.e. $2(2I+1)$ such independent functions. The operator $\hat{H}_{\text{out}}$ contributes only to the diagonal matrix elements V_{kk}. In the operator

$\hat{H}_{hfs}$ the term $\hat{B}$ is zero because the electron orbital with $J = 1/2$ does not interact with the nuclear quadrupole moment and $\hat{H}_{hfs} = A\hat{I}\hat{J}$. In order to find the matrix elements of this operator, we expand it into components over circular unit vectors

$$\hat{H}_{hfs} = A\left[\hat{I}_z\hat{J}_z + \frac{(\hat{J}_x + i\hat{J}_y)(\hat{I}_x - i\hat{I}_y)}{2} + \frac{(\hat{J}_x - i\hat{J}_y)(\hat{I}_x + i\hat{I}_y)}{2} \right]. \tag{5.145}$$

The part of the operator $\hat{H}_{hfs}$ which is equal to $A\hat{I}_z\hat{J}_z$ contributes only to the diagonal matrix elements and is equal to

$$\langle JMIm_I | \hat{H}_{hfs} | JMIm_I \rangle = AMm_I . \tag{5.146}$$

Only the last two terms of the operator $\hat{H}_{hfs}$ (5.145) can contribute to the off-diagonal matrix elements of the operator $\hat{V}$. In accordance with the formulae of the previous section, they give non-zero matrix elements only between the functions Φ_{JMIm_I} and $\Phi_{JM \pm 1Im_I}$. For our special case, the system of equations (5.131) can be rewritten as (here we drop the indices J and I, which are identical for all functions, and take into consideration that $M = 1/2$ and $-1/2$, and that m_I runs through the values from I to $-I$):

$$C_1(\langle M = \tfrac{1}{2}m_I = I | \hat{V} | \tfrac{1}{2} I \rangle - \varepsilon) = 0 ,$$

$$C_2(\langle -\tfrac{1}{2}I | \hat{V} | -\tfrac{1}{2}I \rangle - \varepsilon) + C_3\langle -\tfrac{1}{2}I | \hat{V} | \tfrac{1}{2}I - 1 \rangle = 0 ,$$

$$C_2\langle \tfrac{1}{2}I - 1 | \hat{V} | -\tfrac{1}{2}I \rangle + C_3(\langle \tfrac{1}{2}I - 1 | \hat{V} | \tfrac{1}{2}I - 1 \rangle - \varepsilon) = 0 ,$$

$$\vdots \tag{5.147}$$

$$C_{n-2}(\langle -\tfrac{1}{2} - I + 1 | \hat{V} | -\tfrac{1}{2} - I + 1 \rangle - \varepsilon)$$

$$+ C_{n-1}\langle -\tfrac{1}{2} - I + 1 | \hat{V} | \tfrac{1}{2} - I \rangle = 0 ,$$

$$C_{n-2}\langle \tfrac{1}{2} - I | \hat{V} | -\tfrac{1}{2} - I + 1 \rangle + C_{n-1}(\langle \tfrac{1}{2} - I | \hat{V} | \tfrac{1}{2} - I \rangle - \varepsilon) = 0 ,$$

$$C_n(\langle -\tfrac{1}{2} - I | \hat{V} | -\tfrac{1}{2} - I \rangle - \varepsilon) = 0 .$$

As can be inferred, the system (5.131) decomposes into $2I$ subsystems of "blocks" containing two equations with two unknowns, and two independent equations in which the off-diagonal matrix elements $V_{ik \pm i}$ are absent. These two single equations determine the energies $\varepsilon_{1,n}$ of the sublevels with maximum and minimum values of the sum $m = M + m_I$.

The diagonal matrix elements of the operator $\hat{V}$ are found using the formulae (5.137, 143–146):

$$\langle \tfrac{1}{2}I | \hat{V} | \tfrac{1}{2}I \rangle = \tfrac{1}{2}AI + \mu_0(g_J\tfrac{1}{2} - g_I I)\mathcal{H} = \varepsilon_1 ,$$

$$\langle -\tfrac{1}{2} - I | \hat{V} | -\tfrac{1}{2} - I \rangle = \tfrac{1}{2}AI - \mu_0(g_J\tfrac{1}{2} - g_I I)\mathcal{H} = \varepsilon_n . \tag{5.148}$$

Equation (5.148) shows that the energies of these states vary linearly with the magnetic field, and indicate, furthermore, that their wavefunctions are not changing.

The determinant of each of the remaining blocks of equations must be equal to zero if the system is to have a solution:

$$\begin{vmatrix} V_{ii} - \varepsilon & V_{ik} \\ V_{ki} & V_{kk} - \varepsilon \end{vmatrix} = 0$$

From this it follows that

$$\varepsilon_{q,p} = \frac{V_{ii} + V_{kk}}{2} \pm \frac{1}{2} \sqrt{(V_{ii} - V_{kk})^2 + 4V_{ik}^2} \ . \tag{5.149}$$

Making use of the notation $m_I + M = m$, the equations (5.147) and (5.149) can be written in a generalized form. Using (5.138, 136, 145) and noting that

$$V_{ii} = (H_{\mathrm{hfs}})_{ii} + (H_{\mathrm{out}})_{ii}; \qquad V_{ik \neq i} = (H_{\mathrm{hfs}})_{ik \neq i} \ , \tag{5.150}$$

setting $M_i = 1/2$ and $M_k = -1/2$, we obtain the following expressions for the matrix elements

$$V_{ii} = -\frac{A}{4} + \frac{Am}{2} + \mu_0 (g_J \tfrac{1}{2} - g_I (m - \tfrac{1}{2})) \mathscr{H} \ ,$$

$$V_{kk} = -\frac{A}{4} - \frac{Am}{2} - \mu_0 (g_J \tfrac{1}{2} + g_I (m + \tfrac{1}{2})) \mathscr{H} \ , \tag{5.151}$$

$$V_{ik} = V_{ki} = \frac{A}{2} \sqrt{(I + \tfrac{1}{2})^2 - m^2} \ .$$

Substituting these expressions for the matrix elements into formula (5.149) for ε and introducing the conventional notation

$$\delta w_0 = \frac{A}{2} (2I + 1) \ ,$$

$$x = \frac{g_J + g_I}{\delta w_0} \mu_0 \mathscr{H} \ , \tag{5.152}$$

we find the formula, known as the Breit–Rabi formula[4]

$$\varepsilon_{q,p} = -\frac{\delta w_0}{2(2I + 1)} - m \mu_0 g_I \mathscr{H} \pm \frac{\delta w_0}{2} \sqrt{1 + \frac{4m}{2I + 1} x + x^2} \ . \tag{5.153}$$

---

[4] In the literature concerning the Breit–Rabi formula, one finds in the definition of the value x both the plus and the minus sign in front of g_I. The consequence of the definition of g_I assumed by us (Sect. 5.2) is a plus sign in front of g_I and a minus sign before the second term in (5.153).

In the operator $\hat{V}$ the operators $\hat{J}$ and $\hat{I}$ enter symmetrically. Therefore, the problem for arbitrary J and for $I = 1/2$ is solved in the same way and the final formula will differ from (5.153) in that here, and also in (5.152), I will replace J, in place of g_I will be $- g_J$ and g_J will become $- g_I$.

The expansion coefficients C of the function Ψ expressed in terms of Φ,

$$\Psi'_p = C_i^p \Phi_i + C_k^p \Phi_k \ ,$$
$$\Psi'_q = C_i^q \Phi_i + C_k^q \Phi_k \ , \tag{5.154}$$

are found by substituting the known roots and matrix elements in the system (5.147), which for coefficients of the function Ψ'_p has the form

$$C_i^p(V_{ii} - \varepsilon_p) + C_k^p V_{ik} = 0 \ ,$$
$$C_i^p V_{ik} + C_k^p(V_{kk} - \varepsilon_p) = 0 \ . \tag{5.155}$$

For the second function Ψ'_q, the quantity ε_p will be replaced by ε_q.

The solutions of the equations (5.155) for the two roots ε_p and ε_q together with the normalization condition

$$(C_i^p)^2 + (C_k^p)^2 = 1 \ ,$$
$$(C_i^q)^2 + (C_k^q)^2 = 1 \ , \tag{5.156}$$

give

$$C_i^q = - C_k^p = \left(\frac{1}{2} + \frac{Am + \mu_0(g_J + g_I)\mathcal{H}}{2\delta w_0 \sqrt{1 + \dfrac{4m}{2I+1}x + x^2}} \right)^{1/2} \ ,$$

$$C_i^p = C_k^q = \left(\frac{1}{2} - \frac{Am + \mu_0(g_J + g_I)\mathcal{H}}{2\delta w_0 \sqrt{1 + \dfrac{4m}{2I+1}x + x^2}} \right)^{1/2} \ . \tag{5.157}$$

5.12 The Paschen–Back Effect

The Paschen–Back effect refers to the splitting of levels in strong magnetic fields, in which the hyperfine splitting is much smaller than the magnetic one. In such a case, one can first solve the problem with Hamiltonian

$$\hat{H} = \hat{H}_0 + \hat{H}_M \ ,$$

and find afterwards the perturbation due to the electron–nucleus interaction. One must, of course, take Φ as the basis functions. In this representation, the operator $\hat{H}_M$ is diagonal and its eigenvalues can be written as

$$\langle JMIm_I|\hat{H}_M|JMIm_I\rangle = \mu_0(g_J M - g_I m_I)\mathscr{H} \ . \tag{5.158}$$

Its eigenfunctions will also be the basis functions Φ. The total number of levels and different wavefunctions will be $(2J + 1)(2I + 1)$. However, since $g_I \ll g_J$, the initial level splits essentially into $2J + 1$ components, which diverge at a "speed" of $\mu_0 g_J$. Each of these splits into a further $2I + 1$ components, which, in a first approximation, because of the smallness of g_I, are assumed to be degenerate, i.e. to coincide in energy. Each such degenerate state can then be treated individually and one can find the change of its structure due to the electron–nucleus interaction, i.e. using the operator $\hat{H}_{hfs}$. With these constraints, in each individual problem the quantum number M is fixed at one value and the matrix elements of the perturbation operator in terms of the functions Φ possess only diagonal terms (Sect. 5.11). This means that the wavefunction does not change and the energy variation is expressed by

$$\Delta\varepsilon_i = V_{ii} = \langle JMIm_I|\hat{A} + \hat{B}|JMIm_I\rangle \ . \tag{5.159}$$

Taking into account the level energy for a given M in a magnetic field, we find the following expression for the state energy with the given values M and m_I using (5.146) and (5.64):

$$\varepsilon_{Mm_I} = AMm_I + B\frac{3(Mm_I)^2 + \frac{3}{2}Mm_I - I(I + 1)J(J + 1)}{2J(2J - 1)I(2I - 1)}$$

$$+ \mu_0(g_J M - g_I m_I)\mathscr{H} \ . \tag{5.160}$$

The basis functions Φ are the eigenfunctions of the electron states in a magnetic field. Therefore, they often go by the name of strong field functions.

5.13 Hyperfine Splitting in a Weak Magnetic Field

A field is considered to be weak if the magnetic interaction is significantly smaller than the hyperfine interaction. This allows one to treat each of the hyperfine levels individually. Let us express the Hamiltonian as

$$\hat{H} = \hat{H}_0 + \hat{H}_{hfs} + \hat{H}_M = \hat{H}'_0 + \hat{H}_M = \hat{H}'_0 + \hat{V} \ . \tag{5.161}$$

In this the hyperfine interaction is included in the main Hamiltonian $\hat{H}'_0$, and the perturbation operator contains only the interaction with the magnetic field. As eigenfunctions one must of course choose the eigenfunctions of the Hamiltonian $\hat{H}'_0$, i.e. the functions Ψ, and from these we select a set with a single value of F. This also implies of course that the quantum numbers J and I will be the same. The perturbation operator has the form

$$\hat{V} = \hat{H}_{\text{M}} = \mu_0(g_J\hat{\boldsymbol{J}} - g_I\hat{\boldsymbol{I}}) \cdot \boldsymbol{\mathscr{H}} = \mu_0 g_F \hat{\boldsymbol{F}} \cdot \boldsymbol{\mathscr{H}} \ . \tag{5.162}$$

The off-diagonal matrix elements of this operator are zero when $\boldsymbol{\mathscr{H}} \| z$ as a consequence of the relationship (5.138). The diagonal elements directly describe the change of energy

$$\varepsilon_{m_F} = \langle JIFm_F | \hat{V} | JIFm_F \rangle = \mu_0 g_F m_F \mathscr{H} \ . \tag{5.163}$$

In weak fields a hyperfine sublevel thus splits into $2F + 1$ sublevels, which diverge linearly with the magnetic field at a "rate" of $\mu_0 g_F$.

Of course, an exact solution of the problem gives correct energies and wavefunctions in arbitrary fields; the approximate solutions in "weak" and "strong" fields attract attention only because of their simplicity.

5.14   The Stark Effect in Atoms with Hyperfine Structure in a Weak Electric Field

Similar to the above treatment of the interaction of an atom with a weak magnetic field (Sect. 5.13), let us examine a single hyperfine sublevel in a weak electric field. The Hamiltonian has the form

$$\hat{H} = \hat{H}_0 + \hat{H}_{\text{hfs}} + \hat{H}_{\text{e}} = \hat{H}_0' + \hat{H}_{\text{e}} = \hat{H}_0' + \hat{V} \ . \tag{5.164}$$

The perturbation operator (5.127)

$$\hat{V} = \alpha \mathscr{E}^2 + \beta \left(\hat{J}_z^2 - \frac{J(J+1)}{3} \right) \mathscr{E}^2$$

is not the characteristic operator of the hyperfine wavefunctions Ψ. Its matrix elements between these functions can be found by expanding the function Ψ_{JIFm_F} in terms of the eigenfunctions of the operator $\hat{V}$, i.e. in terms of the functions Φ_{JMIm_I} in accordance with (5.86, 87). However, one can solve the problem differently. In Sect. 5.8 we derived a formula (5.119a) for the energy level shift due to an electric field

$$\varepsilon_F = \sum_k \frac{V_{kF} V_{Fk}}{E_F - E_k} \ ,$$

where V_{kF} and V_{Fk} are matrix elements of the dipole moment (5.110):

$$V_{kF} = -\langle \Psi_k | \boldsymbol{d} \cdot \boldsymbol{\mathscr{E}} | \Psi_F \rangle \ .$$

The functions Ψ are eigenfunctions of the full Hamiltonian minus the electric field perturbation, i.e. of the Hamiltonian

$$\hat{H}_0' = \hat{H}_0 + \hat{H}_{\text{hfs}}$$

and these are the hyperfine wavefunctions Ψ.

The matrix elements of $\hat{d}$ between the hyperfine functions were defined in Sect. 5.7 by the formulae (5.94, 95). Let us substitute them into the formula for the energy ε_F, noting that, when $\mathscr{E} \parallel z$, the index ν will have only one value, $\nu = 0$; moreover, let us replace the coefficient C^2 by its explicit form and multiply it by $(2F_k + 1)/(2F_k + 1)$ in a similar way as was done in Sect. 5.8. As a result we find

$$\varepsilon_F = \alpha_F \mathscr{E}^2 + \beta_F \left(m_F^2 - \frac{F(F+1)}{3} \right) \mathscr{E}^2 , \tag{5.165}$$

where

$$\alpha_F = \sum_k \Pi_{J_F J_{k'}}^2 \frac{2F_k + 1}{3(E_F - E_k)} \left\{ \begin{matrix} I & J_k & F_k \\ 1 & F & J_F \end{matrix} \right\}^2 , \tag{5.166}$$

$$\beta_F = \sum_k \Pi_{J_F J_{k'}}^2 \frac{2F_k + 1}{E_F - E_k} \left\{ \begin{matrix} I & J_k & F_k \\ 1 & F & J_F \end{matrix} \right\}^2$$

$$\times \left(-\frac{\delta_{F,F_k+1}}{F(2F-1)} + \frac{\delta_{F,F_k}}{F(F+1)} - \frac{\delta_{F,F_k-1}}{(F+1)(2F+3)} \right) . \tag{5.167}$$

It is thereby established that the hyperfine level splits and changes its energy in the same way as a level of the finestructure. The "rate" of energy variation is now given by other coefficients, which we denote by α_F and β_F, in order to emphasize their relationship to the hyperfine structure and to other quantum numbers m_F. In this approximation the wavefunctions will remain unchanged.

The constant α_F differs from α but this difference is quite small. Actually, if one neglects the hyperfine splitting, then α_F can be written as

$$\alpha_F \approx \sum_{k'} \frac{\Pi_{J_F J_{k'}}}{3(E_F - E_{k'})} \sum_k (2F_k + 1) \left\{ \begin{matrix} I & J_{k'} & F_k \\ 1 & F & J_F \end{matrix} \right\}^2 , \tag{5.168}$$

where the index k' labels the levels with different J and the index k labels levels with different quantum numbers F. However, due to the properties of the 6j-symbols the second sum in (5.168) is equal to one. The radial part of the matrix element of the dipole moment is thus identical for all hyperfine sublevels, therefore

$$\alpha_F = \alpha .$$

The relationship between the coefficients β_F and β in this same approximation is rather complex. It is expressed by a coefficient that depends on the quantum numbers F, J, I:

$$\beta_F = \beta \frac{3C(2C - 1) - 2F(F + 1)J(J + 1)}{F(F + 1)(2F + 3)(2F - 1)} , \qquad (5.169)$$

where

$$C = [F(F + 1) - J(J + 1) - I(I + 1)] .$$

5.15   The Stark Effect in Atoms with Hyperfine Structure in Intermediate Fields

An atom with a nonzero nuclear spin in an electric field is described by the Hamiltonian

$$\hat{H} = \hat{H}_0 + \hat{H}_{hfs} + \hat{H}_e = \hat{H}'_0 + \hat{V} ,$$

where, for $\mathscr{E} \parallel z$ (5.117):

$$\hat{H}_e = \alpha \mathscr{E}^2 + \beta \left(\hat{J}_z^2 - \frac{J(J + 1)}{3} \right) \mathscr{E}^2 .$$

As is the case for an atom with zero nuclear spin, the splitting is absent for states with $J = 0$ and $J = 1/2$. Actually, in the former case

$$\hat{H}_e = \alpha \mathscr{E}^2 , \qquad (5.170)$$

i.e. the displacement is identical for all hyperfine sublevels and it can be included in the fundamental operator $\hat{H}'_0$, determining the position of the centre of gravity of the structure. For $J = 1/2$ the eigenvalue of the expression in the bracket (5.127) becomes zero

$$M^2 - \frac{J(J + 1)}{3} = 0 .$$

Furthermore $\beta = 0$, which is related to the fact that the 6j-symbol in the formula (5.126) becomes zero and once again

$$\hat{H}_e = \alpha \mathscr{E}^2 . \qquad (5.170)$$

When $J > 1/2$ the problem can generally be solved by numerical techniques according to the usual rules described in Sect. 5.9. However, if the nuclear spin $I = 1/2$, then an analytical solution is possible. In this case $F_1 = J + 1/2$ and $F_2 = J - 1/2$. In the same way as for the state with $J = 1/2$ in a magnetic field, the secular equation (5.132) decomposes into 2×2 blocks, which lead to second-order equations for ε. Their solutions are given by the formula (5.149). In order to determine the explicit form of ε, one needs to find the matrix elements V_{ii}, V_{kk} and $V_{ik} = V_{ki}$. As basis functions, we take the strong field functions Φ_{JMIm_I}. Then the indices i and k will be given identically for both sublevels by the quantum number $m = M + m_I$ and by different numbers m_I, taking only two values $\pm 1/2$.

Let the index i correspond to $m_I = 1/2$ and let k correspond to $m_I = -1/2$. Then

$$V_{ii} = \langle JM_i Im_I | \hat{V} | JM_i Im_I \rangle \ . \tag{5.171}$$

When $I = 1/2$ the operator $\hat{B} = 0$ and $\hat{H}_{hfs} = \hat{A}$. Noting also that $M = m \pm 1/2$, we obtain the following expressions

$$V_{ii} = -\frac{A}{4} + \frac{Am}{2} + \beta \left(m^2 - m + \frac{1}{4} - \frac{J(J+1)}{3} \right) \mathscr{E}^2 \ ,$$

$$V_{kk} = -\frac{A}{4} - \frac{Am}{2} + \beta \left(m^2 + m + \frac{1}{4} - \frac{J(J+1)}{3} \right) \mathscr{E}^2 \ , \tag{5.172}$$

$$V_{ik} = V_{ki} = \frac{A}{2} \sqrt{ \left(J + \frac{1}{2} \right)^2 - m^2 }$$

and finally, for $\varepsilon_{q,p}$ we get

$$\varepsilon_{q,p} = \frac{A}{4} + \beta \left(m^2 + \frac{1}{4} - \frac{J(J+1)}{3} \right) \mathscr{E}^2$$

$$\pm \frac{A}{2} \sqrt{ \left(\frac{2\beta \mathscr{E}^2 m}{A} \right)^2 - \frac{4\beta \mathscr{E}^2 m^2}{A} + \left(J + \frac{1}{2} \right)^2 } \ . \tag{5.173}$$

This formula shows that the separation of the levels $|q\rangle$ and $|p\rangle$, decoupled in a zero field by the hyperfine splitting

$$\delta w_{hfs} = \delta w_0 = A \frac{2J+1}{2} \ ,$$

varies with the electric field as

$$\Delta w_{q,p} = \sqrt{ (2\beta \mathscr{E}^2 m)^2 - 4\beta \mathscr{E}^2 m^2 A + A^2 (J + \tfrac{1}{2})^2 } \ . \tag{5.174}$$

The dependence of $\varepsilon_{q,p}$ on $\mathscr{E}^2$ describes the splitting of the hyperfine structures into $F + 1$ components for the integer values of F and into $(2F + 1)/2$ components for the half-integral values since the number of values $|m|$ is determined in exactly this way by the quantum numbers F.

Let us introduce the notation

$$x_e = \frac{2\beta \mathscr{E}^2}{\delta w_0} \tag{5.175}$$

and rewrite (5.173) in the form

$$\varepsilon_{q,p} = -\frac{\delta w_0}{2(2J+1)} + \beta \left(m^2 + \frac{1}{4} - \frac{J(J+1)}{3} \right) \mathscr{E}^2$$

$$\pm \frac{\delta w_0}{2} \sqrt{x_e^2 m^2 - 2Am^2 x + 1} \ . \tag{5.176}$$

In this form, the expression is reminiscent of the Breit–Rabi formula. Using the same scheme as in Sect. 5.9, we find the expansion coefficients of the wavefunctions Ψ' that describe the atomic system in an electric field

$$C_i^q = -C_k^p = \left(\frac{1}{2} + \frac{(A - 2\beta\mathscr{E}^2)m}{2\delta w_0 \sqrt{m^2 x_e^2 - 2Am^2 x_e + 1}} \right)^{1/2} ,$$

$$\tag{5.177}$$

$$C_i^p = C_k^q = \left(\frac{1}{2} - \frac{(A - 2\beta\mathscr{E}^2)m}{2\delta w_0 \sqrt{m^2 x_e^2 - 2Am^2 x_e + 1}} \right)^{1/2} .$$

For the two states with $m = M + m_I = \pm(J + 1/2)$ the matrix elements $V_{ik} = V_{ki} = 0$ and the energy is determined from the linear equations

$$\varepsilon_p = \varepsilon_q = V_{ii} = V_{kk}$$

and expressed by

$$\varepsilon = \beta \left(\frac{2J(J+1)}{3} + \frac{1}{4} \right) \mathscr{E}^2 \ . \tag{5.178}$$

The wavefunctions do not depend on the electric field strength $\mathscr{E}$.

5.16 Splitting of Atomic Levels with Hyperfine Structure in a Strong Electric Field

Let us define the strong electric field as we did the strong magnetic field, i.e. it is a field whose perturbation is much greater than the hyperfine splitting. We now include the action of the electric field in the fundamental Hamiltonian

$$\hat{H}_0' = \hat{H}_0 + \hat{H}_e$$

and the interaction with the atomic nucleus becomes the perturbation operator

$$\hat{V} = \hat{H}_{hfs} \ .$$

As the basis functions we take the broken link functions Φ_{JMIm_I}, the eigenfunctions of the operator H_0'. The electric field does not interact with the nuclear spin of the atom and therefore the operator H_e does not act on the nuclear component of the strong field function, i.e. on the function Y_{Im_I} (5.87) and the residual energy of the free atom in the state under discussion is determined by the formula (5.124). In strong fields one can ignore the interaction between different hyperfine states, except the two special cases. As follows from Sect. 5.10, the matrix elements of the hyperfine interaction operator $\hat{H}_{hfs}$ differ from zero only between states with $\Delta M = \pm 1, 0$; $\Delta m_I = \pm 1.0$, which is attributed to the operator $\hat{A}$ and between states with $\Delta M = \pm 2$ as a result of the structure of the operator $\hat{B}$. The electric field causes an identical shift of the levels with the same $|M|$, in particular, the states $M = \pm 1/2$ and $M = \pm 1$. There is a hyperfine interaction between these states and the matrix elements

$$\langle J \tfrac{1}{2} I m_I | A \hat{I} \hat{J} | J - \tfrac{1}{2} I m_I + 1 \rangle \tag{5.179}$$

and

$$\langle J \, 1 \, I \, m_I | \hat{B} | J - 1 \, I \, m_I + 2 \rangle$$

cannot be neglected. However, for states with $|M| \neq 1/2$ one can neglect in the secular equation the terms $V_{ik \neq i}$. Then the secular equation (5.132) decomposes into a series of independent linear equations and blocks, leading to second-order equations.

For states with $|M| \neq 1/2$, the energy variation is described simply by the matrix element (5.159):

$$\varepsilon = \langle JIMm_I | \hat{A} + \hat{B} | JIMm_I \rangle \, ,$$

which has already been derived in Sect. 5.12, and the total energy can be written

$$E_{Mm_I} = E_0 + \alpha \mathscr{E}^2 + \beta \left(M^2 - \frac{J(J+1)}{3} \right) \mathscr{E}^2 + AMm_I$$

$$+ B \frac{3(Mm_I)^2 + \tfrac{3}{2}Mm_I - I(I+1)J(J+1)}{2J(2J-1)I(2I-1)} \, , \tag{5.180}$$

i.e. each energy level splits with respect to M and m_I. The wavefunctions do not change.

From the second-order equations we find an expression for the energy states $|q\rangle$ and $|p\rangle$ formed upon mixing the states with $M = \pm 1/2$:

$$\varepsilon_{q,p} = \alpha \mathscr{E}^2 + \beta \left(\frac{1}{4} - \frac{J(J+1)}{3} \right) \mathscr{E}^2 - \frac{A}{4}$$

$$+ B \frac{\tfrac{3}{4}(m^2 - \tfrac{1}{4}) - I(I+1)J(J+1)}{2J(2J-1)I(2I-1)}$$

$$\pm \frac{A}{2} \sqrt{(J+\tfrac{1}{2})^2 [(I+\tfrac{1}{2})^2 - m^2] + m^2} \, , \tag{5.181}$$

and for the coefficients in the expansion of the unknown wavefunctions over the basis functions

$$C^q_{+1/2} = -C^p_{-1/2} = \left(\frac{1}{2} + \frac{m}{\sqrt{(J + \frac{1}{2})^2[(I + \frac{1}{2})^2 - m^2] + m^2}}\right)^{1/2},$$

$$C^q_{-1/2} = C^p_{1/2} = \left(\frac{1}{2} - \frac{m}{\sqrt{(J + \frac{1}{2})^2[(I + \frac{1}{2})^2 - m^2] + m^2}}\right)^{1/2}.$$

(5.182)

The action of the electric field on an atom in the state $J = 1/2, I = 1/2, m = 0$ can be described by the formula (5.181) previously derived, and as well by the formula (5.176), obtained when describing the action of intermediate fields, since this state simultaneously satisfies the conditions under which these formulae are valid. It is not difficult to confirm that, by substituting these quantum numbers into both formulae, one will obtain the same expression.

If, however, a state is characterized by a quantum number $M = \pm 1$ and $\hat{B} \ne 0$ (i.e. $I > 1/2$), then

$$\varepsilon_{q, p} = \alpha \mathscr{E}^2 + \beta\left(1 - \frac{J(J + 1)}{3}\right)\mathscr{E}^2 + Am$$

$$+ B\frac{3m(m + \frac{3}{2}) - I(I + 1)J(J + 1)}{2J(2J - 1)I(2I - 1)} \pm y,$$

(5.183)

$$C^q_{+1} = -C^p_{-1} = \left[\frac{1}{2} - \frac{1}{y}\left(2A + 3B\frac{4m - 1}{2J(2J - 1)I(2I - 1)}\right)\right]^{1/2},$$

$$C^p_{+1} = C^q_{-1} = \left[\frac{1}{2} + \frac{1}{y}\left(2A + 3B\frac{4m - 1}{2J(2J - 1)I(2I - 1)}\right)\right]^{1/2},$$

(5.184)

where

$$y = \frac{1}{2}\left[\left(2A + 3B\frac{4m - 1}{2J(2J - 1)I(2I - 1)}\right)^2\right.$$

$$\left. + B^2\frac{J^2(J + 1)^2[(I + 1)^2 - m^2](I^2 - m^2)}{[2J(2J - 1)I(2I - 1)]^2}\right]^{1/2}.$$

5.17 Behaviour of Atoms in Combined Fields

The term "combined fields" covers a whole multitude of situations. Firstly one can have a magnetic and an electric field in the same direction, perpendicular to one another or directed arbitrarily. One can also consider combined fields that are two components of the magnetic or the electric field. Such an approach is useful when one wants to examine the response of an atomic system to the variation of one of the components.

We begin our discussion with the simplest case of parallel electric and magnetic fields. The Hamiltonian for the atom has the form

$$\hat{H} = \hat{H}_0 + \hat{H}_{hfs} + \hat{H}_M + \hat{H}_e \ . \tag{5.185}$$

As usual, one must choose the z-axis along the direction of the field vectors in order to simplify the form of operators. In a general case, as holds also for the action of one field, the problem can be solved by numerical techniques in accordance with the scheme of Sect. 5.9, with the only difference that the perturbation operator $\hat{v}$ now includes $\hat{H}_{hfs}$, H_M and H_e. If one adopts the broken link functions as the basis functions, then the matrix elements entering in the secular equation will contain the very same expressions as found in the cases of magnetic and electric fields (Sects. 5.12, 16).

If both the fields are weak, then one can assume that the wavefunctions will remain the same, as in the absence of fields, i.e. one assumes the hyperfine wavefunctions. The energy change is described (4.163, 165) by

$$\varepsilon = \alpha_F \mathscr{E}^2 + \beta_F \left(m_F^2 - \frac{F(F+1)}{3} \right) \mathscr{E}^2 + \mu_0 g_F m_F \mathscr{H} \tag{5.186}$$

i.e. in a fixed magnetic field, the levels shift in proportion to the square of the electric field strength, $\mathscr{E}$, and to the square of the quantum number, m_F, whereas in a fixed electric field they split into two sublevels and shift further, increasing linearly with the magnetic field strength.

If one of the fields is stronger than the other, then one can neglect the interaction between the hyperfine components, i.e. the off-diagonal matrix elements, in the secular equation (except $M = \pm 1/2$ and ± 1), and the energy shift is given approximately by

$$\varepsilon = AMm_I + \alpha \mathscr{E}^2 + \beta \left(M^2 - \frac{J(J+1)}{3} \right) \mathscr{E}^2$$

$$+ \mu_0 (g_J M - g_I m_I) \mathscr{H} \ . \tag{5.187}$$

A rather complex case arises when the fields are mutually perpendicular; this problem cannot be reduced to the above discussed case, which contained only the projection of the operator $\hat{J}$ on the z-axis. Let us assume that only a magnetic field is present. We direct the quantization axes perpendicular to it, and we direct the field, for definiteness, along the x-axis. Then the interaction operator with the magnetic field (Sect. 5.10) has the form

$$\hat{H}_M = \mu_0 (g_J \hat{J}_x - g_I \hat{I}_x) \mathscr{H} \ . \tag{5.188}$$

The matrix elements of the operator $\hat{J}_x$ are given in Sect. 5.10. This operator couples the states with $\Delta M = \pm 1$ when $\Delta m_I = 0$ and those with $\Delta m_I = \pm 1$ when $\Delta M = 0$.

In weak fields one can treat each of the hyperfine structures separately. The interaction operator has the form

$$\hat{H}_{\mathrm{M}} = \mu_0 g_F \hat{F}_x \mathcal{H} \ . \tag{5.189}$$

There then remain only the matrix elements with $\Delta m_F = \pm 1$. For $F = 1/2$ and $F = 1$ the problem can be solved analytically.

In the former case $m_F = \pm 1/2$ and the secular equation takes the form

$$\begin{vmatrix} -\varepsilon & V_{1/2\,-1/2} \\ V_{-1/2\,1/2} & -\varepsilon \end{vmatrix} = 0 \ , \tag{5.190}$$

whence

$$\varepsilon = \pm V_{1/2\,-1/2} = \pm V_{-1/2\,1/2} \ .$$

The explicit form of the matrix element, according to (5.138) is given by

$$V_{1/2\,-1/2} = \mu_0 g_F \langle \tfrac{1}{2}\tfrac{1}{2} | F_x | \tfrac{1}{2} - \tfrac{1}{2} \rangle \mathcal{H} = \tfrac{1}{2} \mu_0 g_F \mathcal{H} \tag{5.191}$$

and then

$$\varepsilon = \pm \tfrac{1}{2} \mu_0 g_F \mathcal{H} \ . \tag{5.192}$$

Thus, as expected, the energy structure is described in the same way as for $\mathcal{H} \parallel z$ with

$$\varepsilon = \pm \tfrac{1}{2} \mu_0 g_F \mathcal{H} \ .$$

Let us now find the expansion coefficients of the wavefunctions over the hyperfine functions. For the state $|p\rangle$ with energy $\varepsilon = -\tfrac{1}{2} \mu_0 g_F \mathcal{H}$ we have

$$- C_{1/2}^p \varepsilon + C_{-1/2}^p V_{1/2\,-1/2} = + \tfrac{1}{2} \mu_0 g_F \mathcal{H} C_{1/2}^p + \tfrac{1}{2} \mu_0 g_F \mathcal{H} C_{-1/2}^p = 0 \ ,$$

which yields

$$+ C_{1/2}^p + C_{-1/2}^p = 0 \quad \text{or} \quad - C_{1/2}^p = C_{-1/2}^p \ .$$

Because

$$(C_{1/2}^p)^2 + (C_{-1/2}^p)^2 = 1 \ ,$$

we have

$$C_{1/2}^p = - C_{-1/2}^p = 1/\sqrt{2} \ .$$

For the other state

$$C_{1/2}^q = C_{-1/2}^q = 1/\sqrt{2} \ .$$

The new wavefunctions consist of equal contributions of the eigenfunctions $\Psi(J, I, F, m_F)$ of the operators $\hat{F}_z$:

$$\Psi_p = [\Psi(J I \tfrac{1}{2} \tfrac{1}{2}) - \Psi(J I \tfrac{1}{2} - \tfrac{1}{2})]/\sqrt{2} ,$$

$$\Psi_q = [\Psi(J I \tfrac{1}{2} \tfrac{1}{2}) + \Psi(J I \tfrac{1}{2} - \tfrac{1}{2})]/\sqrt{2} . \tag{5.193}$$

Now let $F = 1$. This state is characterized by three eigenfunctions of the operator $\hat{F}_z$ with the magnetic quantum numbers $m_F = \pm 1, 0$. The secular equation is

$$\begin{vmatrix} -\varepsilon & V_{10} & 0 \\ V_{01} & -\varepsilon & V_{0-1} \\ 0 & V_{-10} & -\varepsilon \end{vmatrix} = 0 . \tag{5.194}$$

Due to equality of the matrix elements

$$\langle J I 1 m_F | \hat{F}_x | J I 1 m_F \pm 1 \rangle = V_{\pm 10} = V_{0 \pm 1} = \sqrt{2}\,\mu_0 g_F \mathcal{H}/2 \tag{5.195}$$

and one can write

$$\varepsilon(\varepsilon^2 - \tfrac{1}{2}\mu_0^2 g_F^2 \mathcal{H}^2) - \tfrac{1}{2}\varepsilon\mu_0^2 g_F^2 \mathcal{H}^2 = 0 .$$

From this we find

$$\varepsilon_p = 0, \qquad \varepsilon_{q,r} = \pm \mu_0 g_F \mathcal{H} .$$

Once more, it is quite natural that the energy structure is the same as in the case $\mathcal{H} \parallel z$. The wavefunctions are given by

$$\Psi_q = \frac{1}{2}\Psi(J I 1\, 1) + \frac{1}{\sqrt{2}}\Psi(J I 1\, 0) + \frac{1}{2}\Psi(J I 1\, -1) ,$$

$$\Psi_p = \frac{1}{\sqrt{2}}\Psi(J I 1\, 1) - \frac{1}{\sqrt{2}}\Psi(J I 1\, -1) ,$$

$$\Psi_r = \frac{1}{2}\Psi(J I 1\, 1) - \frac{1}{\sqrt{2}}\Psi(J I 1\, 0) + \frac{1}{2}\Psi(J I 1\, -1) .$$

Any problem in which the magnetic field is arbitrarily directed relative to the axis of quantization, which is conveniently related to other physical parameters, can be reduced to the problem of two fields. These fields are mutually perpendicular to one another. The axis of quantization z is parallel to one of them, and the x-axis is directed along the other. Thus, the problem is one of two magnetic fields. The Hamiltonian which is the starting point for all problems of this type has the form

$$\hat{H} = \hat{H}_0 + \hat{H}_{\text{hfs}} + (\hat{H}_M)_x + (\hat{H}_M)_z . \tag{5.196}$$

Upon solving this problem in a general form, the secular equation will contain the matrix elements of the operators $(\hat{H}_M)_z$, as was the case when treating an atom in a "longitudinal" magnetic field (Sect. 5.11), and $(\hat{H}_M)_x$, as in the case that has just been examined. The nonzero terms in the secular equation for this case of two magnetic field components are large, as a result of which the problem turns out to be rather cumbersome.

In order to demonstrate the expected properties of the solution, let us consider a special case that has an analytical solution, namely, the behaviour of a single hyperfine sublevel with $F = 1$ in a weak magnetic field. The secular equation is similar to (5.194) and has the form

$$
\begin{vmatrix}
V_{11} - \varepsilon & V_{10} & 0 \\
V_{01} & V_{00} - \varepsilon & V_{0-1} \\
0 & V_{-10} & V_{-1-1} - \varepsilon
\end{vmatrix} = 0 .
$$

In this, $V_{11} = -V_{-1-1} = \mu_0 g_F \mathscr{H}_z$; $V_{00} = 0$ and the remaining matrix elements have just been determined (5.195). The roots characterizing the energy of the sublevels are

$$
\varepsilon_p = 0 ,
$$

$$
\varepsilon_q = -\varepsilon_r = \mu_0 g_F \sqrt{\mathscr{H}_x^2 + \mathscr{H}_z^2} = \mu_0 g_F \mathscr{H} ,
$$

indicating that the additional energy is determined by the strength of the total magnetic field $\mathscr{H}$.

The expansion coefficients (5.129) calculated with the help of these roots, together with the normalization conditions, determine the form of the wavefunctions

$$
\Psi_q = \frac{\mathscr{H} + \mathscr{H}_z}{2\mathscr{H}} \Psi(F = 1\, m_F = 1) + \frac{\mathscr{H}_x}{\sqrt{2}\mathscr{H}} \Psi(10) + \frac{\mathscr{H} - \mathscr{H}_z}{2\mathscr{H}} \Psi(1-1) ,
$$

$$
\Psi_p = \frac{-\mathscr{H}_x}{\sqrt{2}\mathscr{H}} \Psi(11) + \frac{\mathscr{H}_z}{\mathscr{H}} \Psi(10) + \frac{\mathscr{H}_x}{\sqrt{2}\mathscr{H}} \Psi(1-1) ,
$$

$$
\Psi_r = \frac{\mathscr{H} - \mathscr{H}_z}{2\mathscr{H}} \Psi(11) - \frac{\mathscr{H}_x}{\sqrt{2}\mathscr{H}} \Psi(10) + \frac{\mathscr{H} + \mathscr{H}_z}{2\mathscr{H}} \Psi(1-1) .
$$

Let the field along the x-axis be fixed, whereas the second one is variable. This formulation of the problem corresponds to the real situation of the presence of a fixed laboratory or earth field in the experimental measurements of the radiation intensity as a function of a controlled magnetic field. We must put $\mathscr{H}_x = \text{const}$, and variation of $\mathscr{H}_z$ determines the change of the coefficients in the expansion. The additional energy dependence on $\mathscr{H}_z$ is depicted on Fig. 5.2. The curves q and r behave in exactly the same manner as anti-crossing levels. The

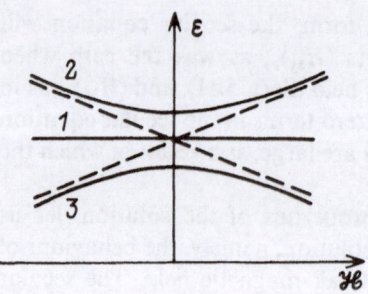

Fig. 5.2. Zeeman structure for the case of two mutually perpendicular external magnetic fields

whole problem can be treated as anti-crossing in which a transverse magnetic field acts as a perturbation which couples the states with $m_F = \pm 1$.

A somewhat simpler case appears to be the solution of a problem for a moment of 1/2. This moment can be the quantum number F for an atom with a hyperfine structure, or J for atoms with a simple nucleus, i.e. without moments. The secular equation has the form

$$\begin{vmatrix} \frac{1}{2}\mu_0 g\mathcal{H}_z - \varepsilon & \frac{1}{2}\mu_0 g\mathcal{H}_x \\ \frac{1}{2}\mu_0 g\mathcal{H}_x & -\frac{1}{2}\mu_0 g\mathcal{H}_x - \varepsilon \end{vmatrix} = 0 \; . \tag{5.197}$$

Its roots are the energies of the Zeeman sublevels:

$$\varepsilon_{q,\,p} = +\frac{1}{2}\mu_0 g\sqrt{\mathcal{H}_x^2 + \mathcal{H}_z^2} = \pm\frac{1}{2}\mu_0 g\mathcal{H}$$

and the wavefunctions are

$$\Psi_q = \left(\frac{\mathcal{H} + \mathcal{H}_z}{2\mathcal{H}}\right)^{1/2} \Psi(\tfrac{1}{2}\tfrac{1}{2}) + \left(\frac{\mathcal{H} - \mathcal{H}_z}{2\mathcal{H}}\right)^{1/2} \Psi(\tfrac{1}{2}\,-\tfrac{1}{2}) \; ,$$

$$\Psi_p = -\left(\frac{\mathcal{H} - \mathcal{H}_z}{2\mathcal{H}}\right)^{1/2} \Psi(\tfrac{1}{2}\tfrac{1}{2}) + \left(\frac{\mathcal{H} + \mathcal{H}_z}{2\mathcal{H}}\right)^{1/2} \Psi(\tfrac{1}{2}\,-\tfrac{1}{2}) \; . \tag{5.198}$$

Hence once more, if one chooses the magnetic field component to be along the x-axis, then the dependence of the energy splitting on the magnetic field $\mathcal{H}_z$ and the behaviour of the wavefunctions constitute precisely the level anti-crossing effect.

We should add that the presence of a transverse magnetic or electric field, or the existence of any other level structure, transforms any kind of level crossing into anticrossing. This is apparent from the presence of the off-diagonal matrix elements in the secular equation, which lead to "mixing" of the wavefunctions.

Perpendicular Electric and Magnetic Fields. If one directs the quantization axis along the magnetic field, then the Hamiltonian of the atom takes the form

$$\hat{H} = \hat{H}_0 + \hat{H}_{\mathrm{hfs}} + \hat{H}_\mathrm{M} + \hat{H}_e \; .$$

Let the x-axis be in the direction of the electric field, then $\mathscr{E} = \mathscr{E}_x$. The electric field perturbation operator is given by

$$\hat{H}_e = \alpha \mathscr{E}^2 + \beta\left(\hat{J}_x^2 - \frac{J(J+1)}{3}\right)\mathscr{E}^2$$

$$= \alpha \mathscr{E}^2 + \beta\frac{\hat{J}_+^2 + \hat{J}_-^2}{4}\mathscr{E}^2 - \frac{\beta}{2}\left(\hat{J}_z^2 - \frac{J(J+1)}{3}\right)\mathscr{E}^2 \ . \tag{5.199}$$

The latter expression is obtained by replacing $\hat{J}_x$ by $\frac{1}{2}(\hat{J}_+ + \hat{J}_-)$.

In terms of the broken link basis functions Φ the operator $\hat{H}_e$ containing the operators $\hat{J}_+^2$, $\hat{J}_-^2$ and $\hat{J}_z^2$ contributes to the diagonal matrix elements with $\Delta m = 0$, and to the off-diagonal elements coupling the states with m and $m \pm 2$. It is obvious that, compared to the problem of parallel fields $\mathscr{E}$ and $\mathscr{H}$, the number of terms in the secular equation increases and the solution becomes complex.

When solving the problem with the perturbation operator $\hat{V} = \hat{H}_{\text{hfs}} + \hat{H}_{\text{M}}$, the secular equation decomposed into blocks of equal m values. In the present problem, the electric field $\mathscr{E}_x$ couples states with $\Delta m = \pm 2$, and thus all the blocks with even values of m are coupled among themselves, likewise the odd ones. The secular equation therefore decomposes into two blocks and leads to two algebraic equations, whose degree is less than $2J + 1$.

One may also express the Hamiltonian in another coordinate system where $z \parallel \mathscr{E}$ and $x \parallel \mathscr{H}$:

$$\hat{H} = \hat{H}_0 + \hat{H}_{\text{hfs}} + \mu_0(g_J\hat{J}_x - g_I\hat{I}_x)\mathscr{H} + \alpha\mathscr{E}^2$$

$$+ \beta\left(\hat{J}_z^2 - \frac{J(J+1)}{3}\right)\mathscr{E}^2 \ . \tag{5.200}$$

In this representation, the perturbation operator contributes to even more of the matrix elements, since the operators $\hat{J}_x$ and $\hat{I}_x$ couple states with $\Delta m = 1$. In connection with this, the secular equation will not decompose into blocks and the degree of the equation that must be solved to calculate the eigenvalues increases in comparison to the previously chosen coordinate system, which is preferable to the latter.

Solutions can be found by analytical means only in special cases in which the problem reduces to a two-level system with $\Delta m = \pm 2$.

Such a two-level system can be isolated in an atom with a complex hyperfine structure placed in a magnetic field in the crossing region of the two levels with $\Delta m = \pm 2$. If there are no other crossings in the vicinity, then this means that there are no sublevels in the region that are coupled to any of those isolated by the perpendicular electric field $\mathscr{E}_x$.

It is convenient to take as basis functions, the eigenfunctions of the operator

$$\hat{H}_0' = \hat{H}_0 + \hat{H}_{\text{hfs}} + \hat{H}_{\text{M}}^{\text{cr}}$$

where the operator $\hat{H}_M^{cr}$ is the perturbation operator of the magnetic field $\mathcal{H}_{cr}$ in which the levels become degenerate. We call these functions Ψ_p and Ψ_q. We will restrict ourselves to a very small range of variation of the magnetic field $\mathcal{H}$, so that these functions can be assumed to be fixed.

The perturbation Hamiltonian of the atom is given by

$$\hat{H} = \hat{H}_0' + \hat{H}_e + \Delta\hat{H}_M . \tag{5.201}$$

The operator $\Delta\hat{H}_M$ describes the change of the Hamiltonian upon detuning the magnetic field from $\mathcal{H}_{cr}$, i.e.

$$\Delta\hat{H}_M = \mu_0(g_J\hat{J}_z - g_I\hat{I}_z)(\mathcal{H} - \mathcal{H}_{cr})$$

or, to shorten the notation

$$\Delta\hat{H}_M = \mu_0(g_J\hat{J}_z - g_I\hat{I}_z)\Delta\mathcal{H} . \tag{5.202}$$

The operator $\Delta\hat{H}_M$ does not couple the functions Ψ_p and Ψ_q; in this basis it is diagonal and in our approximation, as will be seen later on, the energy gap is proportional to $\Delta\mathcal{H}$.

The diagonal elements of the secular equation

$$\begin{vmatrix} V_{pp} - \varepsilon & V_{pq} \\ V_{qp} & V_{qq} - \varepsilon \end{vmatrix} = 0$$

have the form

$$V_{pp} - \varepsilon = \langle \Psi_p | \hat{H}_e | \Psi_p \rangle + \langle \Psi_p | \Delta\hat{H}_M | \Psi_p \rangle - \varepsilon . \tag{5.203}$$

Expanding the function Ψ_p over the broken link functions

$$\Psi_p = \sum_i C_i^p \Phi_i$$

we rewrite the second term as

$$\langle \Psi_p | \Delta\hat{H}_M | \Psi_p \rangle = \Delta\mathcal{H} \sum_i (C_i^p)^2 \langle \Phi_i | g_J M_i - g_I m_I | \Phi_i \rangle \equiv \Delta\mathcal{H} g_p . \tag{5.204}$$

The electric field contribution to the diagonal matrix elements is given by

$$\langle \Psi_p | \hat{H}_e | \Psi_p \rangle = \alpha\mathscr{E}^2 - \beta\frac{\mathscr{E}^2}{2}\sum_i (C_i^p)^2 \left(M_i^2 - \frac{J(J+1)}{3} \right) \equiv \beta_p\mathscr{E}^2 . \tag{5.205}$$

In the last two expressions g_p and β_p are constants that do not depend on the fields.

Let us now consider the off-diagonal elements:

$$\langle \Psi_p | \hat{H}_e | \Psi_q \rangle = \langle \Psi_p | \hat{J}_+^2 + \hat{J}_-^2 | \Psi_q \rangle \beta\mathscr{E}^2/4 . \tag{5.206}$$

The form of matrix elements of the operators $\hat{J}_+^2$ and $\hat{J}_-^2$ is familiar to us for the function Φ (5.137). Considering this, we find

$$V_{pq} = \sum_{i,k} \langle C_i^p \Phi_i(JM_i Im_I^i) | \hat{J}_+^2 + \hat{J}_-^2 | C_k^q \Phi_k(JM_k Im_I^k) \rangle \ . \tag{5.207}$$

In this sum only terms with $M_k = M_i + 2$ and $m_I^i = m_I^k$ are different from zero. We have already seen that in the expansion of the wavefunctions corresponding to the intermediate magnetic fields over the broken link functions, only functions with the same value of the quantum number $m = M + m_I$ enter (Sect. 5.11). Therefore, $M_i + m_I^i = m_p$ and $M_k + m_I^k = m_q = m_p \pm 2$. Let $m_q = m_p + 2$. Then

$$V_{pq} = \frac{\beta \mathscr{E}^2}{4} \sum_{i,k} C_i^p C_k^q \sqrt{(J - M_i)(J + M_i + 1)(J - M_i - 1)(J + M_i + 2)}$$

$$\equiv \lambda \beta \mathscr{E}^2 \ . \tag{5.208}$$

Here the index k is fully determined by the index i in accordance with the relationship between the quantum numbers. It is easy to verify that $V_{pq} = V_{qp}$. If however $m_q = m_p - 2$, then it is simple to transfer to the previous case by exchanging the indices p and q. Since $V_{pq} = V_{qp}$ one can use the very same formula (5.208) with the very same dependence between i and k. Solving the secular equation, we find the already familiar formula

$$\varepsilon_{q,p} = \frac{V_{qq} + V_{pp}}{2} \pm \frac{1}{2} \sqrt{(V_{qq} - V_{pp})^2 + 4V_{pq}^2} \ .$$

Substituting the matrix elements by their expressions in terms of $\Delta \mathscr{H}$ and $\mathscr{E}$, we find

$$\Delta \varepsilon = \varepsilon_q - \varepsilon_p = \sqrt{(\Delta \mathscr{H} g' + \mathscr{E}^2 \beta')^2 + 4\lambda^2 \beta^2 \mathscr{E}^4} \ . \tag{5.209}$$

Here $g' = g_q - g_p$, $\beta' = \beta_q - \beta_p$ and β retain their previous values. The last formula enables one to trace easily the dependence of the energies on $\mathscr{H}$ and $\mathscr{E}$ (Fig. 5.3). The smallest separation between levels is given by

$$\Delta \varepsilon_{\min} = 2V_{pq} = 2\lambda \beta \mathscr{E}^2 \ .$$

On the magnetic field scale this point lags behind the level crossing points of the unperturbed states (see Sect. 3.13) according to

$$\Delta \mathscr{H}_{cr} = \frac{\beta_p - \beta_q}{g_q - g_p} \mathscr{E}^2 \ .$$

If this point lags far enough behind the crossing point of the unperturbed states, then one cannot assume that the functions Ψ_p and Ψ_q are independent (a consequence of this assumption turned out to be a linear dependence of $\Delta \mathscr{H}_{cr}$ on $\mathscr{E}^2$); nonetheless, one can deduce a very accurate solution of the problem by

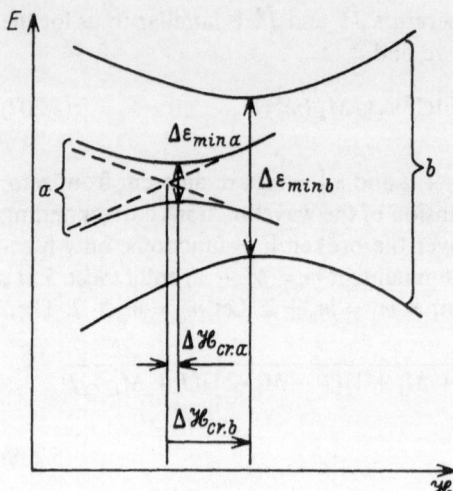

Fig. 5.3. Energy level structure when $\mathscr{H} \perp \mathscr{E}$. The cases a and b correspond to different strengths of $\mathscr{E}$

taking as the basis functions the eigenfunctions of the operator

$$\hat{H} = \hat{H}_0 + \hat{H}_{\text{hfs}} + \hat{H}_{\text{M}}^{\text{cr}} - \frac{\beta}{2}\left(\hat{J}_z^2 - \frac{J(J+1)}{3}\right)\mathscr{E}^2 . \tag{5.210}$$

The above-discussed problems are of sufficient generality to describe practically all other combinations of time independent fields. Thus, for example, if the electric and magnetic field are directed at an arbitrary angle to each other, then it is reasonable to choose the quantization axis along the electric field, because the advantages of the coordinate system with $z \parallel \mathscr{H}$ exist only when $\mathscr{H} \perp \mathscr{E}$; in the coordinate system with $z \parallel \mathscr{E}$ the electric field operator has a simple form and the magnetic field operator decomposes into a sum of two operators: $(\hat{H}_{\text{M}})_x$ and $(\hat{H}_{\text{M}})_z$.

References

Chapter 1

1.1 W. Hanle: Über magnetische Beeinflussung der Polarisation der resonanz Fluoreszens. Z. Phys. **30**, 93 (1924)
1.2 J.N. Dodd, R.D. Kaul, D.M. Warrington: The modulation of resonance fluorescence excited by pulsed light. Proc. Phys. Soc. **84**, 176 (1964)
1.3 E.B. Aleksandrov: Luminescence beats indced by pulse excitation of coherent states. Opt. Spectrosc. (USSR) **17**, 522 (1964)
1.4 M.I. Podgoretskii, O.A. Khrustalev: Interference effects in quantum transitions. Sov. Phys. - Usp (USA) **6**, 682 (1964)
1.5 E.B. Alexandrov: Quantum beats. 6th Int'l Conf. Atom. Phys. Proc., Riga (1978) pp.521–534
1.6 P.A. Franken: Interference effects in the resonance fluorescence of "crossed" excited atomic states. Phys. Rev. **121**, 508 (1961)
1.7 M.P. Chaika: *Interference of Degenerate Atomic States* (University Press, Leningrad 1975)

Chapter 2

2.1 W. Hanle: Über magnetische Beeinflussung der Polarisation der resonanz Fluoreszens. Z. Phys. **30**, 93 (1924)
2.2 L.D. Landau, E.M. Lifshitz: *The Classical Theory of Fields* (Pergamon, Oxford 1962)
2.3 W. Heitler: *The Quantum Theory of Radiation* (Oxford Univ. Press, London 1954)
2.4 R.W. Wood, A. Ellett: Polarized resonance radiation in weak magnetic fields. Phys. Rev. **24**, 243 (1924)
2.5 E.B. Alexandrov: Luminescence beats induced by pulse excitation of coherent states. Opt. Spectrosc. (USSR) **17**, 522 (1964)
2.6 J.N. Dodd, R.D. Kaul, D.M. Warrington: The modulation of resonance fluorescence excited by pulsed light. Proc. Phys. Soc. (London) **84**, 176 (1964)
2.7 E.B. Alexandrov: Optical manifestations of the interference of nondegenerate atomic states. Sov. Phys. - Usp (USA) **15**, 436 (1973)

Chapter 3

3.1 L.D. Landau, E.M. Lifshitz: *Quantum Mechanics* (Pergamon, New York 1977)
3.2 W. Heitler: *The Quantum Theory of Radiation* (Oxford Univ. Press, London 1954)
3.3 A.S. Davydov: *Quantum Mechanics* (Fizmatgiz, Moscow 1973)
3.4 A.I. Achiezer, V.B. Berestetzkii: *Quantum Electrodynamics* (Fizmatgiz, Moscow 1959)

240   References

3.5   A.A. Socolov: *Introduction into Quantum Electrodynamics* (Fizmatgiz, Moscow 1968)
3.6   J. Weisskopf: Zur Theorie der Resonanz Fluoreszens. Ann. Phys. **9**, 23 (1931)
3.7   P.P. Shorygin, T.M. Ivanova: Phenomenon of resonance Raman scattering. Sov. Phys. Docl. **34**, 201 (1952)
3.8   I.M. Beterov, Yu. A. Matyugin, V.P. Chebotayev: Measurement of the level relaxation constants by the method of three-level laser spectroscopy. JETP Lett. (USA) **12**, No.4, 120 (1970)
3.9   V.S. Letokhov, V.P. Chebotayev: *Nonlinear Laser Spectroscopy*, Springer Ser. Opt. Sci., Vol.4 (Springer, Berlin, Heidelberg 1977)
3.10 E.B. Alexandrov: Luminescence beats induced by pulse excitation of coherent states. Opt. Spectrosc. (USSR) **17**, 522 (1964)
3.11 M.I. Podgoretskii, O.A. Krustalev: Interference effects in quantum transitions. Sov. Phys. - Usp. (USA) **6**, 682 (1964)
3.12 L. Allen, J.H. Eberley: *Optical Resonance and Two-Level Atoms* (Wiley, New York 1975)
3.13 E.B. Alexandrov: Quantum beats of resonance luminescence under modulated light excitation. Opt. Spectrosc. (USSR) **14**, 233 (1963)
      A. Corney, G.W. Series: Double resonance excited by modulated light. Proc. Phys. Soc. (London) **83**, 213 (1964)
3.14 W. Demtröder: Bestimmung von Oszillatorenstärken durch Lebensdauermessungen der ersten angeregten Niveaus. Z. Phys. **166**, 42 (1962)
3.15 D.A. Varshalovich, A.N. Moskalev, V.K. Khersonskii: *Quantum Theory of Angular Monentum* (Nauka, Leningrad 1975) [Engl. transl. World Scientific, Singapore (1988)]
3.16 G.I. Khvostenko, M.P. Chaika: States with identical angular-momentum characteristics do not interfere. Opt. Spectrosc. (USSR) **53**, 581 (1982)
3.17 G.I. Khvostenko, M.P. Chaika: Redistribution of radiation under the interference of quantum states. Opt. Spectrosc. (USSR) **59**, 714 (1985)
3.18 G.I. Khvolstenko: Absence of beats during interference of states with the same angular-momentum quantum numbers. Opt. Spectrosc. (USSR) **71**, 509 (1991)
3.19 E.U. Condon, G.H. Shortley: *Theory of Atomic Spectra* (Cambridge Univ. Press, London 1935)
3.20 A.J. Ferguson: *Angular Correlation Methods in Gamma-Ray Spectroscopy* (North-Holland, Amsterdam 1965)
3.21 M.I. D'yakonov: Theory of resonance scattering of light by gas in the presence of a magnetic field. Sov. Phys. JETP **20**, 1484 (1965)
3.22 M.P. Chaika: *Interference of Degenerate Atomic States* (University Press, Leningrad 1975)
3.23 C. Cohen-Tannoudji: Theorie quantique du cycle de pompage optique. Ann. Phys. **7**, 423 (1962)
3.24 L.D. Landau, E.M. Lifshitz: *The Classical Theory of Fields* (Pergamon, Oxford 1962)
3.25 E.L. O'Neill: *Introduction to Statistical Optics* (Addison-Wesley, Reading, MA 1963)
3.26 T.G. Eck, L.L. Foldy, H. Wieder: Observation of anticrossings in optical resonance fluorescence. Phys. Rev. Lett. **10**, 239 (1963)
3.27 G.W. Series: Some remarks on anticrossing in optical resonance fluorescence. Phys. Rev. Lett. **11**, 13 (1963)
3.28 I.C. Lehman: Etude de l'orientation nucleaire par pompage optique des isotopes stable du cadmium. Ann. Phys. **2**, 345 (1967)
3.29 T.G. Eck: Level crossings and anticrossings. Physica **33**, 157 (1967)
3.30 B. Budick, H. Snirz: Coherent anticrossings signals. Phys. Rev. Lett. **20**, 177 (1968)

3.31   G.I. Khvostenko: Coherent anticrossing signals. Opt. Spectrosc. (USSR) **26**, 352 (1969)

3.32   G.M. Strakhovskii, A.V. Uspenskii: *Principles of Quantum Electronics* (Higher School Press, Moscow 1973) Chap.3

Chapter 4

4.1   A. Kastler: Quelques suggestions concernant la production optique et la detection optique du inegalité de population des niveaux de quantification spatiale des atomes. Application a l'experience de Stern et Gerlach et a la resonance magnitique. J. Phys. Rad. **11**, 255 (1950)

4.2   A. Kastler: Les methodes optique d'orientation atomique et leur applications. Proc. Phys. Soc. **A67**, 853 (1954)

4.3   W. Hanle: Über magnetische Beeinflussung der Polarisation der Resonanz-Fluoreszens. Z. Phys. **30**, 93 (1924)

4.4   R.W. Wood: *Physical Optics* (Mcmillan, New York 1934)

4.5   S.A. Kasantsev, E.D. Mishchenko, P.K. Telbisov, M.P. Chaika: Decay Constant of the $^3P_1$ State of Ne. Opt. Spectrosc. (USSR) **45**, 711 (1978)

4.6   M. Shull, J.D. Winefordner: Polarization of atomic fluorescence in flames. Appl. Spectrosc. **25**, 97 (1971)

4.7   Kh. Kallas, M. Chaika: Alignment of Excited States of Neon in a Direct Current Discharge. Opt. Spectrosc. (USSR) **27**, 376 (1969)

4.8   C.G. Carrington, A. Corny: Hanle effect in a neon discharge. Opt. Commun. **1**, 115 (1969)

4.9   M. Lombardi: Note sur la possibilité d'orienter un atom par superposition de deux interactions separement non orientates en particulier par alignement electronique et relaxation anisotrope. C.R. Acad. Sci. **265**, 191 (1967)

4.10   M. Lombardi: Orientation by electric field of atoms excited in a radiofrequency discharge. Proc. Int'l Conf on Opt. Pumping, Warsaw (1969) p.69

4.11   M. Lombardi: Alignment and orientation production measurement and conversion. Beam-Foil Spectros. **2**, 731 (1977)

4.12   U. Fano: Precession equation of a spinning particle in non-uniform fields. Phys. Rev. **133**, 828 (1964)

4.13   U. Fano: J.H. Macek: Impact excitation and polarization of the emitted light. Rev. Mod. Phys. **45**, 553 (1973)

4.14   W.S. Buckel: Modulation spectroscopy using the beam-foil light spectroscopy. Surf. Sci. **37**, 971 (1973)

4.15   W.L. Wiese: Atomic transition probabilities, a survey of our present knowledge and future needs. Nucl. Instrum. Methods **90**, 25 (1970)

4.16   O. Nedelec: Phenomenes de coherence Zeeman daus une vapeur atomique excitee par des electrons lents: Theses L'universite de Grenoble (1965)

4.17   J.P. Descombes, J.C. Pebay-Peyroula: Sur la resonance magnetiques des niveaux atomiques due mercure excites par bombardement electronique. C.R. Acad. Sci. **247**, 2330 (1958)

4.18   J. Buchhaupt: Level-Crossing Experiment an den Helium Zuständen 3^3D and 4^3D bei Ionenstossanregung. Z. Naturf. A **24**, 1058 (1969)

4.19   R. Muller, G. von Oppen, W.D. Perschmann: Analysis of level crossing signals after impact excitation of $1snd^3D$ levels of He I by He^+-He collision. Z. Phys. **A315**, 41 (1984)

4.20   E.B. Aleksandrov, A.M. Bonch-Bruevich, V.A. Khodovoy: Weak Magnetic Field Measurements using Methods of Optical Pumping. Opt. Spectrosc. (USSR) **23**, 282 (1967)

242   References

4.21   J. Dupont-Roc, S. Haroche, J.C. Cohen Tannoudji: Detection of very weak
        magnetic fields (10^{-9} gauss) by ^{87}Rb zero field level crossing resonances. Phys.
        Lett. 25, 638 (1969)
4.22   J.C. Cohen-Tannoudji, J. Dupon-Roc, S. Haroche, et al.: Diverses resonances de
        croisement de niveaux sur des atomes pompes optiquement en champ fables.
        Rev. Phys. Appl. 5, 102 (1970)
4.23   N.M. Pomerantzev, V.M. Ryzhkov, G.V. Skrotzkii: Principles of Quantum Mag-
        netometry (Nauka, Moscow 1972)
4.24   R.E. Slocum, B.I. Marton: Measurement of weak magnetic fields using zero
        field parametric resonance in optically pumped He. IEEE Trans. MAG-9, 221
        (1973)
4.25   E.I. Dashevskaya, E.E. Nikitin, S.Ya. Umanskii: Alignment of atoms due to col-
        lisions. Prof. 9th All-Union Conf. on Electronic and Atomic Collisions, Riga
        (1984) Vol.1, p.77
4.26   M.P. Chaika: Self-alignment of molecular beam. Opt. Spectrosc. (USSR) 60, 681
        (1986)
4.27   H.G. Weber, F. Bylicki, G. Miksch: Inversion of polarization by light induced
        stabilization. Phys. Rev. 30, 270 (1984)
4.28   P. Toschek: Stossausrichtung in Molekularstrahlen. Z. Phys. 187, 56 (1965)
4.29   G.N. Nicolaev, G.D. Rodionov, S.G. Rautian et al.:Observation of the influence
        of anisotropic collisions upon light absorption in NE. Preprint No.238, Inst. of
        Automatics and Electrometry, Siberian Branch, USSR Academy of Sci. (1985)
4.30   G.N. Nikolaev, S.G. Rautian, G.D. Rodionov, E.G. Saprykin: Magnetooptic
        resonance in linear absorption caused by anisotropic collisions. Opt. Spectrosc.
        (USSR) 60, 149 (1986)
4.31   P. Toschek: Anomale Anisotropie der zwischenmolekularen Kraft. Z. Physik
        187, 52 (1965)
4.32   V.N. Rebane: Depolarization of Resonance Fluorescence during Anistropic Col-
        lisions. Opt. Spectrosc. (USSR) 24, 309 (1968)
4.33   P. Manabet, T. Yabusaki, T. Ogawa: Observation of collisional transfer from
        alignment to orienation of atoms excited by a single-mode laser. Phys. Rev. Lett.
        46, 637 (1981)
4.34   M.I. D'yakonov, V.I. Perel': Coherence Relaxation of Excited Atoms in Colli-
        sion. Sov. Phys. JETP 21, 227 (1966)
4.35   W.D. Ewans, W. Gough: Anomalous Hanle effect signal in sensitised fluores-
        cence. J. Phys. B12, 2271 (1979)
4.36   E.B. Aleksandrov: Luminescence Beats Induced by Pulse Excitation of Coherent
        States. Opt. Spectrosc. (USSR) 17, 522 (1964)
4.37   J.N. Dodd, R.D. Kaul, D.M. Warrington: The modulation of resonance fluores-
        cence excited by pulsed light. Proc. Phys. Soc. 84, 176 (1964)
4.38   S.A. Bagaev, V.N. Smirnov, L.P. Kantserova, V.Ju. Cherepanov: Quantum-beat
        study of the splitting of the magnetic sublevels of n ^{3}D (n = 3÷6) state of He I.
        Opt. Spectrosc. (USSR) 57, 592 (1984)
        V.B. Smirnov, V.Yu. Cherepanov: Quantum beats under pulsed electron excita-
        tion of He I D states: I Singlet states, II triplet states. Opt. Spectrosc. (USSR) 72,
        164-168 (1992)
4.39   H.J. Andrä: Zero-field quantum beats subsequent to beam foil excitation. Phys.
        Rev. Lett. 25, 325 (1970)
4.40   H.J. Andrä: Quantum beats and laser excitation in fast beam spectroscopy.
        Atomic Physics IVB (Plenum, New York 1974) p.635
        J. Carlsson, P. Johnsson, L. Stukesson: Accurate time-resolved laser spectros-
        copy of silver atoms. Z. Physik D 15, 87 (1990)
        J. Carlsson, L. Stukesson: Accurate time-resolved spectroscopy of lithium atoms.
        Z. Physik D 14, 281 (1989)

4.41 W. Gornik, D. Kaiser, W. Lange, et al.: Quantum beats induced by laser excitation. Opt. Commun. **6**, 327 (1972)

4.42 C. Fabre, M. Gross, S. Haroche: Determination by quantum beats spectroscopy of fine structure intervals in a series of highly excited sodium D-states. Opt. Commun. **13**, 393 (1975)

4.43 T.W. Ducas, M.G. Littman, M.L. Zimmerman: Observation of oscillation in resonance absorption from a coherent superposition of atomic states. Phys. Rev. Lett. **35**, 1752 (1975)

4.44 F.A. Hopf, R.L. Shoemaker: Observation of quantum beats in photon echoes and optical nutation. Phys. Rev. Lett. **33**, 1527 (1974)

4.45 E.B. Alexandrov: Quantum beats of resonance luminescence under modulated light excitation. Opt. Spectrosc. (USSR) **14**, 233 (1963)

4.46 A. Corney, G.W. Beries: Double resonance excited by modulated light. Proc. Phys. Soc. **83**, 213 (1964)

4.47 W.E. Bell, A.L. Bloom: Optically excited spin precession. Phys. Rev. Lett. **6**, 280 (1962)

4.48 G. Guedard, J.C. Lehmann: Effect Hanle et resonancs en lumiere modulee sur le niveau Bl_u, $v = 0$, $J = 105$ de molecule $(^{80}Se)_2$ excitee par la raie 4727 Å d'une laser a argon ionise. C.R. Acad. Sci. **280**, 471 (1975)

4.49 E.I. Ivanov, M.P. Chaika: Observation of interference light beats in spontaneous emission of He-Ne laser. Opt. Spectrosc. (USSR) **29**, 67 (1967)

4.50 H.R. Schlossberg, A. Javan: Saturation behavior of Doppler broadened transition involving levels with closely spaced structure. Phys. Rev. **150**, 267 (1966)

4.51 M. Kaivola, N. Bjerre, O. Poulsen, et al.: Observation of populating trapping in a two-photon resonant three-level atom. Opt. Commun. **49**, 418 (1984)

4.52 O.V. Konstantinov, V.I. Perel': Coherence of States in the scattering of modulated light. Sov. Phys. JETP **18**, 195 (1964)

4.53 E.B. Alexandrov: Pulsations in luminescence due to phase modulation of excited states. Opt. Spectrosc. (USSR) **19**, 452 (1965)

4.54 A.I. Okunevich: Parametric relaxation resonance of optically oriented Atoms in a transverse magnetic Field. Sov. Phys. JETP **39**, 773 (1974)

4.55 E.B. Aleksandrov, O.V. Konstantinov, V.I. Perel', V.A. Khodovoi: Modulation of scattered light by parametric resonance. Sov. Phys. JETP **18**, 346 (1964)

4.56 L.N. Novikov, G.V. Skrotzkii: Nonlinear and parametric effects in atomic RF spectroscopy. Sov. Phys. - Usp. (USA) **21**, 589 (1978)

4.57 S.A. Kazantzev: Astrophysical and laboratory application of self-alignment. Sov. Phys. - Usp. (USA) **26**, 328 (1983)

4.58 D.Z. Zhechev, M.P. Chaika: Radiation from a hollow cathode discharge in a weak magnetic field. Opt. Spectrosc. (USSR) **43**, 352 (1977)

4.59 S.A. Kazantzev: Alignment of highly excited states of rare gases in a discharge. Bull. Leningrad University, Phys. and Chem. Sects., No.14 (1980) p.52

4.60 M.R. Atadzhanov, E.N. Kotlikov, M.P. Chaika: Investigation of the alignment of the $^2D_{5/2}$ state of Cd II state in a gas discharge. Opt. Spectrosc. (USSR) **53**, 378 (1982)

4.61 S.A. Kazantsev, A.G. Rys', M.P. Chaika: Alignment of excited atoms by electron collision in a discharge. Opt. Spectrosc. (USSR) **54**, 124 (1983)

4.62 M.P. Chaika: Polarization of spontaneous emission of atoms. Autometry **1**, 101 (1979)

4.63 M.R. Atadzhanov, E.N. Kotlikov, M.P. Chaika: Alignment of metastable state of Cd I in a gas discharge. Opt. Spectrosc. (USSR) **50**, 447 (1981)

4.64 M. Chaika: Latent alignment of excited state of gas atoms due to isotropic irradiation. Opt. Spectrosc. (USSR) **30**, 443 (1971)

4.65 M.P. Chaika: Alignment of excited atoms in a gas discharge: Proc. 6th Int'l Conf on Atom Phys., Riga (1978) p.423

4.66   A.V. Pavlov, V.A. Polishchuk, M.P. Chaika: Dichroism in DC discharge in Ne. Opt. Spectrosc. (USSR) **49**, 544 (1980)

4.67   N.G. Lukomsky, V.A. Polischuk, M.P. Chaika: Conversion of latent alignment into orientation in a low-pressure plasma. Opt. Spectrosc. (USSR) **59**, 606 (1985)

4.68   N.G. Lukomsky, V.A. Polischuk, M.P. Chaika: "Hidden" anistropy of collisions in a low-temperature plasma. Opt. Spectrosc. (USSR) **58**, 284 (1985)

4.69   N.G. Lukomsky, V.A. Polischuk, M.P. Chaika: Experiments on the observation of orientation in a plasma. Opt. Spectrosc. (USSR) **60**, 11 (1986)

4.70   C.L. Hyder: The polarization of emission lines in astronomy. The polarization of coronal emission lines. Astrophys. J. **141**, 1382 (1965)

4.71   C.L. Hyder: Polarized light magnetographs and solar magnetic fields. Solar Phys. **5**, 29 (1968)

4.72   V.E. Stepanov, A.B. Severnyi: Photoelectric method for measurement of a magnetic field strength and direction on solar surface. Proc. Astrophysical Observatory of Crimea **28**, 166 (1962)

4.73   L.L. House: Coronal emission line polarization. Solar Phys. **23**, 103 (1972)

4.74   S. Sahal-Brechot, V. Bommier, J.L. Leroy: The Hanle effect and the determination of magnetic fields in solar prominences. Astron. Astrophys. **59**, 223 (1977)

4.75   V. Bommier: Quantum theory of the Hanle effect. Astron. Astrophys. **87**, 109 (1980)

4.76   I.O. Stenflo: The Zeeman effect for weak magnetic fields. Solar. Phys. **8**, 260 (1969)

4.77   F.K. Lamb: Line formation in magnetic fields. Solar Phys. **12**, 186 (1970)

4.78   V. Bommier, S. Sahal-Brechot: Quantum theory of the Hanle effect: Calculations of the Stokes parameters of the helium line for quiescent prominences. Astron. Astrophys. **69**, 57 (1978)

4.79   A.A. Ruzmaikin: Possible evaluation of the magnetic field near pulsars from beat frequency of atomic transition. Sov. Astron. (USA) **19**, 702 (1975)

4.80   A.C.G. Mitchell, M.W. Zemanski: *Resonance Radiation and Excited Atoms* (Cambridge Univ. Press, London 1961)

4.81   W. Hanle, E.F. Richter: Polarization Erscheinungen bei der stufenweisen Anregung der Fluoresenz. Z. Phys. **54**, 811 (1929)

4.82   B.P. Kibble, S. Pancharatham: Interference effects in stepwise fluorescence due to level crossing. Proc. Phys. Soc. **86**, 1351 (1965)

4.83   Kh. Kallas, M. Chaika: Polarization of the spontaneous emission of a gas laser. Opt. Spectrosc. (USSR) **22**, 283 (1967)

4.84   M. Ducloy, B. Decomps: Etude experimental du transfert de population et d'alignement par emission spontanee. C.R. Acad. Sci. **266**, 412 (1968)

4.85   M. Ducloy, M. Dumont: Transfert de population et de coherence hertzeinne par emission spontanee. C.R. Acad. Sci. B **266**, 340 (1968)

4.86   N.D. Bhaskar, A, Lurio: Lifetime of the $1s_2(^1P_1)$ and $1s_4(^3P_1)$ levels of neon by the cascade Hanle effect. Phys. Rev. A **13**, 1484 (1976)

4.87   C.G. Carrington: Cascade effects in neon lifetime measurements. J. Phys. B **5**, 1572 (1972)

4.88   S.A. Kazantzev, A.G. Rys', M.P. Chaika: Alignment of kripton atoms in gas discharge. Opt. Spectrosc. (USSR) **44**, 249 (1978)

4.89   T. Holstein: Imprisonment of resonance radiation in gases. Phys. Rev. **72**, 1212 (1947)

4.90   J.P. Barrat: Etude de la diffusion multiple coherent de la lumieve de resonance optique. J. Phys. Rad. **20**, 541 (1959)

4.91   M.I. D'yakonov, V.I. Perel': Coherence relaxation during diffusion of resonance radiation. Sov. Phys. JETP **20**, 997 (1965)

4.92   W. Happer, D.B. Saloman: Observation of different lifetime for atomic states, excited with linearly and circulary polarized light. Phys. Rev. Lett. **15**, 441 (1965)

4.93   B. Decomps, M. Dumont: Effect Hanle d'atomes de neon irradies par un laser. C.R. Acad. Sci. B **262**, 1695 (1966)

4.94   C.G. Carrington: Cascade effects in Ne lifetimes measurements. J. Phys. B **5**, 80 (1972)

4.95   S.A. Kazantsev, A. Kisling, M.P. Chaika: Alignment of excited argon atoms in a positive column of direct current discharge. Opt. Spectrosc. (USSR) **36**, 606 (1974)

4.96   E.B. Saloman, W. Happer: Lifetime coherence narrowing and HFS of the $^3P_1{}^0$ state in lead. Phys. Rev. **144**, 7 (1966)

4.97   V.I. Perel', I.V. Rogova: Alignment of excited atoms in a gas discharge. Sov. Phys. JETP **38**, 501 (1974)

4.98   M.P. Chaika, E.N. Kotlikov, G. Ts. Todorov, M.R. Atadzhanov: Alignment pseudosignals in magnetic fields. Opt. Spectrosc. (USSR) **51**, 32 (1981)

4.99   Kh. Kallas, G. Markova, G. Khvostenko, M. Chaika: Determination of the hyperfine-structure constants of cesium from the intersection of magnetic sublevels. Opt. Spectrosc. (USSR) **19**, 173 (1965)

4.100 G. Markova, G. Khvostenko: Lifetime of $5^2P_{3/2}$ and $7^2P_{3/2}$ states of cesium. Opt. Spectrosc. (USSR) **23**, 456 (1967)

4.101 A.L. Mashinsky: Precise measurement of atomic constants using level-crossing signals of the $3^2P_{3/2}$ state of Na. Opt. Spectrosc. (USSR) **28**, 1 (1969)

4.102 V.N. Grigorieva: Atomic Constants of $6^2P_{3/2}$ state of RB. Bull. Leningrad University, Phys. Chem. Sects. **16**, 44 (1973)

4.103 E. Kotlikov, M. Chaika: Radiation lifetime of $3p_4$ state of neon. Opt. Spectrosc. (USSR) **27**, 281 (1969)

4.104 Y.D. Flichtner, J.H. Gallagher, Mizushima Masatake: Lifetime of the first excited atomic states of Rb^{87}. Phys. Rev. **164**, 44 (1967)

4.105 D. Lecler: Effect de la dispersion anomal sur les formes des combes d'effect Hanle. J. Physique **29**, 739 (1968)

1.106 M. Chaika: Level-crossover signal in the case of arbitrary escitation-versus-time dependence. Opt. Spectrosc. (USSR) **34**, 108 (1973)

1.107 M. Chenevier, M. Lombardi: Magnetic depolarization of atomic fluorescence in flames. Chem. Phys. Lett. **16**, 154 (1972)

4.108 J. Bonn, C. Huber, H.J. Kluge et al.: Orientation of $^{199\,m}$Hg by optical pumping detected by γ-radiation anisotropy. Z. Physik A **272**, 375 (1975)

4.109 H. Schweickert, J. Dietrich, R. Neugart, et al.: Nuclear spins and magnetic moments of ^{20}Na and ^{36}K by β-radiation detected optical pumping. Nucl. Phys. A **246**, 187 (1975)

4.110 M. Diemling, R. Neugart, H. Schweikart: Spin and magnetic moment of ^{25}Na by β-radiation detected optical pumping. Z. Phys. A **273**, 15 (1975)

4.111 G. Huber, J. Bonn, H.J. Kluge, et al.: Nuclear radiation detected optical pumping of neutron deficient Hg isotopes. Z. Phys. A **276**, 189 (1976)

4.112 E.W. Otten: Determination of nuclear gound state properties far from stability by optical pumping. J. Physique **34**, 63 (1973)

4.113 J.E.M. Goldsmith, E.W. Weber, T.W. Hansch: New measurement of the Rydberg constant using polarization spectroscopy of H. Phys. Rev. Lett. **41**, 1525 (1978)

4.114 E.B. Aleksandrov: Optical manifestations of the interference of nondegenerate atomic states. Sov. Phys. - Usp. (USA) **15**, 436 (1973)

4.115 M.P. Chaika: *Interference of Degenerate Atomic States* (University Press, Leningrad 19755)

4.116 C. Cohen-Tannoudji: Theory quantique du cycle du pompage optique. Ann. Phys. **7**, 423 (1962)

4.117 E.B. Aleksandrov, M.P. Chaika: Optical self-pumping. Sov. Sci. Rev. 6, 3 (1985)
4.118 W. Happer: The Hanle effect. Beam-Foil Spectros. 1, 305 (1968)
4.119 A. Omont: Irreducible components of the density matrix. Applications to optical pumping. Prog. Quant. Electron 5, 69 (1977)
4.120 V.G. Pokazan'ev, G.V. Skrotskii: Crossing and anticrossing of atomic levels and their use in atomic spectroscopy. Sov. Phys. - Usp (USA) 15, 452 (1973)
4.121 G. Hermam, G. Lasnitschka, A. Schazmann et al.: A factor of the ^{111}Cd and ^{113}Cd isotopes by hyperfine level crossing of the Cd II $^2D_{5/2}$-state in stimulated emission. Phys. Lett. A 74, 55 (1979)
4.122 M.S. Mathur, F.M. Kelly: Comparison of the density of the dependence of the Hanle effect in the resonance line of atomic barium and magnesium. J. Phys. Chem. 84, 1783 (1980)
4.123 D.W. Fabey, W.F. Parks, L.D. Schearer: Hanle lifetime measurements of Sr I $^1P_1$ and Ca I $^1P_1$ levels excited by a neutral beam of $^1S_0$ helium atoms. Phys. Lett. A 74, 405 (1979)
4.124 M.S. Mathur, F.H. Kelly: Hanle effect of the $5p^1P_1{}^0$ level of calcium and its density dependence. Cdn. J. Phys. 60, 1237 (1982)
4.125 V.P. Kafandjian, V. Doppzanowski, L. Klein et al.: Optical Hanle effect of Doppler broadened transitions. Phys. Rev. A 28, 1173 (1983)
4.126 M. Baumann, H. Lundel, B. Lindenberger: Investigation of the HFS in the $4d^5 5pz^7P_{3,4}$ states of ^{95}Mo and ^{97}Mo by level-crossing spectroscopy. Opt. Commun. 50, 353 (1984)
4.127 M.A. Bouehiat, J. Guen, L. Pottier: Absolute polarization measurement and natural lifetime in the $7S_{1/2}$ state of Cs. J. Phys. Lett. 45, 523 (1984)
4.128 M. Soltanolkotabi, R. Gupta: Hyperfine structure and lifetime measurement in the $5^2P_{3/2}$ state of ^{107}Ag. Physica C 123, 386 (1984)
4.129 L.R. Hunter, G.M. Watson, D.S. Weiss, et al.: High precision measurement of lifetime and collisional decay parameters nCa^{1}D states using the two photon Hanle effect. Phys. Rev. A 31, 2268 (1985)
4.130 M. Bauer, L. Baumann, H.A. Liening: A level crossing experiment in $6s6p^1P_1$ state of Yb^{173} with additional optical pumping of the ground state. Phys. Lett. A 60, 101 (1977)
4.131 J. Mlynek, W. Lange: A simple method of observing coherent ground state transitions. Opt. Commun. 30, 337 (1979)
4.132 E. Bernalean, A. Cavero: Multiple anticrossing of magnetic hyperfine atomic sublevels. Spectrosc. Lett. 12, 609 (1979)
4.133 H.J. Beyer, K.J. Kolath: Electric field induced singlet-triplet anticrossing in helium. J. Phys. B 10, 15 (1977)
     L. Windholz: The sodium D2 line in crossed electric and magnetic fields. Z. Physik D 15, 87 (1990)
4.134 Th. Krist, P. Kuske, A. Gaupp, et al.: Improved ^{23}Na($3^2P_{3/2}$) HFS measurement beyond the natural linewidth by beam laser quantum beats. Phys. Lett. A 61, 94 (1977)
4.135 D.J. Pegg, P.M. Griffin, B.M. Jonson, et al.: Intensity modulations in the decay of the $3^2P_{1/2}{}^0$ level in the sodiumlike ion Cu^{18}. Phys. Rev. Lett. 38, 1471 (1977)
4.136 P.B. Kramer: Spectroscopy on coherent populations using quantum beats. Phys. Rev. Lett. 38, 1021 (1977)
4.137 P. Grundevik, H. Lundberg, A.M. Martenson et al.: Hyperfine structure study in the P-sequence of ^{23}Na using quantum beat spectroscopy. J. Phys. B12, 2645 (1979)
     A.I. Alekseev, O.V. Zhemerdeev: Optical free-induction-decay quantum beats in multilevel systems. J. Opt. Soc. Am. B 4, 30 (1987)

4.138 P. Kondo Koji: Synchronized quantum beat spectroscopy and its application to the measurement of pressure shifts of the hyperfine splitting in $3^2S_{1/2}$ of sodium atoms. J. Phys. Soc. Jpn. 52, 2340 (1983)

4.139 A. Lurio: Lifetime of the $1s_2(^1P_1)$ and $1s_4(^3P_1)$ levels of neon by cascade Hanle effect. Phys. Rev. A 13, 1484 (1976)

4.140 M. Druetta, A. Denis: Measurement of g-factors and polarisation for levels in neon. Nucl. Instrum. Meth. 110, 291 (1973)

4.141 R. Luypaert, J. van Craen: Cascade quantum beats: The calculation of intensity profiles. J. Phys. B 12, 2613 (1979)

4.142 B.J. Dalton: Theory of cascade effects of quantum beats. J. Phys. B 12, 2625 (1979)

4.143 H. Brand, K.H. Drake, W. Lange, et al.: Level crossing in forward scattering by laser-polarization spectroscopy. Phys. Lett. A 75, 345 (1980)

4.144 V.N. Ochkin, N.G. Preobrazhenskii, N.N. Sobolev, N.Ya. Shaparev: Optogalvanic effect in plasmas and gases. Sov. Phys. - Usp. (USA) 29, 260 (1986)

4.145 C. Cohen-Tannoudji, V. Dupont-Roc, S. Haroche et al.: Diverses resonances de croisement de niveaux sur des atomes pompes optiquenment en champ nul. Rev. Phys. Appl. 5, 95 (1970)

4.146 S. Garpman, S. Svanberg: Investigation of the hyperfine structure of the Te[125] in the $5p^3 6s^5 S_2$ level of the Te I spectrum by optical double resonance and level crossing spectroscopy. Phys. Soc. 5, 213 (1972)

4.147 M. Ducloy, M. Dumont: Etude de transfert d'exitation par emission spontanee. I. Analyse theorique. J. Physique 31, 419 (1970

4.148 M.A. Bouchiat, J. Guena, L. Poffier: Optical orientation, spin-orbit coupling in the final state of the Stark induced 6s-7s cesium transition. Opt. Commun. 37, 265 (1981)

4.149 G. Borghs, P. Bichop, H. van Hove, et al.: Time differential ground-state Hanle effect in fast-beam laser spectroscopy. Phys. Rev. Lett. 52, 2030 (1984)

4.150 D.W. Fahey, W.F. Parks, L.D. Schearer: The Hanle effect in Penning excited ions. J. Phys. B 12, 619 (1979)

4.151 W.D. Evans, W. Gough: Anomaleous Hanle effect signals in sensitized fluorescence. J. Phys. B 12, 2271 (1979)

4.152 M.A. Rebolledo: Level-crossing detection by photon statistics. Opt. Eng. 22, 583 (1983)

4.153 B. Decomps, M. Dumont: Une nouvelle verification de la theorie de la diffusion multiple coherente. C.R. Acad. Sci. B 266, 1272 (1968)

4.154 J. Ryde Nils: *Atoms and molecules in electric fields* (Almquist and Wiksel, Stockholm 1976)

4.155 O.V. Khait, G.I. Khvostenko: Antilevel crossing of an rf field-dressed atom in an electric field. Opt. Spectrosc. (USSR) 67, 727 (1989)

4.156 S.B. Smirnov, O.V. Khait, G.I. Khvostenko: Characteristic properties of Stark interaction anisotropy in an alternating magnetic field. Opt. Spectrosc. (USSR) 70, 703 (1991)

4.157 P. Cacciani, S. Libermas, T. Luc-Koenig, J. Pinard, C. Thomas: Anticrossing effects in Rydberg states of lithium in the presence of parallel magnetic and electric fields. Phys. Rev. A 40, 3026 (1989)

4.158 P.J.O. Teubner, R.E. Scholten, G.F. Shen: Orientation and alignment in electron collisions with sodium. Proc. Int'l Symp. on Correlation and Polarization in Electronic and Atomic Collisions, Hoboken, NJ (1989) (NIST, Gaithersburg, MD 1990) NIST SP 789, p45

Chapter 5

5.1   D.A. Varshalovich, A.H. Moskalev, V.K. Khersonskii: *Quantum Theory of Angular Momentum* (Nauka, Leningrad 1975) [Engl. transl. World Scientific, Singapore (1988)]
5.2   E.U. Condon, Shortley: *Theory of Atomic Spectra* (Cambridge Univ. Press, London 1935)
5.3   L.D. Landau, E.M. Lifshitz: *Quantum Mechanics* (Pergamon, Oxford 1977)
5.4   L.D. Landau, E.M. Lifshitz: *The Classical Theory of Fields* (Pergamon, Oxford 1962)
5.5   H. Kopfermann: *Kernmomente* (Akadem. Verlagsgesllschaft, Frankfurt/M 1956)
5.6   A.S. Davydov: *Quantum Mechanics* (Fizmatgiz, Moscow 1973)
5.7   I.I. Sobelman: *Atomic Spectra and Radiative Transitions*, 2nd edn., Springer Ser. Atom. Plasma Phys., Vol.12 (Springer, Berlin, Heidelberg 1992)

Subject Index